NIKOLA TESLA'S ELECTRICITY UNPLUGGED

Edited by Tom Valone, Ph.D.

Adventures Unlimited Press

Nikola Tesla's Wireless Unplugged

Wireless Transmission of Power as the Master of Lightning Intended

Edited by Tom Valone, Ph.D.

ISBN: 978-1-939149-57-2

Special thanks and acknowledgement to: Nikola Tesla, William Terbo, Jacqueline Panting, Stephan Hall, Sava Kosanovic, James Rybak, Marc Seifer, A. S. Marancic, Nick Simos, Oliver Nichelson, Andre Waser, Kurt Van Voorhies, Jim and Ken Corum, Mike Gamble, Roy Davis, Konstantin Meyl, Koen Van Vlaenderen, and Gary Peterson

Published by:
Adventures Unlimited Press
One Adventure Place
Kempton, Illinois 60946 USA
auphq@frontiernet.net

http://AdventuresUnlimitedPress.com
http://integrityresearchinstitute.org

A portion of the proceeds from the sale of this book benefits Integrity Research Institute, a nonprofit, charitable 501(c)3 organization.

NIKOLA TESLA'S ELECTRICITY UNPLUGGED

Wireless Transmission of Power as the Master of Lightning Intended

Edited by Tom Valone, Ph.D.

Other Books in the Lost Science Series:

Harnessing the Wheelwork of Nature
The Fantastic Inventions of Nikola Tesla
The Tesla Papers
Prodigal Genius: The Life of Nikola Tesla
The Free-Energy Device Handbook
The Time-Travel Handbook

Other Books From Adventures Unlimited Press:

Hitler's Suppressed and Still-Secret Weapons
Mind Control and UFOs
Dark Star
Hitler's Flying Saucers
Covert Wars and the Clash of Civilizations
Saucers, Swastikas and Psyops
Covert Wars and Breakaway Civilizations
LBJ and the Conspiracy to Kill Kennedy
Roswell and the Reich
Nazi International
Reich of the Black Sun
The S.S. Brotherhood of the Bell
Secrets of the Unified Field
The Cosmic War
The Giza Death Star
The Giza Death Star Deployed
The Giza Death Star Destroyed

www.AdventuresUnlimitedPress.com

NIKOLA TESLA'S ELECTRICITY UNPLUGGED

Wireless Transmission of Power as the Master of Lightning Intended

By
Thomas F. Valone, Ph.D., P.E.

ADVENTURES UNLIMITED PRESS

OTHER BOOKS BY THOMAS VALONE

Zero Point Energy: The Fuel of the Future
Bioelectromagnetic Healing: A Rationale for Its Use
Homopolar Handbook: Definitive Guide to Faraday Disk & N-Machine Technologies
Electrogravitics Systems: A New Propulsion Methodology, Editor
Electrogravitics II: Validation of a New Propulsion Methodology, Editor
Energy Crisis: The Failure of the Comprehensive Energy Strategy
Bush-Cheney Energy Study: Analysis of the National Energy Policy
Energetic Processes, Volume I and II, Editor
Modern Meditation: Science & Shortcuts
The Future of Energy: An Emerging Science
Proceedings of the Conference on Future Energy, COFE1-7, 1999-2016, Editor
The Zinsser Effect: Cumulative Electrogravity Invention of Rudolph G. Zinsser
Proceedings of the First International Tesla Science Conference & Exposition
The Eric Laithwaite Report: Gyromagnetic Engineering Genius, Editor
Practical Conversion of Zero-Point Energy: Feasibility Study of Energy from Vacuum
Harnessing the Wheelwork of Nature: Tesla's Science of Energy, Editor

ABOUT THE AUTHOR

Thomas F. Valone has a Ph.D. in General Engineering from Kennedy-Western University, a Masters in Physics from SUNY at Buffalo, and Professional Engineering license. Dr. Valone is President of Integrity Research Institute. The Institute is dedicated to scientific integrity in the areas of energy, propulsion, and bioenergetics. It is a membership-driven, nonprofit, 501(c)3 charitable organization offering books, DVDs, CDs, and electronic downloads. Donations are tax deductible (www.IntegrityResearchInstitute.org).

Thomas Valone is a retired US Patent Examiner, previously specializing in Class 324 Physics, Measuring, and Testing apparatus. He was elected a Board Member of the Patent & Trademark Office Society in 1999. He taught engineering, physics and electronics at the SUNY-accredited, Erie Community College, where he also designed the Instrumentation & Process Control Curriculum and managed engineers and technicians in lab projects. At Scott Aviation, he was a research scientist and Director of R&D. He was responsible for numerous sensor circuit design, instrument design and testing projects.

Memberships include: Institute of Electrical and Electronic Engineers (IEEE), National Space Society (NSS), Union of Concerned Scientists (UCS.org), American Institute of Aeronautics and Astronautics (AIAA), New York State Society of Professional Engineers, and formerly, the U.S. Energy Association and the Bioelectromagnetics Society. He is also a Fellow of the World Innovation Foundation.

He is listed in *Who's Who in U.S. Writers, Editors & Poets, 1988*

Table of Contents

Foreword – William Terbo   11
1. Introduction to Electricity Unplugged – Thomas Valone   17

SECTION I – Tesla's Life Events Leading to Wireless Power

2. Visit With Tesla – Kosanovic   29
3. Look Back at Niagara Falls – Panting   31
4. Tesla: Scientific Saint, Wizard, Carnival Sideman – Hall   37
5. Nikola Tesla: Scientific Savant – Rybak   49
6. Secret History of Wireless – Seifer   63
7. Tesla's Manifesto – Tesla   93
8. The True Wireless – Tesla   97

SECTION II – Popular Tesla Wireless Power Concepts

9. Wireless Transmission of Energy – Marincic   115
10. Tesla Unplugged – Simos   125
11. Tesla's Self-Sustaining Generator – Nichelson   161
12. Nikola Tesla's Wireless Systems – Waser   169
13. Worldwide Wireless Power Prospects – Van Voorhies   187
14. Distribution of Electric Power: Cavity Resonator Modes – Corum   203
15. The Real Tesla Electric Car – Gamble   213
16. High Q Resonant Wireless Power Transfer – Davis   239
17. Wireless Energy Transfer with Efficient Scalar Waves – Meyl   333
18. Electrodynamics with the Scalar Field – Van Vlaenderen   341

SECTION III – Zenneck Surface Wave Theory

19. Wardenclyffe: How Does Wireless Work – Peterson   355
20. Bell Labs: Radio Surface Wave Experiment – Corum   375
    Index   451

Artist's conception of the completed Wardenclyffe Generator powering flying craft

Foreword

William H. Terbo

On Wednesday, March 10, 2010, Nikola Tesla was inducted into the Long Island Technology Hall of Fame. I was privileged to receive this award on his behalf, being his next-of-kin, great-nephew who spoke of his famous ancestor's impact on the world. I was introduced by Jane Alcorn, president of Tesla Science Center at Wardenclyffe, who noted the familial trend in engineering and technology: Tesla was an electrical engineer; his nephew, my father, was an engineer; and I turned out to be a mechanical engineer. Tesla's popularity has grown since 1979, when I was fortunate to become the Honorary Chairman of the Tesla Memorial Society, Inc., in Lackawanna, NY, near Niagara Falls where Tesla masterminded the first grand-scale hydropower system and helped promote Tesla's name ever since, until today we have a car named after him.

In 1984, at the first International Tesla Society conference, I recounted some of the important events in the professional life of Nikola Tesla as he recounted them in 1916 for the AIEE, the predecessor of the IEEE. Significant among them was his arrival in the United States in 1884 and his subsequent citizenship. He recognized the opportunity living and working in the United States brought to him—and he respected it by becoming a citizen, while maintaining a clear appreciation for his roots in his native land. In that way, Tesla was an inspiration to the many immigrants who also came to these shores. An inspiration that being born in a small town in a faraway country was not a barrier to reaching fame and fortune in America.

In 1986, I recounted the growing list of honors commemorating Tesla in the past several years, such as a U.S. commemorative postage stamp and early induction into the Inventors Hall of Fame. And I concluded that Tesla was worthy of the honors not only because of his accomplishments, but because of his character.

Well, the recognition continues to grow. The drama of his life and the burst of creative energy that changed the world and that foretold developments that are still unfolding have become the basis for articles, books, plays and movies. The momentum is building. My hope is that one day every school child will know that when a light is turned on, that Edison may have invented the bulb, but it was Tesla who brought the electricity that lit it.

Getting recognition for Nikola Tesla has been a difficult process. His name finally does appear on a brand of car you drive, but not the TV you watch, the food on your breakfast table—or on the electrical utility bill you pay each month. In addition to reminding the general public of his accomplishments, it has been necessary to overcome some reticence on a part of the academic and professional cadre who harbor some misgivings about a man who was not specifically "one of them." Fortunately, these misgivings have been recognized as ill-founded, and are less and less an impediment to the proper recognition of the man who gave us our modern version of electricity.

Tesla was an "outsider" to an influential part of the academic and professional society in which he moved, and which sometimes chose to judge him in a manner unworthy of his accomplishments or of their intellectual calling. Tesla didn't strive for acceptance but gained it through performance. He didn't follow the norm of the academic world with its "publish or die" attitude which gluts the professional journals.

His regular use of the popular press, rather than dealing exclusively with professional journals, came back to haunt him in later years. In his later years, he held his annual "news conferences" for the popular press in which he proposed scientific developments which, while often prescient, were beyond his ability to prove. This gave substance to that part of the scientific community which wished to consider him a "crank" and to denigrate his earlier accomplishments. They equated personal eccentricities with less than scholarly work.

While no knowledgeable person disputes his preeminent position in the development of concepts and technologies in electrical power, there has been precious little recognition in other disciplines where his work was equally fundamental. An example of this was his long battle with Marconi over priority in radio art in the first decade of the century.

Tesla's well documented research and experiments in the wireless transmission of electrical energy (which we so appropriately commemorate with the publication of this book) was only the preparation for a system to be installed at Wardenclyffe on New York's Long Island. Tesla's proposal to illuminate the Paris exposition with power generated in Long Island was pure fantasy to the scientific community of the day. The concept was too grand for the time, the cost too great and the practical applications too remote. The project never came to completion.

That Tesla's vision of wireless power transcended wireless radio was completely ignored until Marconi's "original" work became public. The ensuing lawsuits, originally decided in Marconi's favor, gave more substance to the attitude of some in the scientific community towards this man who had already earned an exalted place in the annals of world science. Of course, the final irony was the overturning of Marconi's patents by the U.S. Supreme Court—but 30 years too late.

The episode with Marconi is instructive on many levels. It was startling that the "Father of Alternating Current Technology" could have seminal concepts in such diverse technologies as radio, lasers, x-rays, turbines and the like. It shows the dynamism of his thought. It was disappointing and unworthy that what was his truly original work with wireless power transmission could reinforce negative regard among those naysayers in his profession—both then and later. It was distressing that the economic penalty exacted on him by the elaborate and costly construction at Wardenclyffe would thence forward foreclose to Tesla the ability to bring his future innovations to public use and benefit. It is ironic that the name of his student, by virtue of this disputed claim, by a company that commercialized the technology and by the riches that flowed from it, is better known than the teacher.

This irony was further amplified when Tesla, Marconi, Samuel F.B. Morse and the Wright Brothers were inducted into the Inventor's Hall of Fame in 1975 in Washington DC. Tesla's contribution honored at the affair was his patent for the inductive motor, the foundation of the world 's modem electrical technology. Marconi's honored patent was for transmitting electrical signals, the basis for which was overturned 32 years earlier. At least Tesla laid no claim to the Wright Brothers flying machine.

It might seem that Tesla was a contrary sort of person—suing Marconi and engaging Thomas Edison in the decade long "Battle of the Currents." (Of course, Tesla, Westinghouse, and his alternating current won that battle, but popular history has certainly ignored the outcome.)

However, Tesla was not a contrary sort. His relationship with George Westinghouse was warm and mutually beneficial. It resulted in the transition of Westinghouse Electric from a striving competitor to a recognized world leader in electrical power generating equipment through the use of the "Tesla" patents. The relationship also caused the epic harnessing of Niagara Falls, which changed forever every aspect of what we so casually call "modern living."

Tesla was a cultured and educated person. One who took the considerable effort to produce polished English translations of poems and stories from his Serbian ethnicity. Tesla was not the lifelong recluse he was often pictured in later life. Anyone who included Mark Twain among his personal friends would not fit that category. He simply outlived most of his contemporaries and outlived his finances. His personality didn't change as much as his circumstances.

Let me digress with an anecdote told to me by my father. One time in the 1930s, Tesla visited my father in Detroit. They met downtown in the afternoon and went to the Book-Cadillac Hotel, Detroit's finest, for a chat and a late bite to eat. They were about to enter the dining room at about five minutes to two when the maitre d' suggested they wait the five minutes, when the five dollar cover charge would be lifted. Tesla would hear none of it and marched right in. Remember, this was the 1930s, when $500 would buy a new car.

If you 're curious about what Tesla ate, he ordered bread, milk and a chafing dish and prepared the warm dish himself at the table—going through the ritual of wiping the silverware (for which he was so famous) to the chagrin of the headwaiter. And talking about Tesla's personal eccentricities, his legendary phobia about germs doesn't ring particularly true to me, based on my meeting with the old gentleman.

My mother and I were concluding a summer vacation at the Jersey Shore with a trip to New York City, before returning to Detroit. I was 8 or 10 at the time and Tesla was over 80. When we met, I was totally intimidated by this very tall, very, very old man, who proceeded to hug me, kiss me on each cheek and pat my head (as one would pat a good puppy).

Now, I kissed my dad in the regular European family manner, but my British mother (and her brother) took great pains to teach me the "American" way of greeting adults—male and female, family members or not; a firm handshake with heels together. (This suited me fine, but some female adults refused to play by the "rules.")

So Tesla's greeting came as quite a surprise to me. If Tesla lived in fear of germs, I hardly think he would consider touching—much less kissing—an 8 year old boy as promoting good health. Everyone knows that 8 year old boys are magnets for every category of dirt and germs.

Another aspect of Tesla's character that generates a certain amount of controversy was his altruism. In today's world of self-centeredness, it seems almost inconceivable that a person would relinquish great riches—as Tesla did in giving up his Westinghouse royalties—to make sure his inventions were exploited to the benefit of society, by a man he trusted. To understand this altruism, one must understand Tesla's upbringing: the son of a minister in a very strict church. The Golden Rule wasn't just mouthed, it was lived. My father, who was also the son of a minister in that same church, Serbian Orthodox, understood that. If one wishes to call that a character flaw—then, so be it—but certainly it is a flaw practiced rarely by a rare breed of Christians.

I recall, as a young boy, after hearing about the value of the Westinghouse royalty agreement, that if Tesla had died rich, my father would have been a principal beneficiary and we would have been rich. My father said that he never considered it a possibility, because Tesla had said, "I'll never die rich unless the money comes in the door faster than I can shovel it out the window."

But Tesla was an innovator who worked independently. And working independently brings into play a man's complete resources. This was his life's work—not a sabbatical from some more structured existence.

It was not enough for Tesla to develop concepts and invent methods of harnessing them. He had to raise money, design equipment, build laboratories, make contracts, sell ideas and follow through to completion, all the while maintaining an atmosphere conducive to new ideas. This takes character—and to do it alone takes innate character. As it has been said: "You cannot build up a character in solitude; you need a formed character to stand solitude."

The accomplishments of the man we honor, with the publication of *Nikola Tesla's Electricity Unplugged* centered on his greatest work in wireless power, arouse in us a great admiration for the combination of character and intellect that brought so many discoveries to the world and its people. It is sad to say there were some, during his life and after, in whom a certain amount of envy was also aroused, and this has had much to do with the delay in reviving the proper acknowledgment of his gifts and recognition of his name.

Tesla considered himself a student because he never stopped inquiring into the mysteries of knowledge. But, in truth, he was a teacher because he shared his knowledge with us all. It is my hope and belief that this book will finally serve to teach the public what wireless power transmission can accomplish if it is designed with Tesla's intentions and blueprints in mind. Let me emphasize what this book can do with a quotation by Henry Ward Beecher that expresses Tesla's essence: "Every man should use his intellect, not as he uses his lamp in the study, only for his own seeing, but as a lighthouse uses its lamps, that those afar off on the sea may see the shining, and learn their way."

About The Author:

Born April 10, 1930, to Nicholas J. and Alice H. Terbo in Detroit, Michigan, William H. Terbo is the Chairman of the Executive Board of the Tesla Memorial Society, Inc. His father Nicholas Terbo (Nikola Trbojevich), was a world-known research engineer, mathematician and inventor - and nephew and friend of Nikola Tesla. Trbojevich held nearly 200 US and foreign patents, principally in the field of gear design, including the basic patent for the Hypoid Gear – used on nearly every rear drive automobile in the world since 1931.

William Terbo's father modeled his professional life after Nikola Tesla, a man 30 years his senior. He was the only family member to join Tesla in the United States. With such a family history in science and engineering, William Terbo's higher education was a matter of "which engineering school" rather than "what area of concentration." William Terbo graduated from Purdue University, with a degree in Mechanical Engineering – an area closer to his father's specialty.

1 Introduction to Tesla's Unplugged Electrical Power

Thomas F. Valone, PhD., P.E.

This book serves to provide the world with perhaps the only update available today on the status of realizing Nikola Tesla's greatest dream from the past century, over one hundred years after he invented it. However, before introducing that vital topic, it would be prudent to set the record straight about the continued misunderstanding about the first person in the world to transmit wireless signals. For example, even though **Marconi** is often publicized as the originator with his backyard spark transmission experiments in 1895 (most recently on a PBS NOVA science program in March, 2016), **Nikola Tesla** was already in the public's eye two years before, at the 1893 Columbian Exposition in Chicago where he was able to wirelessly power lamps from across the stage and out into the room.[1] By 1895, a famous scientist named **Jagdish Chandra Bose** was also demonstrating in Kalkata India, the transmission of signals over a distance of nearly a mile with microwaves.[2] Therefore, the history is clear regarding the third place that Marconi should be relegated to when discussing the inception of wireless transmission of electromagnetic waves.

Wireless transmission and reception of electrical power *across the earth* without power lines, the subject of this book, was first demonstrated by Tesla in 1899 at Colorado Springs. In recalling that experience, Tesla (Dec., 1904, *L.A. Times*) stated, "With my transmitter I actually sent electrical vibrations around the world and received them again, and I then went on to develop my machinery."

This topic was also introduced in the "first book" in this series, *Harnessing the Wheelwork of Nature*[3] by this author, and is recommended reading since only two short chapters from that book are reprinted in this "second book," *Nikola Tesla's Electricity Unplugged* (credit is given to Dr. Nick Simos for inspiring the title), on Tesla technology. Needless to say, when implemented, such a technology will drastically reduce the use of fossil fuels for electrical power generation by boosting the energy

---

[1] "Electricity at the Columbian Exposition" By John Patrick Barrett. 1894. Page 168–169. Also see Rybak's Chapter 5 in this book, section entitled, "Tesla Demonstrates Wireless".
[2] "Jagadish Chandra Bose". Pursuit and Promotion of Science: The Indian Experience (Chapter 2). Indian National Science Academy. 2001. pp. 22–25.

THOMAS VALONE, PH.D.
AUTHOR, HARNESSING THE WHEELWORK
OF NATURE: TESLA'S SCIENCE OF ENERGY

conversion and transmission efficiency and make energy available to remote locations of the globe, which will also bring up the standard of living in all third world countries. Tesla's superior technology, carefully discerned by the world's Tesla experts in this book, answers the present energy and climate crisis worldwide, saves electrical conversion losses, and provides a real alternative to transmission lines and inefficient generation which tend to degrade and attenuate the transmitted electrical power, until today, only a third of the power is received by the consumer, according the **US Energy Association**, a think-tank based in Washington DC.

Nikola Tesla stated the remarkable difference of his transmission method, "When there is no receiver there is no energy consumption anywhere. When the receiver is put on, it draws power. That is the exact opposite of the Hertz-wave system...radiating all the time whether the energy is received or not." Thus, with Tesla's futuristic and efficient transmission of power, source dissipation will only be experienced when a load is engaged in a tuned, resonant receiver somewhere on the earth. This fact alone represents a major leap forward in electrical transmission efficiency, even now, one hundred years later.[3]

This second book in the Tesla series of edited anthologies about Nikola Tesla accomplishments focuses mainly on his *wireless electricity concepts* with articles by the world's experts and serves to be a companion to the first one that was published for the **Wardenclyffe Tower Centennial**[4]. It also expands the theories published in the first book to show how much has changed since then with the scientific understanding of Tesla's wireless power technique. Is this topic being pursued commercially today? The answer is "yes" where **Global Energy Transmission Corp.** (http://getcorp.com) and **Texzon** (www.texzont.com) are prime examples, using the "conductive crust of the earth" to transmit useful power with prototype demonstrations and citing the earth-ionosphere resonant cavity as well. Both corporations are seriously pursuing the Wardenclyffe-style tower transmission of electricity and tested successful prototypes.

---

[3] Please note that health effects have been fully considered and are negligible. See the first book for details of how low the energy density is even for an atmospheric, earth-ionosphere transmission. – Ed. Note

[4] Valone, Thomas, *Harnessing the Wheelwork of Nature*, Adventures Unlimited Press, 2002. Also see www.IntegrityResearchInstitute.org for the *Proceedings of the Tesla Science Conference*, 2003 and DVDs of the speakers, many of whom contributed to the *Harnessing* book and *Proceedings*.

Chapter Arrangement in This Book

The book is organized in the following manner. The *first several articles are historical* and introduce the reader to the life and times of Nikola Tesla, with an emphasis on those stories of his transmission of power endeavors, including Niagara Falls, Colorado Springs, and Long Island. An effort has been made to make the succession of articles proceed, from the front to the back of this book, *in the order of:* 1) increasing technical detail and 2) more recent and modern interpretations. I grew up in Buffalo New York, which was the *first city in the world* to receive transmitted power from twenty miles away in 1896 from Tesla's invention of AC power generators driven by the Niagara Falls in 1895. My research at the **Niagara Falls Library** uncovered many original articles at the time that Niagara electric power was eminent, as well as locating the site of the three **Adams Plant** buildings. I reported on that history in the first book and promoted Tesla's history during interviews on the History H2 channel and Science channel, as has other contributors to this book, like Dr. Marc Seifer. As to the segment with me that was included on the H2 channel on Tesla's wireless power, I have made that portion of the interview available on our IRI[5] organization's website under "Videos."

Before we get into the summary of the amazing contributions to this book, it is only proper to give a hearty endorsement to the recent Tesla-related activities on Long Island. On Wednesday, March 10, 2010, Nikola Tesla was inducted into the **Long Island Technology Hall of Fame**. The award was given on his behalf to his next-of-kin, great-nephew William Terbo (see the Foreword to this book), who spoke of his famous ancestor's impact on the world. Terbo was introduced by Jane Alcorn, president of **Tesla Science Center** at Wardenclyffe, who noted the familial trend in engineering and technology: Tesla was an electrical engineer; his nephew, Terbo's father, was an engineer; and Terbo is a mechanical engineer. Those interested in the gallant effort to create a museum and learning institute at the Tesla Science Center can visit the website for more information and to offer their support http://www.teslasciencecenter.org/wardenclyffe/. Another recent Tesla event was the promotion of a crowdfunded "Tesla feature film" that failed in 2014 to raise even $20,000 for the project, though the remaining webpage is very exciting to review https://www.indiegogo.com/projects/tesla-feature-film-final-campaign#/ . However, the PBS television special, **"Tesla: Master of Lightning"**, which inspired the subtitle of this book, aired back in 2000 and then again in 2004. Details are still online at http://www.pbs.org/tesla/ and a DVD of the show is available for $19.99 from www.shoppbs.org .

---

[5] IRI = Integrity Research Institute and www.IntegrityResearchInstitute.org refer to the nonprofit 501(c)3 charitable organization, registered with the IRS with the published mission of researching scientific integrity of energy, propulsion, and bioenergetics. A portion of the proceeds from this book benefits the organization. Subscribe to the monthly Future Energy eNews for free.

Back to the subject of this anthology, Dr. Elizabeth Rauscher states in the first book[6] that the earth's magnetosphere is a source of electrical energy for the ionosphere, as Tesla emphasized. She points out that the relatively small longitudinal impulses that the Tesla Tower supplies is a trigger for the magnetosphere-ionosphere oscillations to take place so the receivers can tap the earth's atmospheric cavity electrical energy. I also found confirmation since then that the ionosphere couples and pumps energy into the magnetosphere by an over-reflection of Alfven waves, in an unstable mode of feedback, which also feeds energy back into the ionosphere.[7] Tesla estimated the available energy of the earth-ionosphere cavity at 7.5 gigawatts whereas Dr. Rauscher updates this and shows that it is closer to 3 terawatts (3 billion kW), while the US only consumes about 360 million kW today for electrical needs (at 27% of the world usage). Therefore, the earth has several times the capacity available for electrical consumption than the entire world presently utilizes everyday, if the earth-ionosphere cavity with its magnetosphere energy can be utilized by a Wardenclyffe tower stimulus package. However, there is much more information available to expand on the wireless story, which is the main reason for this second book.

Why wasn't the prototype of Wardenclyffe finished in 1903, besides the narrow mindedness of J. P. Morgan? Tesla offered this visionary conclusion: "The world was not prepared for it. It was too far ahead of time. But the same laws will prevail in the end and make it a triumphal success... Let the future tell the truth and evaluate each one according to their work and accomplishments. The present is theirs; the future, for which I really worked, is mine."

Some suggest that Tesla was not practical about his Wardenclyffe Tower (Wardenclyffe Centennial 1903-2003). Research by the scientists in this book actually proves he was very practical but more than a century ahead of his time. Nikola Tesla's discovery of pulsed propagation of energy does not resemble the standard transverse electromagnetic waves so familiar to electrical engineers everywhere. Many engineers and physicists have dismissed Tesla's wireless energy transmission as unscientific without examining the unusual characteristics and benefits of longitudinal waves.[8] In recognition of the Centennial of the Wardenclyffe Tower (1903-2003) and the IRI-sponsored Tesla Science Conference (Nov. 8-9, 2003), a detailed explanation of Tesla's superior energy

---

[6] Rauscher, Elizabeth et al., "Fundamental Excitatory Modes of the Earth and Earth-Ionosphere Resonant Cavity", in *Harnessing the Wheelwork of Nature*, Thomas Valone, Adventures Unlimited Press, 2002, p. 233

[7] Wang, Xu-Yu, et al., "Ionospheric feedback instability in the coupling of the magnetosphere-ionosphere", *Chin. Phys. Lett.*, vol. 20, No. 8, 2003, p. 1:101

[8] Carlson, W. Bernard, *Tesla: Inventor of the Electrical Age*, 2013, Princeton University Press – quoted in interviews on radio and in magazines alleging that "no one in the world" understands Tesla's wireless power, while promoting his biographical book.

transmission discovery is warranted and obviously needed by the public to understand the inspired genius behind this ecological and futuristic technology.

The first book in this series contains several papers from prominent physicists detailing the unusual method of pulsing a broadband Tesla coil with a deep grounding tube at a repetition rate of 7.5 Hz to resonate with the Earth's Schumann cavity. In this *second book*, we find that a discovery of the harmonic of 11.7 Hz seems to be more theoretically and experimentally supported by the present experts, including Dr. Nick Simos, and Dr. James Corum. As Corum explains, in one of his papers in the first book (p. 198) entitled, "Tesla & the Magnifying Transmitter: A Popular Study for Engineers," a mechanical analog of the lumped circuit Tesla coil is an easier model for engineers to understand. From mechanical engineering, the "magnifying factor" can be successfully applied to such a circuit. "All that is necessary," says Corum, "is that his transmitter power and carrier frequency be capable of round-the-world propagation." In fact, Tesla (in L.A. Times, Dec., 1904) stated, "With my transmitter I actually sent electrical vibrations around the world and received them again, and I then went on to develop my machinery." Corum actually reiterates that same basic philosophy in his latest contribution to this second book regarding the Tesla theory of wireless propagation of power, emphasizing the proper frequency of transmission to reach the necessary distance with his recently uncovered Zenneck waves. The Corums' new Zenneck wave discovery involves the conductive ground wave transmission which also is the subject of patent applications by his company. For example, see the polyphase waveguide probe photo (Fig. 14A of the new 2016 patent application entitled, EXCITATION AND USE OF GUIDED SURFACE WAVE MODES ON LOSSY MEDIA, http://www.patentsencyclopedia.com/imgfull/20160072300_14).[9]

The power loss experienced by Tesla's pulsed, electrostatic discharge mode of propagation was less than 5% over 25,000 miles as recorded in his lab notebooks circa 1900 and rediscovered by Dr. Corum. Dr. Van Voorhies says (first book, p. 151), "...path losses are 0.25 dB/Mm at 10 Hz", which often is difficult for engineers to believe, who are used to transverse waves, a lossy resistive medium, and line-of-sight propagation modes.[10]

The capacitive dome of the Wardenclyffe Tower (Shoreham, Long Island) is perhaps a key to the understanding of longitudinal waves which are also called "scalar waves." Dr. Rauscher quotes Tesla (first book, p. 236), "Later he compared it to a Van de Graaff generator. He also explained the purpose of Wardenclyffe...'one does not need to be an

---

[9] See also Corum US patent apps 20160079643, 2016007944, 20160079645, 20160080034, 20140252886, 20140252865 and Corum patents 4622558, 4751515, 5442978, 5654723 related to Tesla wireless transmission.

[10] Note: a "dB" or decibel is a unit of attenuation where 0.25 dB is about a 5% decrease but notice the distance quoted by Dr. Van Voorhies is "Mm" which is a megameter or a million meters!

expert to understand that a device of this kind is not a producer of electricity like a dynamo, but merely a receiver or collector with amplifying qualities.'" Such a perspective undoubtedly contributed to the concept of the *Tesla Magnifying Transmitter* which is described in the first book in detail, as well as in Chapters 11 and 12 of this book. Since then, other authors have also presented theories to explain the concept as well. Dr. Paul LaViolette for example suggests the presence of phase conjugate waves being created from the "repeating series of very high-voltage shock fronts" emanating from the Wardenclyffe dome.[11]

Summary of Chapters

In this second book, I have collected some amazing updates since the publication of the first book. For example, Dr. Konstantin Meyl contributes a longitudinal wave explanation from his 650-page opus on *Scalar Waves*[12] that was published in English in time for the 2003 Centennial but never included in the first book. Dr. Nick Simos also has a great 2015 conference slide presentation[13] on "Tesla Unplugged" that has been edited especially for this book. Simos explains nature's available longitudinal wave coupling between the ionosphere and the earth using what are known as "cavity modes" (earth-ionosphere cavity) based on Maxwell's Equations and derives the 11.78 Hz resonant frequency for transmission, independently from Corum's work in the last chapter.

Beginning **Section I**, we have a sentimental story about a personal meeting with Tesla. Mr. Kosanovic gives us a glimpse of what Tesla was like in Chapter 2. Dr. Panting continues our historical section with a candid summary of Tesla's contribution to Niagara Falls electrical generation in Chapter 3. Thanks to Stephan Hall, we find that history has not been kind to the showy inventor of alternating current motors and more in Chapter 4, but the tide is at last turning. Though it is a brief article from the *Smithsonian* magazine, it gives the reader a great review of all of the electrical generation projects during Tesla's life. Continuing with a similar review, James Rybak gives us more details of Tesla's professional life history in Chapter 5, including the 1893 event mentioned earlier when Tesla demonstrated wireless transmission of electrical energy.

A friend and colleague, Dr. Marc Seifer, has contributed the most comprehensive article on Tesla's secret history of wireless in Chapter 6 and perhaps the best introduction to the later chapters in Section II. In preparation for Chapters 8, 19, 20, notice the letter on the first page of Dr. Seifer's article which emphasizes that Tesla spoke at a meeting about

---

[11] LaViolette, Paul, *Secrets of Antigravity Propulsion*, Bear & Co., 2008, p. 246-248
[12] Meyl, Konstantin, *Scalar Waves*, Indel, GmbH, Verlagsabteilung, 2003, www.k-meyl.de or www.meyl.eu – also distributed in the US by Integrity Research Institute
[13] Simos, Nick, "Wireless Energy Transmission: Nikola Tesla Unplugged", Seventh Conference on Future Energy (COFE7), DVD available from www.futurenergy.org and *Proceedings of COFE7* from www.proceedings.com and www.KnowledgeE.com .

"messages he sends out are conducted along the earth." This reflects the latest interpretation, also seen in Tesla's "True Wireless" article, that the ground conduction is an important dominant avenue for wireless transmission. Of course, the rest of the article contains lots of secret intrigue and suspense which reflects more information about Tesla's battle over wireless transmission than previously published anywhere else.

With Chapters 7 and 8, we have a new window on Tesla's frame of mind in 1904 compared with 1919. In the latter chapter, Tesla reveals many more of his details concerning "The True Wireless" (except for some that are still classified in the National Archives even today). We can see how Tesla was determined to contrast his transmission antenna design with the Hertz transverse wave style of radio waves that we emerging at that time, due to Marconi's monopole antennas.

In **Section II**, we move from the narrative to the technical. To begin with, Chapter 9 features an older article by Prof. A. S. Marincic from the University of Belgrade who gives us an electrical engineering review of the wireless transmission of energy from the archives of the Nikola Tesla Museum in Belgrade. The existence of stationary waves is one of the discoveries that Marincic found in Tesla's notes, as well as a graphical confirmation from a 1978 book of Tesla's claims of amplitude attenuation with distance claims. Marincic also reproduces a very intriguing drawing of alternative antenna designs by Tesla and mentions how Tesla's claim for energy flow through the earth "was peculiar" which it was, decades ago. In Chapter 10, friend and colleague Dr. Nick Simos presents a nicely illustrated review of a Maxwell equation analysis of Tesla's wireless transmission capacitor concepts. Nichelson's article in Chapter 11 is one of only two reprinted from the first book since it includes an interesting analysis that defends the elusive Tesla Magnifying Transmitter concept. In Chapter 12, Andre Waser offers us a more technical analysis of Tesla's patented transmitter designs, including a better explanation of the recently acknowledged ground conduction.

Dr. Van Voorhies presents in Chapter 13 the second reprinted article from the first book on prospects from Tesla's worldwide wireless power. It is a worthwhile addition to this book since it offers a short but technical summary of Tesla's "proof of concept" and explains the attenuation at a particular frequency per distance in megameters (Mm) where a fraction of a decibel (dB) per Mm is quite unequalled by any other means. Van Voorhies also gives credit to the Corums for waveguide coupling and source/transmitter concepts. In Chapter 14, friend and colleague Dr. Jim Corum gives us the conventional view of the earth-ionosphere terrestrial cavity resonator modes, a great comparison of physical parameters that Tesla accurately disclosed, as well as the *quality factor* Q. The last quantity also helps introduce the Roy Davis article of Chapter 16 which shows how Q is being practically used today with inductive, wireless power transfer for vehicles, in an exciting slideshow presented in 2015 at the US Patent and Trademark Office. Davis also includes a wonderful tutorial he calls a "History Lesson" to give budding engineers the tools to understand how his company accomplishes the high Q circuit design, even with a Design Example, to show how he maximizes the power transfer inductively. Davis also has a nice Tesla wireless review, including a slide which he found that shows an original vehicle powered wirelessly.

Nikola Tesla's Viewpoint

Brooklyn Eagle (July 10. 1932)

I have harnessed the cosmic rays and caused them to operate a motive device. Cosmic ray investigation is a subject that is very close to me. I was the first to discover these rays and I naturally feel toward them as I would toward my own flesh and blood. I have advanced a theory of the cosmic rays and at every step of my investigations I have found it completely justified.

The attractive features of the cosmic rays is their constancy. They shower down on us throughout the whole 24 hours. and if a plant is developed to use their power it will not require devices for storing energy as would be necessary with devices using wind. tide or sunlight.

All of my investigations seem to point to the conclusion that they are small particles, each carrying so small a charge that we are justified in calling them neutrons. They move with great velocity, exceeding that of light.

More than 25 years ago I began my efforts to harness the cosmic rays and I can now state that I have succeeded in operating a motive device by means of them. I will tell you in the most general way. the cosmic ray ionizes the air. setting free many charges– ions and electrons. These charges are captured in a condenser which is made to discharge through the circuit of the motor. I have hopes of building my motor on a large scale. but circumstances have not been favorable to carry-ing out my plan.

In Chapter 15, friend and colleague Mike Gamble offers us his research results into the famed automobile that many believe was designed by Nikola Tesla and powered wirelessly. It is perhaps the most convincing presentation to date since it includes his engineering analysis that starts to make sense when compared to Roy Davis' practical engineering work on the same subject. Also reprinted on this page is an early news article out of our IRI archives from 1932 that also lends credence to Tesla's auto.

It seemed that this anthology would not be complete unless some credit was given to Prof. Konstantin Meyl's tireless work with demonstrating a miniature Tesla tower transmitter and receiver at the IRI 2003 Tesla Science Conference and his many scientific presentations on scalar waves which is summarized in Chapter 17. All of Dr. Meyl's numerous DVDs and books are available from his website, www.k-meyl.de (click on the US or British flag for English).

To finish up Section II, we are pleased to include Koen Van Vlaenderen's technical article in Chapter 18 that expands upon Meyl's work, with Andre Waser's contribution, to give a rigorous derivation of scalar waves based on standard electrodynamics. It has been included mainly to satisfy those physicists who always doubt that longitudinal or scalar waves can by physically proven to exist. The beauty of this article is the proof, for the first time, of longitudinal scalar wave generation from a dynamic charge density distribution, which is what many describe the Wardenclyffe tower antenna as generating.

In **Section III**, we reach the most modern development in the evolution of scientific analysis of Tesla's wireless transmission of power. Gary Peterson kindly rewrote his original article from the first *Tesla* magazine (V. 1, No. 1, July 10, 2013)[14] just for Chapter 19 of this book to address the Corum discovery of the Zenneck surface wave and put it in perspective for us. Finally, in Chapter 20, we are honored to have the Corum brothers' contribution on the history of the surface wave theory and experiment which includes the famed Bell Telephone Laboratory experiment. After studying the graph of Figure B1 it occurs to anyone of ordinary skill that as the frequency of transmission is *lowered*, the range for the Zenneck wave *increases*, which is in keeping with other research reported in this book. This insight into the value of this discovery relates directly to this available manipulation of attenuation if worldwide transmission is desired, with minimum attenuation. The Corum US patent application, 20140252886 also confirms the observation made above:

> "Another way to state that is that the higher the frequency, the smaller the region over which the energy is spread, so the greater the energy density. Thus, the 'knee' of the Zenneck surface wave shrinks in range as the frequency is increased. Alternatively, the lower the frequency, the less the propagation attenuation and the greater the field strength of the Zenneck surface wave at very large distances from the site of transmission..." (par. 144)

This insight is brought to an exciting conclusion in the same patent application with the Corum observation, "Note that if the frequency is low enough, it may be possible to transmit a guided surface wave around the entire Earth. It is believed that such frequencies may be at or below *approximately 20-25 kilohertz*" (par. 148). With that ideal frequency range, the flat surface model can be modified to become a sphere, just like the earth, "where the propagation distances approach the size of the terrestrial medium" (par. 148). Needless to say, such a clear breakthrough in both the theoretical and experimental confirmation of Tesla's original wireless transmission records, leading to a practical conclusion with much more precision than ever before, should expedite the implementation of wireless energy for mankind. Just the number of patent applications that the Corums have filed (see footnote 8) is a good indication of intellectual property that we all hope will lead to a more secure and stable electrical transmission virtual "grid" for this 21st century. Contrast this picture with the present aging and problematic national electric grid we currently struggle with, which has trouble accepting large renewal energy input surges, lest a grid section becomes unbalanced. Our hope also is

---

[14] Visit www.teslainfo.org to obtain this collector's edition of *Tesla* magazine or to subscribe. On one of their webpages http://www.teslainfo.org/view_tesla_magazine.html is posted a free PDF of a few articles, including one on Tesla's wireless energy.

that renewable, carbon-free energy can be exclusively used to generate the electrical input. Then the true, wirelessly powered Tesla automobile may quite possibly be resurrected, besides benefiting stationary electrical receivers all over the globe.

Speaking of wireless transmitters and receivers, it is worth mentioning the Leyh and Kennan's *Lightning On Demand* company article on efficient wireless transmission of power with two large Tesla coils which was published in an IEEE Power Symposium proceedings in 2008.[15] They are also building the world's largest lightning generator pair with ten story high Tesla coils, which is also on their website.

Also interesting information for wireless transmission designers is the fact that more than just another Tesla coil is recommended for good reception. Tesla has disclosed some information about the receiver design "synchronizing" with the transmitter, in his US patent 685,954 (see below), which is available through www.uspto.gov and elsewhere. However, apparently *there is more* to the specific design that is surprisingly still classified. For those with clearance, this editor has learned that the Tesla classified records are stored and available at the National Archives Library in College Park MD.

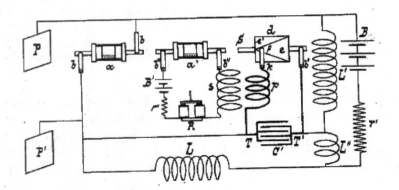

All of us who are part of the Tesla community hope that we will see Tesla's dream of wireless electrical power manifest within our lifetime, so that we may truly be "unplugged" from the umbilical cord that J. P. Morgan forced upon us, through the electrical utility grid, when he denied Tesla's request for further funds to finish the Wardenclyffe tower. Perhaps this book will help that process along so wireless electricity will be as commonplace as wireless telephones and television have become, about a century, more or less, after their inception.

---

[15] Leyh and Kennan article http://ieeexplore.ieee.org/xpl/abstractKeywords.jsp?arnumber=5307364 but is free from the authors' website http://lod.org/misc/leyh/papers/naps2008final.pdf

Beginning: "My Inventions," by Nikola Tesla

FEB.
1919
20 CTS.

OVER
175
ILLUST.

ELECTRICAL EXPERIMENTER

SCIENCE AND INVENTION

THE TESLA
WIRELESS LIGHT
SEE PAGE 692

2 A Visit with Nikola Tesla

SAVA N. KOSANOVIC, TESLA'S NEPHEW, FIRST VISIT in 1926.
Translator - Nicholas Kosanovich, TMS, Executive Secretary/Treasurer

This article was printed in a reader for high school students for literature prior to WWII.

On the 7th Avenue, in the largest and most comfortable hotel, lives a great inventor, our Nikola Tesla. In his rooms on the 15th floor of this mammoth structure - a skyscraper - above the noise of autos, underground subways, lives this great thinker, engineer, inventor with self-composed peace for meditation. The tumult does not disturb him. Late in the night through his windows can be seen that he is working.

It is a strange feeling that engulfs us when we first meet with Nikola Tesla. Things swirl in one's head all that has been said and read about him. Somehow we fear, not knowing how we will behave before him, how to hold a conversation, how to talk to him. With such a feeling I went into the lobby of Hotel Pennsylvania where Tesla had invited me. I lose myself in the crowd of people, but I didn't wait long. Tesla is prompt. Eye contact is made from a distance. Tesla nods farewell to his friends.

As he is approaching, I study him. A distinguished man, in a dress coat, with a derby hat on his head, carrying his gloves, a thin cane in hand. Wearing no jewelry. Straight, erect like a flower, thin, elastic, without a trace of age, even though he is no longer young, stiff head, as though it was chiseled. A large straight nose with thin lips and a high forehead. But the eyes? When one sees them, he will never forget them. A gentle, gentle glance, but deep and penetrating, as though he was looking into your soul: full of kindness and melancholy.

He recognizes me from a distance. He approaches hurriedly. He shakes hands with a firm grip, athletic hands of a worker. After the first words, in the purest form of our language, my fears left me and we talked warmly and friendly.
In the Hotel Pennsylvania restaurant room, Tesla usually eats with the guests there. He is a most gracious host, Tesla serves you himself. After supper there is talk and his first question was, "How is it at home?" Then he talks about our situation and the future of our people. He often says, "Just work! Work more!" He doesn't drink wine and doesn't smoke and enjoys mostly fruits and vegetables.

After supper Tesla takes his prepared bag with bird food, for the pigeons, and goes for a walk. He stops in front of a huge building, the New York City Public Library, and broadcasts the seeds who gather around just like at Saint Mark's Square in Venice. After his long walk along New York's streets, he arrives home to work, which usually lasts late into the night.

I will never forget the picture of this healthy man, who moves like a young man, even though he was then over 70 years of age. But I shall always remember some words that I heard from the mouth of this scientist, engineer and physicist, who however always reads our folk poetry and enjoys the work of poets and artists.

This is what I heard and remembered from him: "I received a lot of money for my inventions, but I invested everything into further research. I cannot refuse a beggar. I know that at times begging is a form of skill, but I cannot see how someone stands in front of me and begs. Maybe, this one is deserving of assistance. Who am I to judge him? One must distribute to everyone who seeks alms, because we cannot know all of the problems of the beggar."

"I do not care for fame and fortune. I only get satisfaction in my work." "The rich financiers ask me to work with them, but I do not want to. I do not want to work with others capital or money and be dependent."

"I am by birth a Serb, but I am also a Croat, because we are one. "When I complete some projects, I shall go to the fatherland. There is a strong desire to see and visit it. I would like most of all to publish a voluminous work of our folk poetry with pictures of our painters.

"If there were no mirrors, I would not know that I was aging. My hand is now firmer than in my youth. From time to time I exercise.

"I do not want the youth to revere me, but, I understand my homeland. That is why I told them at Belgrade University: "I desire that this spark engenders a flame of inspiration to the youth of Yugoslavia."

3 A Look Back at Nikola Tesla's Accomplishments in the Niagara Falls Region

Jacqueline Panting, N.D.

"It will not be long before we can transmit that power [of Niagara Falls] by means of a wire...over great distances...I believe the time will come when we shall transmit that energy without any wire." – Nikola Tesla

These prophetic words from the great scientist Nikola Tesla stated in 1893, show how he never doubted his ability to transform the world by transmitting electricity over long distances by means of a wire with the awesome power of the Niagara Falls River. He was so far ahead of his time, that his belief in transmitting energy without wires is still not a reality today, but certainly feasible in the near future[1]. One only has to see the ability worldwide wireless communications to see that the next natural progression will be to transmit energy without wires around the whole world. The dream of harnessing the power of Niagara Falls for generating electricity was something Tesla thought of when he

was only a child. Seeing pictures of the awesome power of the Falls, he stated "Someday I shall harness it." He set huge ideas and forces in motion years before this work ever started. With his unique genius and sheer volition, he knew he could accomplish his dream. And what a great accomplishment indeed! Thanks to Tesla's invention of the AC induction motor, which allowed electricity to be transmitted long distances (DC electricity could be transmitted only for very short distances), Niagara Falls was the first city in the world to have commercial alternating current generation of electricity. The city of Buffalo, New York, was the first city in the world to receive electric power generated from a long distance away (22 miles), in 1896.

Birdseye view of Goat Island & Niagara Falls

Then, power lines were extended to Syracuse in 1905. This application of long distance power transmission eventually led to the present Niagara Mohawk Power Co. that distributes electricity throughout western and central New York. Soon, the whole world was lit by his genius just as he predicted in his statement above.

The *Niagara Gazette* (the foremost paper of the area at the time) had nothing but praise for the man they described as the "greatest living electrician." Scientists, politicians, financiers, men and women alike wanted to meet and hear him speak of his amazing inventions and discoveries. He was one of the most famous celebrities of his time.

Wanting to record all that we could relating to Tesla history in the Niagara/Buffalo area, a group of us researched all the landmarks still existing today that stands as witnesses to his incomparable genius by taking a trip back in time, when the excitement and thrill of Tesla's fame was felt by everyone.

We started our trip from the city of Buffalo, New York and headed to the Niagara Falls region. We decided to first go to Goat Island, a landmass between the Falls, and on the American side. Here is the beautiful Niagara Falls State Park, where a bigger than life-

---

[1] See Tom Valone's book, *Harnessing the Wheelwork of Nature, Tesla's Science of Energy* (Adventures Unlimited, 2002)

size bronze sculpture of Nikola Tesla towers over the plaza. The statue created by a Yugoslavian sculptor, was unveiled on July 25, 1976 commemorating the 120th anniversary of Tesla's birth.

Behind it, is the only remaining part of the original Adams Power Station (Number One), the ornate entrance archway, which has at the top center, a medallion that features a Mohawk Indian traveling the Niagara River by canoe. The Edward Dean Adams Hydro-Electric Power Station Number One was inaugurated on August 26, 1895 by the Niagara Falls Power Company. The building was designed by Mr. Stanford White, one of the foremost architects of the time who used the Richardsonian Romanesque style then favored for churches, universities and public buildings. The Station's nameplate is

Adams Plant Number One Entrance Archway, the only remnant of this building that housed the original Tesla dynamos, is located behind Tesla's statue (credit: J Panting)

currently on this entrance archway along with other commemorative plaques, are described further in this article.

As we walked toward the Niagara Falls River, the thundering power of this rushing water was so loud that we were easily able to understand why Tesla was inspired on tapping some of it for the large-scale generation of electricity. This river has an average of 202,000 cubic feet per second of water flow. The gusting winds along with the heavy mist that the falls produces, visible hundreds of feet away, are really quite a unique experience. This untamed power and energy of Niagara Falls was revered by Tesla even

Adams Plant One. Photo from Niagara Falls Library Archives

as a child and we have to be forever grateful of his tenacity and perseverance in fulfilling his childhood dream of "Someday I shall harness it."

After experiencing the inspiring beauty of the river and falls, we started to look for the original generating Stations. We wanted to find the areas where the original plants we erected, so we headed to the Niagara Falls Library to see what they had in their archives. As we looked at the old Niagara Falls maps, showing the Adams plant site, we realized that we could find them if we would go out and look for them. Talking with the Librarian, she mentioned that one of the plants was still there. Armed with maps and our digital camera, we headed toward the Robert Moses Expressway. Once there, by hiking down the side of it, we reached the ground and then realized that we were now standing on the site of the original Adams Plant Number One.

As we looked around, on the ground still remain large pieces of stone, probably parts of the original Station building. Sadly, there are no plaques or signs commemorating the site. This Plant contained ten 5000 horsepower Tesla AC generators yielding 37,000 kilowatts. The second Adam's Plant (Number Two) doubled that output.

Adams Plant Number 3. The only building left standing today

The original plant at that time was designed for 25Hz. however, the subsequent expansion included conversion to 60 Hz.

As we continued to walk along we saw the Niagara Falls Sewage Treatment facility. Here is where the Adams Building Number Two once stood. Then adjacent to the canal, as we continue to survey the landscape, there it was, the Adams Plant Number Three. Still intact, this building was the original transformer house and is all that remains of the three original buildings. Over 100 years ago, this trio comprised the world's first complex of Alternating Current Power Generating Stations. The Tesla alternating current generators housed in this particular building were still working in 1961. At that time they were shut down, due to the opening of the new Robert Moses Power Plant. We have been told that at least one is stored at the Smithsonian Museum in Washington DC, although not on display. The Moses plant has a capacity of 1,950,000 kilowatts enough to supply a city the size of Chicago. Tesla was right when he foresaw the enormous potential of Niagara Falls!

Sadness and nostalgia came over us, as we saw what was left of Tesla's magnificent achievement in Niagara Falls. As we sat and talked we realized something: Yes, the buildings are gone, the generators shut off, but we had overlooked the most important fact, that is, thanks to Tesla, we have the ability of transmitting electricity anywhere in the world, no matter how far away from the generating plant. Electricity has become necessary for our existence, as Tesla predicted back in 1893 during an interview by the Niagara Gazette: *"Electricity is becoming more and more an important factor in our lives...after a considerable amount of time it will become practically necessary for our existence."*

Also on this archway is a large plaque honoring Tesla and crediting him as the inventor of the AC induction motor which made possible the long distance transmission of electricity that so many had dreamed of up till then, but that only he could accomplish. The plaque was erected and dedicated on July 25, 1976, the same day as his statue was, on occasion of the 120th anniversary of this genius's birth. On the plaque, next to his name, on the right side are etched replicas of the original towering AC generators

35

invented by him and placed in the Adam Plant Stations in 1896. On the left side are etched huge wire tower transmitters, which symbolize the long distance transmission, made possible by these AC generators. How wonderful to see that here in The Niagara Falls State Park on Goat Island, is a monument to Tesla's genius as the inventor of AC electricity. Visited by hundreds of thousands from all over the world, every year, we were very happy to see so many people reading the plaques, climbing the statue and becoming acquainted with how Tesla changed the world.

Leaving nostalgia behind and totally overwhelmed by the magnitude of Tesla' accomplishment here, we decided to walk up once again to his magnificent statue. As we walked toward it, my heart was filled with reverence and gratitude. I thought what a different world this would have been without Tesla's inventions (and he has so many). Most likely, we would have much more expensive DC electricity since Edison fought AC continually. Fortunately for us, Tesla changed forever the way we all live on the Earth.

I decided to climb up the statue as many people do, and feel what its like to be next to him. I was thrilled to be so close to the statue of an incredible genius! I started to think how it must have felt to be next to the real Tesla. I remembered how many celebrities of his time worked with him, Westinghouse, J.P. Morgan,

The author proudly standing at Tesla's feet (photo credit: T. Valone)

his famous friends like Mark Twain, Sara Bernhardt. How wonderful it must have been to be able to be with him, hear him speak of his many new inventions yet to come.

After a long day, filled with the excitement of visiting all the places related to Tesla in this area, we were ready to sit and relax at the Top of the Falls restaurant, on top of the Niagara State Park gift shop, which has a magnificent view of the falls. As we dined, the sun started to set and the falls were no longer visible by its rays. What a shame that no longer we could enjoy the white cascading falls. At least, I thought, I could still hear the mighty rumble. Then, suddenly the beautiful colored beams of light were turned on and the falls were visible again! Beams of pink, blue, red, purple, green and white flashed giving the falls a different and breathtaking look. I smiled and thought to myself: To be able to see these beautiful falls, by AC electricity, even at night is yet another never-ending gift of Tesla to the whole world!

For further information

Tesla's history, biography, magnifying transmitter, wireless electricity, Niagara Falls history....get the book: *Harnessing the Wheelwork of Nature*, available on www.Amazon.com

4   Tesla: A Scientific Saint, Wizard or Carnival Sideman?

Stephen S. Hall, <u>Smithsonian</u>, June, 1986, P. 120 & https://teslauniverse.com/nikola-tesla

History has not been kind to the showy inventor of alternating-current motors and more, but the tide is at last turning

As showman in 1894, Tesla astonished crowds with his cordless, phosphor-coated bulb lit by radio waves.

The laboratory, infamous among neighbors, was located on the fourth floor of a building on south Fifth Avenue, just above Bleecker Street. Strange glows and flashes of blue lightning emanated from it in the middle of the night. In Victorian New York, these were silent, eerie announcements that the age of electricity would soon similarly glow and flash all across America.

Nikola Tesla would usually show up after dinner, around 10 o'clock, and work through until morning. When journalists or friends like Mark Twain dropped by, Tesla would amuse them with electrical tricks. For his pièce de résistance, the tall, immaculately dressed host would leap onto an electrified platform. Ever the picture of erect charm and courtesy, he remained motionless as the voltage steadily rose; soon, two million volts of electricity were coursing over his body. The electricity created an interesting effect: a kind of electrical halo surrounded Tesla as he stood there, unmoving, an object of astonishment and envy, alone in his private universe of electrons and light.

Whether Tesla is the patron saint of modern electricity or some carnival sideman in the history of science is an issue that prompts a good deal of heated disagreement even now, 130 years after his birth. In Yugoslavia he is a national hero. To boosters in the International Tesla Society he is the unacknowledged inspiration for everything from radio to robots to radar. A devoted group of disciples in California is convinced that at his death he was whisked back to Venus, from whence he first came. Among contemporary academics and historians of science one hears more modest assessments of his talents, as well as occasional muttered use of the word "crackpot." Variously described as a "gifted madman" and as "a medieval practitioner of black arts," Tesla remains in death as in life a lightning rod for all manner of praise, criticism, controversy and exaggeration.

His brain, wrote one admirer, was to the intelligence of other inventors "as the dome of Saint Peter's to pepper-pots." Hardly anyone disputes that Tesla possessed a brilliant intellect, and a good thing, too, for his personal idiosyncrasies would have thoroughly undermined the credibility of a lesser talent. The same man who compulsively computed the cubic contents of every meal set before him made alternating current practical for mass use. The neurotic who reacted with phobic horror to pearl earrings on women was one of the pioneers of wireless transmission — what we call radio. The visionary who foresaw plasma physics and the cyclotron had a deathly fear of germs. The eccentric who insisted on having exactly 18 napkins with which he polished the silver and glassware before every dinner, and who tried to stay in hotel rooms whose numbers were divisible by three, received patents for robots and remote-control devices before the turn of the century. Millions scoffed when, almost five decades ago, he spoke of death rays and an electronic "Chinese wall" of defense around America created out of radio waves; now, billions of dollars are being spent on the conceptually similar "Star Wars" defense.

Tesla admirers attribute to him the origins and application of a dizzying array of devices; fluorescent lights, X-rays, the electron microscope, microwave transmission, satellite communications, solar energy, guided missiles, computers, the automobile speedometer, television, vertical-takeoff aircraft and radar. A symposium to be held this summer in Colorado Springs, jointly sponsored by the International Tesla Society and the local section of the Institute of Electrical and Electronics Engineers, attests to the interest Tesla still generates among scholars.

William Terbo, honorary chairman of the Tesla Memorial Society and the inventor's grandnephew, is gratified, of course, but sighs at the lack of credit. "When you're a little ahead of your time, you can be proven right in your lifetime. When you're way ahead of your time, the cause and effect may not be so evident."

In 1900 world's largest Tesla coil, 52 feet in diameter, created ganglia of light during 12-million-volt discharge. Tall, stylish, with smoldering blue eyes and a fine mustache, aloof yet full of charm, Tesla was a master of electricity when electricity transformed American life, a peer and rival of Thomas Alva Edison. Indeed, rarely has the combination of personal grandiosity and technical genius produced such spectacular results. At the height of his powers, Tesla created an earthquake that made a Manhattan building rumble, lightning that rolled across the plains of Colorado and, in a regrettable excursion into man-made pomposity, threats of planetary omnipotence. With his oscillating vibrator, he boasted, he could split the Earth "as a boy would split an apple — and forever end the career of Man."

He arrived in New York City in 1884 with four cents and a few of his own poems in his pocket. Within 20 years his adopted homeland would begin to be wired with a system using alternating current, which evolved from an idea he had brought from Europe to America.

Tesla was born at midnight between July 9 and 10, 1856, in the Croatian village of Smiljan. His father was a clergyman, his mother a strong inventive woman with a prodigious memory. Accounts of his childhood (of which Tesla himself was the primary source) are more inspirationally than factually satisfying. As a youngster, he invented his own waterwheels and motors driven by June bugs; he also took apart watches. By the time he was a teenager, he spoke four languages. As an 11-year-old, one story goes, he would stay up all night reading by candlelight; this being the boy-inventor nonpareil, he of course made the candles himself.

One of Tesla's particular gifts was mathematics. His genius in the subject reportedly stemmed from a more controversial "talent": the ability to experience visions. Tesla's biographers report that he was able to visualize a blackboard and compute upon it, thus performing complicated mathematical calculations almost as soon as they were recited.

As brilliant as the young Tesla was, so too was he chronically infirm and bedridden. An ill-conceived childhood invention (refined some years later by the Wright brothers) had Tesla hyperventilating on a barn roof to lighten himself, and then leaping into the air with an umbrella; the convalescence was long and painful. In his teenage years, he contracted malaria, cholera and unspecified nervous disorders that brought him to the verge of death and may have contributed to his germ phobia.

Soon he was off to Graz, Austria, to study electrical engineering; he later attended the University of Prague. In 1881 he found himself employed by the new Central Telegraph Office in Budapest. Working 19 hours a day, sleeping only two (a pattern he allegedly maintained for the rest of his life), Tesla suffered what we would now call a nervous breakdown.

While recovering, he had a vision that led to the most widely applied of all his ideas: the rotating magnetic field and the alternating-current induction motor. It had all the elements of high drama.

Thomas Alva Edison, shown with phonograph, first employed Tesla, later tried to block the use of his alternating current. Tesla was strolling through a city park with a friend at the time. The sun was setting. Lines from Goethe's *Faust*, committed to memory years earlier, came to mind and Tesla recited: "Ah, that no wing can lift me from the soil / Upon its track to follow, follow soaring!" Suddenly the idea hit him "like a flash of lightning and in an instant the truth was revealed." While his friend worried that the sickly Tesla had been seized by a fit, Tesla grabbed a stick and quickly diagrammed the wiring of an electrical motor in the dust. This drawing represented the rotating magnetic fields of an alternating-current motor.

Tesla realized that by surrounding the armature and drive shaft of a motor with several wire-taped blocks, or field windings, one could electrically feed each winding with alternating current slightly out of phase with the others. By precisely timing these "polyphase" currents, in fact, a north-south magnetic field could be made to circle around the drive shaft and induce it to follow magnetically. Goethe's "follow, follow," it has been suggested, put him in mind of a rotating magnetic field and armature following it.

In 1909 Guglielmo Marconi shared a Nobel Prize for contributions to "wireless," but Tesla had also transmitted wireless signals. What he had hit upon was a successful solution to problems (such as sparking) inherent in early direct-current motors. His concept — the possibility of using alternating current — without exaggeration revolutionized the use of electricity.

Alternating current was his entrée to the front ranks of electrical engineering. The Continental Edison Company, based in Paris, hired the 26-year-old Tesla as a troubleshooter. Although no one expressed much interest in his AC motor design, one of his bosses, Charles Batchelor, who had worked with Thomas A. Edison, sent Tesla off to New York with one of the pithiest and most prophetic letters of introduction ever penned. Wrote Batchelor to Edison: "I know two great men and you are one of them; the other is this young man." America's reigning inventive genius hired him on the spot.

J. P. Morgan's daughter Anne loved Tesla, but wore pearls, which he abominated. They were two extremely bright, willful and egocentric men, destined to clash. Tesla lasted less than a year in Edison's Manhattan workshop, but the rivalry went on for a lifetime. The two inventors parted company in 1885 on less than felicitous terms. Tesla claimed Edison had reneged on a promise to pay him $50,000 after he had designed a complete line of new dynamos and that Edison lamely explained away the dispute by saying, "You are still a Parisian. When you become a full-fledged American, you will appreciate an American joke."

The hostilities assumed gigantic proportions when Tesla's AC system came to the attention of George Westinghouse, the inventor whose growing industrial empire was based in Pittsburgh. Westinghouse immediately saw the practical advantages offered by Tesla's AC motors and an alliance was forged with the newly formed Tesla Electric Company in 1888. What was at stake was nothing less than the power system which would drive America's booming industrial sector, feed its subways, power its streetlights, and light homes in every state in the Union.

George Westinghouse backed Tesla in the struggle over alternating current. This was the ultimate power struggle, the War of the Currents: Edison's DC versus Tesla's AC. Direct current could be delivered economically only within a few miles' radius of its source. A national power grid of Edisonian stamp would have littered the land with nearly as many power stations as fire hydrants.

Tesla's polyphase system, by contrast, allowed power to be transmitted hundreds of miles because alternating current could be economically transmitted at high voltages, then reduced, via transformers, to lower voltages for household use. The economic stakes were enormous. For the rights to use his patents, Westinghouse was willing to pay Tesla $70,000 in cash and notes, plus a fabulous royalty of $2.50 per each horsepower produced.

The propaganda that flew back and forth had the voltage of a religious schism. According to historian Thomas Hughes, Edison's campaign to discredit AC reflected "one private enterprise's endeavor, through political power and legislation, to outlaw the technological advantage of another." Edison and his allies had huge capital investments tied up in direct current, so they argued that alternating current was dangerous and potentially fatal to consumers. Edison forces even conspired to place a Westinghouse AC generator in New York's Auburn State Prison to power its electric chair: after the first prisoner was electrocuted in 1890, they gleefully campaigned against "the executioner's current."

Fan, built in 1892, ran on Tesla's alternating-current induction motor. But Westinghouse proved a formidable ally for Tesla. His engineers modified the

Tesla system for practical use, and bit by bit it gained adherents. Edison's outrageous slanders notwithstanding, AC was perfectly safe, if properly handled, and finally made long-distance transmission of power economically feasible. In 1892, Westinghouse underbid Edison and won the right to provide the power system to light the 1893 World's Columbian Exposition in Chicago. Adoption of alternating current for the 1896 Niagara Falls hydroelectric project (p. 128), first of its kind in the country, signaled America's inevitable conversion to the Tesla system.

Yet Tesla's victory bad a bittersweet economic outcome. Early on, when the issue was still in doubt and the Westinghouse Company found itself in severe financial difficulty, George Westinghouse faced intense pressure from bankers to revoke the lucrative royalty agreement with Tesla. Tesla, never especially shrewd in financial matters, agreed to accept a $216,000 cash settlement. It was a tragic miscalculation, for at the stroke of a pen he surrendered a fortune — estimated as high as $12 million in only four years. After one decade of glory, his scientific work would forever suffer for lack of cash.

Poverty was hardly a concern to Tesla in the 1890s. He had become such a celebrity that at the 1893 Fair he starred in his own exhibition. He cut a dashing figure as he strode onstage in white tie and tails. Before a gasping crowd, he calmly stood in his cork-bottomed shoes as several hundred thousand volts of electricity from it nearby Tesla coil coursed over his body. He would pick up light bulbs or long glass tubes. Unconnected to any wire or any source of power, the bulbs bloomed with light in his hand, the glass tubes flared into luminous swords. Sometimes he even struck a Statue of Liberty pose to heighten the effect.

He lived at the Waldorf-Astoria and had a regular table at Delmonico's. Even his

love life — or, more accurately, his resolute bachelorhood — became the subject of comment. The *Electrical Review* observed in 1896, "We are certain that science in general, and Mr. Tesla in particular, will be all the richer when he gets married." Tesla, the celibate loner, revealed his own views around 1897 when asked if he believed in marriage for people of "artistic temperament."

"For an artist, yes; for a musician, yes; for a writer, yes; but for an inventor, no," he replied. "The first three must gain inspiration from a woman's influence and be led by their love to finer achievement, but ... I do not think you can name many great inventions that have been made by married men. It's a pity, too, for sometimes we feel so lonely."

First great application of Tesla's alternating-current system was with 5,000-horsepower generators of Niagara Falls Power Company. His strange phobias were never cited, of course. Anne Morgan, the handsome daughter of financier J. Pierpont Morgan, reportedly fell in love with Tesla; she made the unforgivable mistake, however, of wearing pearl earrings at their first meeting.

If your definition of radio is the wireless communication of signals from a transmitter to a receiver, then Tesla demonstrated something very much like this in 1893. Not only did he propose the technical means of transmitting signals with special high-frequency transformers he'd invented in 1890 known as "Tesla coils"; he later had the vision to predict their ultimate application. "A cheap and simple receiving device," he wrote in 1904, "which might be carried in one's pocket may then be set up anywhere on sea or land, and it will record the world's news or such special messages as may be intended for it."

Tesla's concept of "wireless intelligence" was, as usual, part of a larger technical cosmology. He believed either radio signals or pure electrical power could be transmitted by electromagnetic vibrations throughout Earth. He devoted two years to a variety of experiments until his work was tragically interrupted in March 1895. The building housing his laboratory burned to the ground. Tesla lost everything — machinery, equipment, papers, mementos.

It took months to find a new lab, on East Houston Street, and another two years to get it outfitted. This delay may be considered something like a gestation period for one of science's great controversies. Later in 1895 Guglielmo Marconi (SMITHSONIAN, March 1982) first demonstrated the wireless transmission of signals in Italy. In June 1897 Tesla had accomplished much the same thing: he had broadcast signals from his Houston Street laboratory to a boat 20 miles up the Hudson River. Patents were filed by Tesla the following September.

It is Marconi and not Tesla, of course, who is generally accorded the honor of having invented the radio, although Tesla sarcastically remarked on the very day in 1901 that Marconi sent his famous "S" signal across the Atlantic, "Let him continue. He is using 17 of my patents." Tesla devotees argue that, beginning with the invention of the Tesla coil in 1890 and culminating with the 1893 lectures, Tesla had briefly outlined basic principles of wireless transmission. They are not alone. Haraden Pratt, a respected historian of electrical science, has observed that Tesla, though mistaken for a dreamer by his contemporaries, "stands out as not only a great inventor but, particularly in the field of radio, as the great teacher."

Even today the debate rages on. In one of those selections that does not so much resolve controversy as preserve it forever in amber, the Nobel committee awarded a shared prize in physics to Marconi and German scientist Karl Braun in 1909. It wasn't until 1943, on the other hand, nearly six months after Tesla's death, that the U.S. Supreme Court concluded a long-running patent dispute by ruling that three turn-of-the-century inventors — Sir Oliver Lodge, John Stone Stone and Tesla — appeared to have priority over Marconi in radio-tuning circuits. Tesla was usually gracious about credit, but he let down his guard in a 1927 conversation with a fellow Yugoslav. "Mr. Marconi," he opined, "is a donkey."

Many would agree that Tesla was much the more gifted inventor. During the incredibly fertile decade between 1890 and 1900, his inspired tinkering anticipated a rich catalog of modern devices. He apparently conceived of something like fluorescent lighting, "black light" and neon. In experimenting with his so-called "button lamps," Tesla predicted the existence of cosmic rays; Nobel laureates Robert Millikan and Arthur Compton both acknowledged Tesla's early speculations in this field. In his work and writings, one sees the germ of such seminal technical developments as the cyclotron and the electron microscope. In a tour de force of electrical engineering that opened at Madison Square Garden in 1898, Tesla guided the movements of a scale-model boat through a large tank of water by remote-control radio waves.

Small oscillation transformer stepped up voltage and converted DC to AC. And yet, in a reversal of fortunes perceptible only in retrospect, Tesla's career soon began an irreversible decline. Perhaps as a backlash from the academic community (it is said that he never submitted a single article to an academic publication), perhaps because there was a professional jealousy about his public grandstanding, Tesla found credit harder to come by, credibility harder to maintain. No one rushed to support his studies anymore. His mind remained fertile, but then as now, science is an expensive undertaking; those who can't afford to do experiments cannot reduce theoretical ruminations to practical application.

In 1899, Tesla set up a laboratory in Colorado Springs and essentially built a barn around, and flagpole above, the world's largest Tesla coil, 52 feet in diameter. In the course of experiments, he sent bolts of man-made lightning soaring some 135 feet into the air, created thunder heard 15 miles away. He was also among the first to invent the blackout; the power demands of one experiment burned out Colorado Springs' electric plant, plunging the town into darkness. Tesla believed he had demonstrated the wireless transmission of power, but the experiments have never been considered an unqualified success.

There was one last hurrah for this ambitious concept. Tesla had grand plans to construct a huge radio transmitter on Long Island called, with romantic panache, Wardenclyffe. He got J. P. Morgan to fund it and noted architect Stanford White to design the building. Then a string of scientific and psychological setbacks ensued. Morgan withdrew support and in 1906 work ceased. In 1917, the wooden tower, by then rumored to be a hideout for German spies, was blown up.

Huge transmitting tower was part of an unsuccessful worldwide wireless system. Dangerously close to a parody of the lone-wolf inventor, Tesla dunned former patrons for research support. This time the Westinghouses and the Morgans had no interest. The inventor who once exuded charm and suavity now seemed a pest. His scientific ideas, so often visionary, began to take on a reactionary cast. The emerging age of quantum physics passed him by, and Tesla spent many ill-used hours trying to disprove Einstein's theory of relativity.

His contributions did not go totally neglected, however. In 1916, the American Institute of Electrical Engineers, the premier professional organization in the field, voted to honor Tesla with its prestigious Edison Medal. Tesla at first balked, and his bitterness spilled out. "You would bestow an outward semblance of honoring me," he wrote the organizers, "but you would decorate my body and continue to let starve, for failure to supply recognition, my mind and its creative products which have supplied the foundation upon which the major portion of your Institute exists."

Ultimately Tesla relented, although his sensitivity was not entirely misplaced. H. Otis Pond, an engineer who worked with both Edison and Tesla, in a way seconded Tesla's discontent. Edison, he believed, was "the greatest experimenter and researcher this country has produced — but I wouldn't rate him as much of an originator"; Tesla, he

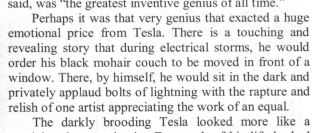

said, was "the greatest inventive genius of all time."

Perhaps it was that very genius that exacted a huge emotional price from Tesla. There is a touching and revealing story that during electrical storms, he would order his black mohair couch to be moved in front of a window. There, by himself, he would sit in the dark and privately applaud bolts of lightning with the rapture and relish of one artist appreciating the work of an equal.

The darkly brooding Tesla looked more like a magician than a scientist. For much of his life he had been ahead of his time; by 1929, with the Great Depression beginning to unfold, he finally was of his time — his major contributions behind him and able to earn little as a consultant. His personal eccentricities now inspired pity. He was a familiar figure at Bryant Park, behind the main branch of the New York Public Library, where he fed the pigeons, whom he called "my sincere friends." He brought ailing birds back to his hotel room for convalescence and on at least one occasion was asked to leave a hotel because of the avian infirmaries he maintained.

Tesla fully expected to live 125 years, and had planned all along to set his life story and scientific principles on paper one day. But by 1943 he was suffering from heart trouble and fainting spells. Weak and apparently confused, he dispatched a messenger from his room in the Hotel New Yorker to carry a loan of $100 to his friend Mark Twain, dead 33 years. He was furious when told the envelope could not be delivered. On January 4 he complained of chest pains during an experiment and returned to his room. The last person to see him alive was the hotel maid on January 5. Alone, as always without a home, literally and figuratively heartsick, Tesla fended off human aid and comfort to the very end, this time with a "Do Not Disturb" sign. His body was discovered on January 8. He was 86.

More than 2,000 people attended the funeral at the Cathedral of St. John the Divine in Manhattan. A testimonial arrived from President Roosevelt; the Mayor of New York, Fiorello La Guardia, read a eulogy on the radio; Nobel laureates sent accolades. Perhaps the most fitting epitaph was something Tesla himself had said, apropos of the radio controversy, many years earlier. "Let the future tell the truth and evaluate each one according to his work and accomplishments," he told a friend. "The present is theirs, the future, for which I really worked, is mine."

Modern-day experimenters like this one by researcher Robert Golka, studying high-voltage discharge and ball lightning, use Tesla-style devices.

5 NIKOLA TESLA: SCIENTIFIC SAVANT

James P. Rybak, <u>Popular Electronics</u>, Nov., 1999, p. 40 & https://teslauniverse.com

Calling Tesla merely "an inventor" would be like referring to Frederic Chopin as simply "a piano player."

What would cause a person to refuse a Nobel Prize? A 1915 Reuters dispatch from London stated, albeit unofficially, that Nikola Tesla and Thomas Edison had been chosen to share that year's Nobel Prize in physics. Numerous magazines and newspapers throughout the world published this report as fact. However, the awards were never made either to Tesla or to Edison. The complete story is not known, but many believe that Nicola (sic) Tesla may have refused to accept the award.

Tesla was very much in need of the $20,000, which would have been his half of the cash award accompanying the Nobel Prize. His work had resulted in the creation of fortunes for many others; but he, himself, lived much of his later life in near poverty and he would die penniless. If Tesla did decline the Nobel Prize, it likely was a matter of principle that precipitated his decision to refuse this prestigious honor. From Tesla's perspective, Edison was merely an "inventor" who devised useful applications of science. Tesla, meanwhile, considered himself a "discoverer" of new scientific principles and only, incidentally, an inventor. In Tesla's mind, the importance of the discoverer far outweighed that of the inventor.

Others believe that it was Edison who refused the award. Perhaps he was still angry that Tesla had quit working for the Edison Company and had aligned himself with Edison's arch competitor, Westinghouse.

Two years later, Tesla initially refused to accept the Edison Medal that the American Institute of Electrical Engineers (AIEE) planned to award him for his outstanding work in the development of alternating current theory and applications. Perhaps Tesla was miffed because it had taken the AIEE almost thirty years to recognize the significance of his work. Perhaps, too, Tesla was insulted to be given an award named after the person who had so strongly opposed the adoption of alternating current power distribution systems and who, Tesla believed, had reneged on a promise to pay him a large amount of money for solving important technical problems. In any event, it was clear that awards meant nothing to Tesla. It took an AIEE official several visits together with much coaxing and pleading before he could get Tesla to agree reluctantly to accept the Edison Medal. Tesla then almost failed to appear at the award ceremony.

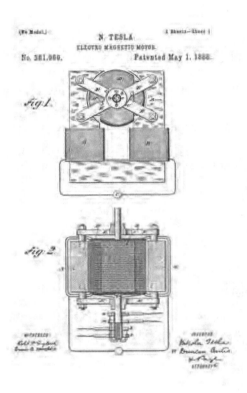

N. TESLA
ELECTRO MAGNETIC MOTOR.
No. 381,968.   Patented May 1, 1888.

Fig.1.

Fig.2.

Who was this man, this pioneer whom many obviously have admired for the past century? Join us for a look at this savant's life and his accomplishments. As you'll see, forgetting Tesla is unforgivable for electronics activists. We owe him much.

Tesla's second AC patent, granted in 1888, was for a then-groundbreaking induction motor.

Early Days. Nikola Tesla was born in 1856 to Serbian parents who lived in a Croatian village in the southern part of the Austro-Hungarian Empire. His father had abandoned a military career to become a priest in the Serbian Orthodox Church. Although Nikola's mother had received no formal education, she was bright and had an exceptional memory. Tesla always credited his mother as the source of his intellectual abilities.

Mathematics was Nikola Tesla's favorite subject in school and the one in which he truly excelled. When given a problem to solve, he needed neither a blackboard nor a sheet of paper. Tesla had the extraordinary ability to visualize in his mind all the steps needed to solve the problem, just as though he had written them down. The ability to visualize mathematical problems and engineering designs clearly in his mind was of great value to Tesla throughout his life.

As a child, Nikola loved to read and retained knowledge extremely well. He also learned several foreign languages. This enabled him to read far more than what was written in his native Serbo-Croatian. Young Tesla loved to devise complex mechanical devices in his mind and then build them from whatever materials were at hand.

While in what we would call high school and the first years of college, Tesla studied so intensely that his health was seriously affected. His father feared that engineering, which required many years of intense study and to which young Nikola aspired, would further jeopardize his son's well-being. The elder Tesla urged the boy to enter the ministry because that profession required a less demanding program of study. Periodic episodes of severe illness due to overwork would plague Tesla throughout his life.

Although further weakened when he contracted malaria, Nikola successfully completed the four-year Higher Real Gymnasium (college) program in Croatia in only three years. It was during this time that Tesla become unalterably determined that he wanted to devote his life to electrical experimentation.

Upon his return home, Tesla's parents were alarmed by their son's demanding career choice and by the unrelenting pace at which he continued to pursue his studies. His health was still very much at risk, and they again insisted that he become a priest, not an engineer. Very quickly, Nikola experienced the additional discouragement of contacting cholera and receiving his draft notice for army service.

Tesla now became despondent and was almost at death's door. He knew that if he survived the cholera, he would have to serve time in the army and then study for the priesthood. He felt he had nothing to live for. Upon realizing this, Nikola's father relented and gave permission for his son to study electrical engineering.

Tesla slowly began to regain his health. His father then sent him off for a year's rest to further recover his health. During this time, the elder Tesla used the influence of relatives to get his son's military obligation cancelled.

Committed to Developing AC. In 1875, Tesla went to the city of Gratz, Austria to study electrical engineering. There he still continued to overwork himself and again jeopardized his health. It was at this time that Tesla realized the inherent limitations of DC motors and generators due to the sparking associated with commutator action (the switching of current polarity in a motor to keep the armature coil moving). This discovery clearly convinced Tesla of the need to develop alternating current motors and generators that would not need commutators.

Developing the details of how this goal could be accomplished occupied much of Tesla's time for the next several years. He rejected the claim of his professor who taught the courses on motors and generators that the development of AC motors and generators was an "impossible idea."

It was "instinct" that told Tesla his professor was wrong. Tesla's instincts were almost always correct when it came to solving scientific problems. Nonetheless, progress toward his goal did not come quickly.

Following some additional engineering study in Prague, Tesla went to Budapest in 1881 where a family friend had offered him a job at the new telephone central station that was being started. Tesla's design, computational, and estimating abilities soon attracted the attention of his supervisors. When the telephone exchange was completed, Tesla was placed in charge of its operation. Once again, Tesla worked excessively and his health rapidly declined. Exhaustion soon forced him to quit his job.

The Key AC Concept. In February of 1882, shortly after recovering his health, the solution to the alternating current problem came to Tesla. He now could visualize clearly in his mind how he would use alternating currents to create a rotating magnetic field. This was the key concept needed to produce a practical AC motor.

Tesla created the rotating magnetic field by using two circuits in which the currents were out of phase with each other. Others had tried to develop AC motors using only one circuit, but their approach could not produce continuous rotation of the motor. Nikola Tesla's two-phase system successfully eliminated the need for a commutator.

The work Tesla had begun was far from completed, however. He now developed designs for dynamos (generators), motors, transformers, and the other devices needed for alternating-current power systems. Tesla extended his rotating magnetic field idea to include currents of three, four, and six different phases. Nikola Tesla had developed in his mind a true polyphase power system. He also believed that he could even build a successful single-phase AC motor.

The telephone company in Budapest where Tesla had worked prior to his illness and to which he had hoped to return was sold. The same family friend who had helped Nikola get his job in Budapest now helped him obtain a job in Paris with the Continental Edison Company, which was licensed to make DC motors, generators, and lighting equipment under Edison's patents.

Tesla tried to interest every likely person he could find in Paris to help him develop his polyphase AC system. He did not have to worry about people stealing his ideas as no one showed any real interest in them at all.

Tesla was assigned to a special project in Germany. Here he used his spare time to build a two-phase generator and a two-phase motor. Tesla did the close tolerance machining work himself. There were no working drawings on paper. Tesla had all the details clearly fixed in his mind. When he first tested the AC machines in 1883, they functioned extremely well. His theory was correct.

Upon returning to Paris after successfully completing the project in Germany, Tesla became dissatisfied with his immediate supervisors. He had been promised a generous bonus for his successful work in Germany but that bonus was never given. Indignant, Tesla decided not to show the Company officials his two-phase system in operation and resigned his position.

Tesla Meets Edison. Continental Edison's manager, Charles Batchelor, was an associate and good friend of Thomas A. Edison. Batchelor was impressed with Tesla and urged him to go to the United States to work directly with Edison.

Tesla welcomed this suggestion together with a letter of introduction to Edison that Batchelor is said to have written for him. Batchelor's letter to Edison reportedly stated "I know two great men and you are one of them; the other is this young man." Tesla sold his personal possessions to pay for the train and ship tickets he needed and left for New York in 1884.

Because his wallet and extra clothes were lost during his travels, Tesla arrived in the United States with little more than the four cents in his pocket and the clothes on his back. Fortunately, Tesla had a friend in New York with whom he could stay temporarily.

While Tesla was very favorably impressed with Edison when the two men first met, the reverse was not true. Edison had very little formal education and did his inventing by trial and error experimentation, whereas Tesla solved all his technical problems mentally and did virtually no experimentation. Perhaps an even greater barrier was that Edison was an unshakable proponent of DC power systems and was strongly opposed to the development of AC systems. Tesla was firmly convinced of the superiority of AC. Despite these fundamental differences, Edison gave Tesla a job, likely on the basis of Batchelor's recommendation.

Edison quickly saw that Tesla consistently put in long hours and made many valuable contributions. When Tesla suggested that he could improve the efficiency and lower the operating cost of the DC dynamos the Edison firm manufactured, the plant manager reportedly told him "There's fifty thousand dollars in it for you if you can do it."

During the following months, Tesla designed twenty-four new types of DC dynamos. He replaced the previously used long field magnets with more efficient shorter ones and added some important automatic controls. The machines performed as Tesla had promised and the Edison firm took out numerous new patents.

In the spring of 1885, when Tesla asked for the fifty thousand dollars he believed he had been promised and had earned, Edison's reply was "Tesla, you don't understand our American humor." Furious because he received not an extra dime beyond his $18 per week salary for all the successes he had produced, Tesla immediately quit his job with Edison.

The Tesla Electric Company. Tesla now was unable to find an engineering job and was forced to work as a laborer. In early 1887, Tesla's abilities and stories about his AC developments attracted the attention of the foreman of the labor crew on which Tesla worked. The foreman also was working far below his own level of training and was sympathetic to Tesla's situation. He introduced Tesla to A. K. Brown of the Western Union Telegraph Company. In April of 1887, Brown and a friend provided the money to create the "Tesla Electric Company." By coincidence, Tesla's new laboratory was located within sight of Edison's facility.

Quickly, Tesla built not only the two-phase AC generator and the induction motor he had built in Europe but also the other machines he had designed in his mind while in Budapest. While concentrating on single-phase, two-phase, and three-phase systems, he also experimented with four- and six-phase devices. Tesla also developed the mathematical theory needed to explain the operation of his AC systems so that others could and would both understand and accept his work.

After having proven that his AC systems were practical, Tesla applied for a number of fundamental patents. These were granted to him in 1888. Word of Tesla's accomplishments and genius spread quickly. On May 16, 1888, Tesla was invited to present a lecture entitled "A New System of Alternate Current Motors and Transformers" at an AIEE meeting in New York. He now was a recognized and accepted member of the electrical engineering "establishment."

George Westinghouse was a farsighted individual who already had made a fortune in Pittsburgh manufacturing his air brake for trains as well as a variety of electrical devices he had invented. He recognized the major advantages which AC power systems held over DC and he saw huge commercial potential in the work Tesla had done.

The DC electrical systems favored by Edison could not be used to distribute power farther than about one-half mile from the generator due to the excessive voltage drops that resulted from the resistance of the power lines and the large currents that flowed through the lines. AC voltages, however, are stepped-up at the generator using transformers, thereby reducing both the current and the transmission losses. The result is a substantially increased distribution range. Transformers then convert the AC voltages to safe levels at the point where the power is utilized.

An Alliance with Westinghouse. Shortly after Tesla gave his AIEE lecture, he was contacted by Westinghouse who wished to see the AC equipment in person. The two men had many interests in common and immediately formed a good relationship. Westinghouse quickly offered Tesla one million dollars for his AC patents. He also invited Tesla to come to Pittsburgh for a year at a "high" salary as a consultant. Tesla quickly agreed. Half of the one million dollars went to A. K. Brown and his partner who had financed Tesla's work but Tesla still was now rich beyond his wildest dreams.

Problems developed when Westinghouse's engineers tried to use Tesla's designs to produce small, single-phase motors. In addition, the priorities and urgencies associated with manufacturing AC power systems for sale were different from Tesla's research priorities. Furthermore, Tesla was adamant that his AC machines worked most efficiently at a frequency of 60 Hz (then "cycles per second") while Westinghouse's engineers had been used to working with frequencies of 133 Hz.

Dissatisfied with working for others, Tesla returned to his New York laboratory. He was now independently wealthy and wanted to return to his research. He rejected a very lucrative offer by Westinghouse to remain in Pittsburgh on a permanent basis. Soon after leaving Pittsburgh, Tesla was granted U.S. citizenship.

High-Frequency AC. Aware that the electromagnetic spectrum extends all the way up to visible light and beyond, Tesla now investigated the behavior of his circuits at higher frequencies. Part of his work would result in transformers, which we, today, call "Tesla coils." Another part of this work would result in tuned circuits.

While developing his mathematical AC circuit theory, Tesla became aware of the roles played by inductance and capacitance in producing electrical resonance. He found that he could produce extremely high voltages with frequencies measured in tens or hundreds of kHz by adding the appropriate amount of capacitance to the primary of an air core transformer. (While iron cores make 60-Hz transformers perform well, they severely degrade transformer performance at high frequencies.) A spark gap discharge connected to the transformer's primary winding resulted in on oscillator which produced high-frequency, high-voltage discharges.

As he predicted by theory and confirmed by experiment, Tesla quickly established that high-frequency AC current flows along the surface of the human body rather than through it. Thus, no electrical shock is felt. As early as 1890, he recognized the therapeutic value high-frequency electric fields could produce in the human body. The effect became known as "diathermy."

Tesla gave his first public lecture and demonstration concerning his high-frequency work to the AIEE in May of 1891. In addition to producing long electrical sparks from his fingertips, Tesla created electrical sheets of flame and caused sealed tubes of gas (Geissler tubes) to glow even though there was no direct electrical connection to the tubes. This spectacular demonstration, coupled with his AIEE lecture on polyphase AC power systems three years earlier, established Tesla as a premier scientist and engineer.

Tesla Demonstrates Wireless. At the Spring 1893 meeting of the National Electric Light Association in St. Louis, Tesla gave his first public demonstration of the wireless transmission of electrical energy and, thereby, the feasibility of wireless communication. On one side of the stage, Tesla had a tuned circuit consisting of a bank of Leyden jar capacitors and a coil. The tuned circuit was connected to a spark gap and a 5-kVA power-distribution transformer. A vertical wire (antenna) extended from the coil to the ceiling. This arrangement formed his "transmitter."

On the other side of the stage Tesla had his "receiver," which consisted of another, identical tuned circuit with a vertical wire extending to the ceiling. A gas-filled Geissler tube was connected to this tuned circuit in place of the spark gap used with the transmitter.

No wires connected the transmitter and receiver. When Tesla applied power to the transmitter, the Geissler tube in the receiver glowed brightly. This demonstration occurred two years before Marconi went to London with his wireless telegraphy equipment. Soon, Tesla was routinely causing gas-filled tubes to light in a manner that predicted the development much later by others of neon signs and fluorescent lamps.

Ar this same time, a grand example of the value of Tesla's polyphase AC system was being undertaken. The concept of harnessing the energy of Niagara Falls and using it to generate electricity had been discussed for some time. Now the technology existed to achieve this goal. If the power of Niagara were to be used to generate DC, the area over which this potentially huge amount of electricity could be distributed would be very small. Even Buffalo, only 22 miles away, could not be served if DC were generated.

Both the Westinghouse Electric Company and the General Electric Company (successor to the Edison General Electric Company) submitted proposals in 1893 to install a Tesla polyphase system. GE, now a firm believer in AC since Edison no longer controlled the restructured company, had obtained a license to use Westinghouse's Tesla patents.

Westinghouse won the contract for the generating plant at Niagara while GE was chosen to build the transmission line to, as well as the distribution system within, Buffalo.

The plant was delivering power in 1895, and the transmission line was completed the following year. Tesla's stature as a technological hero was reinforced once again.

From 1891 until 1893, Tesla lived the life of a celebrity. He was in constant demand at scientific and high society gatherings both in the U.S. and abroad.

Lectures and spectacular demonstrations were given in both London and Paris. Now Europe, too, fully appreciated the magnitude of Tesla's accomplishments. Tesla then abandoned the active celebrity life, because it kept him from the research he loved. However, he had become attracted to the trappings of affluence and would endeavor to maintain that image for the rest of his life, even when he clearly could not afford to do so.

The previous successes Tesla had achieved in making sealed tubes of gas glow when in the vicinity of his high-frequency, high-voltage transformers demonstrated that wireless transmission of electrical energy over short distances was possible. Now Tesla wanted to develop that concept further — much further. He envisioned transmitting energy without wires, not only for communicating, but also for powering lights and motors around the world.

During the winter of 1894-95, Tesla built a transmitter at his laboratory together with a portable receiving station to test his latest plan. Successful wireless transmission was achieved over short distances. Then tragedy struck. Just as he was preparing to make the first public demonstration of his wireless transmission system, fire completely destroyed Tesla's laboratory together with all his equipment and records. Tesla was devastated. Virtually all his money had been invested in his work. Nothing had been insured.

With funds personally provided by the man who had organized the Niagara power plant project, Tesla painstakingly reconstructed his laboratory. He resumed the wireless transmission tests with his transmitter and portable receiver in the spring of 1897. The receiver was operated on a boot traveling up the Hudson River, successfully demonstrating the feasibility of wireless transmission at distances of 25 miles. Tesla's two fundamental wireless patents (<u>645,576</u> and <u>649,621</u>) were issued in September of 1897. In 1943, the U.S. Supreme Court would rule that this work of

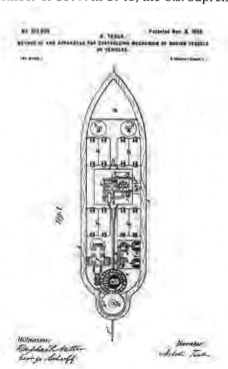

Tesla's, together with related, independent achievements by Oliver Lodge and John Stone, anticipated Marconi's work. As a result, Marconi's important 1904 wireless patent was declared invalid.

In 1898 Tesla demonstrated the world's first radio-controlled boat, and received another patent for a device that would survive him in some form for decades.

First RC Boat. A year later, in September of 1898, Tesla startled visitors to the Electrical Exhibition at New York's Madison Square Garden by demonstrating the world's first radio-controlled boat using what he called his "mind-powered" or "Teleautomatic" system. Tesla remotely controlled a 3-foot long iron-clad boat through a variety of maneuvers in front of large audiences every night for a week. To demonstrate its simplicity of operation, Tesla permitted volunteers from the audience to operate the controls. Patent number 613,809 was awarded to Tesla for this invention. His goal was to sell a similar remotely operated submarine to the U.S. Navy for use in the Spanish-American War. Tesla hated war and felt his invention could save lives. The Navy was not interested.

Tesla had long been the beneficiary of good press coverage concerning his numerous previous inventions. Now he attempted to enlist assistance of the press to create public support with the hope of pressuring the Navy into using his invention. Tesla's written announcement concerning his "mind-powered" submarine together with his responses at a press conference were too fantastic, even for the press of that day which normally thrived on sensationalism. As a result, Tesla found himself

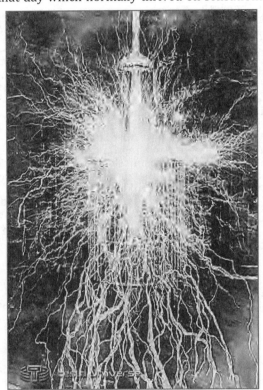

criticized in print by some of the members of the press for, what seemed to them as, his exaggerated claims and blatant attempts at headline-seeking. Nonetheless, the journalists still found Tesla's activities to be of great interest to the public. However, as Tesla continued to announce what seemed to be ever more fantastic plans, the enthusiasm of the press became tempered with skepticism.

Tesla was anxious to proceed with his planned project to beat Marconi in establishing a worldwide wireless communication system, as well as one for the global distribution of electrical power. The problem, again, was money until a wealthy friend loaned him $10,000. Now Tesla built a high-frequency oscillator, which generated 4 million volts, but the sparks produced were too large and violent for his New York City laboratory. More space was needed.

Shown here is the breathtaking discharge from Tesla's "magnifying transmitter" in Colorado Springs.

Off to Colorado. Leonard E. Curtis, a former Westinghouse lawyer who now was associated with the Colorado Springs Electric Company, invited Tesla to move his laboratory to Colorado. Curtis promised Tesla the use of land east of Colorado Springs as well as all the electricity he needed, both free of charge. John Jacob Astor, owner of the Waldorf-Astoria Hotel where Tesla now lived and had dined for years, provided the $30,000 needed to make the move and set up a laboratory.

Tesla arrived in Colorado Springs in May of 1899. Within three months, he built a laboratory complete with a tower and mast topped by a 3-foot copper sphere reaching 200 feet into the sky. A giant high-frequency oscillator, which Tesla called his "magnifying transmitter," also was readied. This magnifying transmitter incorporated a resonant transformer designed to electrically excite the earth and was optimized for maximum wireless transmission of energy. Tesla had carloads of laboratory equipment together with several assistants sent to him from New York.

Using a receiver connected to the earth to monitor the effects of the large number of lightning discharges that occurred in the region daily during the summer, Tesla reached a dramatic conclusion. He now was sure that the earth was filled with fluid electrical charges. Tesla believed that when this electricity is disturbed by repeated electrical discharges occurring at the proper time interval, resonant low frequency electrical waves of tremendous magnitude are produced.

Tesla had produced similar resonance effects in his electrical circuits. He reasoned that he could cause resonant waves in the earth with his high-voltage discharges. Tesla also believed that these waves would provide large amounts of electrical energy that could be tapped throughout the world.

Tesla's nighttime initial test of his new magnifying transmitter went well. Lightning bolts 135 feet in length surged from the top of the mast, and the resulting thunder crashes were heard 15 miles away. Then came silence and darkness.

At first, Tesla thought that his assistant had turned off the power. Finding that not to be the case, he telephoned the power company to demand that his power be restored. The curt reply from the power company was that his experiment had destroyed their generator. All of Colorado Springs was in darkness. A standby generator soon restored power to the city, but Tesla was told that his power would be restored only if and when he repaired their damaged generator.

One evening when the monitoring receiver was connected to the earth to listen for distant thunderstorms, Tesla heard three pulses in quick succession. He

knew that these sounds were not characteristic of thunderstorms and declared that they must be of extraterrestrial origin. Later he concluded that the signals had not come from just any planet but that they had come from Mars!

On January 7, 1900, Tesla left Colorado Springs for New York. He had spent $100,000 in eight months and now was out of money. He intended to return to Colorado to conduct additional experiments once his finances were in better order, but this plan was never realized.

During his time in Colorado, Tesla performed many interesting experiments and learned much from them. However, there is no evidence that he succeeded in transmitting any significant amount of power over long distances without wires.

When he reached New York, Tesla was ridiculed by reporters for his claim of having heard extraterrestrial signals. Shortly afterward, he wrote a seemingly fantastic, metaphysical magazine article entitled "The Problem of Increasing Human Energy," which did little to help his believability in the minds of most people. His own extravagant claims and predictions again were eroding his credibility. Not every prediction in this article was preposterous, however. One prediction described the "radar" systems that were not developed by others until almost 40 years later.

During this same time, Tesla filed for and was granted several patents involving the use of cryogenic techniques for the underground transmission of high voltages. These anticipated similar developments that would later take place in the 1970s, both in the U.S. and abroad.

The Wardenclyffe Project. Tesla needed badly to obtain new financing for what would be his most ambitious project: a giant tower and laboratory with which he planned to establish worldwide wireless communication. There he also expected to refine his plans for wireless electrical power distribution. Neither Westinghouse nor Astor was willing to loan Tesla the money he needed. J. Pierpont Morgan, however, did provide $150,000 to build the tower and other needed facilities at Wardenclyffe, Long Island in exchange for control of some patents Tesla still had.

No doubt Tesla's most ambitious project, the tower at Wardenclyffe was designed to establish worldwide communication — no small feat for the dawn of the 20th century. Unfortunately, Tesla lost the tower to creditors in 1915.

Although Tesla wanted something taller, the finances available limited him to the construction of a tower 187 feet high with a hemispherical dome 68 feet in diameter. The Wardenclyffe project was still under construction in December of 1901 when Marconi succeeded in sending wireless telegraph signals across the Atlantic using much simpler equipment and facilities than what Tesla was proposing. Tesla contended that Marconi's equipment violated many of his (Tesla's) patents. Nonetheless, Tesla's plan was looking more and more extravagant.

Rapidly rising prices together with on overly ambitious design made it impossible for Tesla's Wardenclyffe project to be completed as planned. Creditors constantly hounded him and, despite Tesla's best efforts, additional financing could not be found. Negative rumors concerning the status of Tesla's remaining patents together with growing skepticism concerning his fantastic predictions mode people wary. Tesla become despondent.

By 1906, virtually all construction at the Wardenclyffe site had stopped due to Tesla's inability to pay his bills. The high-voltage oscillator had been completed, but lack of funds made it difficult for Tesla to test it. When he did, however, people throughout Long Island and as far away as Connecticut could observe the bright flashes in the nighttime sky. However, no wireless transmission of messages or electrical power ever occurred from Wardenclyffe.

In 1915, Tesla finally lost Wardenclyffe to creditors. The tower was dynamited for its scrap value in 1917. Despite this huge setback, Tesla never gave up his ideas concerning wireless power transmission and broadcasting. Coincidentally, the years 1915 and 1917 also were the years of Tesla's alleged selection for the Nobel Prize and his receipt of the Edison Award, respectively.

The US Supreme Court in 1943 declared Marconi's important wireless patent of 1904 invalid due, in part, to this earlier patent of Tesla's.

Down, Not Out. While Tesla was dejected due to his inability to complete the Wardenclyffe project, his mind still produced many far-sighted ideas. He designed a VTOL (vertical takeoff and landing) aircraft during 1907-08. A turbine placed at the center of the aircraft had a propeller mounted above, as in a helicopter, for takeoffs and landings. Once airborne, the pilot operated a lever that moved the propeller to the front of the craft, as in a conventional airplane. Tesla did not build a prototype of this VTOL but was awarded patents on its design. In 1908, Tesla publicly described the limitations of propeller driven airplanes and predicted the development of jet aircraft.

Tesla filed for patents in 1909 on a powerful and lightweight "bladeless turbine," which appeared to have the potential to revolutionize the design of prime movers in terms of horsepower produced per pound of weight. Tesla's turbine consisted of a series of horizontally stacked, closely spaced disks attached to a shaft and enclosed in a sealed chamber. A fluid (liquid or gas) under pressure entered the sealed chamber at the periphery of the disks. Viscosity caused the disks to rotate as the fluid moved in circular paths toward the shaft where it exited the turbine.

Successful small models of the turbine were built, but the inadequate materials then available together with Tesla's serious financial problems prevented his development of larger versions. Several firms paid for the rights to refine Tesla's turbine design, but their efforts were largely unsuccessful. The two patents Tesla was awarded on his turbine design in 1909 are still being studied today by engineers trying to develop this far-sighted design.

The Gernsback Connection. Shortly before the Wardenclyffe tower was demolished, a man named Hugo Gernsback renewed on old acquaintanceship with Tesla. Gernsback was editor of the magazine *The Electrical Experimenter* that had evolved from his earlier publication, *Modern Electrics*. Both were similar in many respects to the two magazines, *Popular Electronics* and *Electronics Now*, currently published monthly by Gernsback Publications, Inc. During his lifetime, Gernsback would publish a variety of magazines devoted to electrical technology and related topics.

As a youngster in Luxembourg, Hugo Gernsback had first heard of Tesla and had become fascinated by his accomplishments. Fixed in Gernsback's mind was the photograph of Tesla he had seen which showed high frequency arcs of current passing through the electrical inventor's body. His admiration for Tesla would continue lifelong. Gernsback immigrated to the U.S. in 1903, at the age of 19, after having studied electronics in Europe. The two met briefly in 1908 but Gernsback had followed the press reports of Tesla's activities.

A legitimate scientist and electrical inventor in his own right, Hugo Gernsback was awarded 37 patents during his life. Most people, however, recognize him as the "Father of Modern Science Fiction." Gernsback authored a number of science fiction (or "scientifiction" as he initially called them) stories. He is better known, however, for publishing the science fiction works of many other popular authors in the various magazines he headed between 1910 and his death in 1967.

In 1916, Gernsback asked Tesla to edit a major article on the magnifying transmitter and the Wardenclyffe project. The article was published in the March 1916 issue of *The Electrical Experimenter*. Tesla needed the modest amount of money Gernsback paid him for this work. Gernsback, in turn, was pleased to publish this article about a device and project which, if Tesla had been successful, would have turned many science fiction predictions into reality.

In 1919, Tesla wrote a 6-part series entitled "My Inventions," which Gernsback also published in *The Electrical Experimenter*. Articles by or about Tesla still fascinated Gernsback's readers.

In 1983, the United States honored Tesla with a postage stamp.

Tesla continued to spawn new ideas in his mind. However, as time went on, more and more of these ideas seemed in the realm of science fiction and some appeared even to violate the known laws of nature. Several of his more fantasy-like ideas included a machine for capturing and utilizing the energy of cosmic rays, a technique for communicating with other planets, and a particle-beam weapon for destroying a fleet of 10,000 enemy aircraft at a distance of 250 miles.

Some of Tesla's ideas were more practical, and occasionally he was able to sell to others the rights to develop these concepts. Designs for an automobile speedometer and a locomotive headlight were particularly innovative and practical. These sales provided him with a small amount of money but, due to his many staggering debts, he lived in near-poverty for the rest of his life. Despite his chronic financial problems, Tesla always tried to project a personal image of sophistication and elegance.

Tesla was saved from complete destitution in 1934 when the Westinghouse Corporation agreed to pay his hotel rent together with a monthly stipend to serve as a "consultant." In exchange, Tesla agreed to drop his complaint that Westinghouse had violated his wireless patents. The government of Yugoslavia charitably awarded Tesla a pension of $600 per month in 1937. Anxious creditors eagerly awaited the monthly arrival of these funds.

Tesla become virtually a total recluse and an extreme eccentric in his last years. His main contacts were with the city's pigeons that he cared for and fed. Tesla died alone in a small hotel room on January 7, 1943 at the age of 86.

The New York City cathedral in which Tesla's funeral service was held was packed with over two thousand mourners. Tributes from political and scientific notables, including three Nobel Prize Laureates, poured in from around the world.

Still a great admirer of this world famous scientist and inventor, Hugo Gernsback was among the first to be notified of Tesla's death. Gernsback arranged to have a death mask made and covered with copper. The mask was kept it in Gernsback's office as a personal remembrance of this scientific savant.

Hugo Gernsback firmly believed that Nikola Tesla was the world's greatest inventor of all time — bar none. His admiration for Tesla is best summarized by the following tribute Gernsback wrote in the January 1919 issue of *The Electrical Experimenter*:

"If you mean the man who really invented, in other words, *originated* and discovered — not merely *improved* what had already been invented by others, then without a shade of doubt, Nikola Tesla is the world's greatest inventor, not only at present but in all history…. His basic as well as revolutionary discoveries, for sheer audacity, have no equals in the annals of the intellectual world."

Countless others around the globe shore Gernsback's feelings concerning Nikola Tesla.

ABOUT THE AUTHOR

JAMES RYBAK BSE.E., M.S., Ph.D. Professor of Engineering and Math (2005).

Colorado Mesa University, Prof. Emeritus

Feb 16, 2005 ... Trustee Monfort moved to approve Dr. James Rybak as Emeritus Professor of Engineering and Mathematics; Trustee Wist seconded.

SUGGESTED READINGS

Cheney, Margaret; _Tesla — Man Out of Time_, Dorset Press, New York, NY, 1989.

Gernsback, Hugo; "Nikola Tesla and His Inventions," _The Electrical Experimenter_, January 1919, p. 614.

Glenn, Jim (editor); _The Complete Patents of Nikola Tesla_, Barnes & Noble Books, New York, NY, 1994.

Hunt, Inez and Draper, Wanetta W.; _Lightning in His Hand — The Life Story of Nikola Tesla_, Omni Publications, Hawthorne, CA, 1964.

O'Neil, John J.; _Prodigal Genius — The Life of Nikola Tesla_, Ives Washburn, Inc., New York, NY, 1944.

Secor, H,. Winfield; "The Tesla High Frequency Oscillator," _The Electrical Experimenter_, March 1916, pg. 614.

Seifer, Marc J.; _Wizard: The Life and Times of Nikola Tesla_, Birch Lane Press, Secaucus, NJ, 1996.

Tesla, Nikola; "The Problem of Increasing Human Energy," _The Century Magazine_, June 1900, p. 175.

Tesla, Nikola; "My Inventions," _The Electrical Experimenter_, February 1919, p. 696; March 1919, p. 775; April 1919, p. 864; May 1919, p. 16; June 1919, p. 112; October 1919, p. 506. Also reprinted in edited form with an introduction by Ben Johnston; _My Inventions_, Barnes & Noble Books, New York, NY, 1995.

6 The Secret History of the Wireless

Marc J. Seifer, Ph.D.

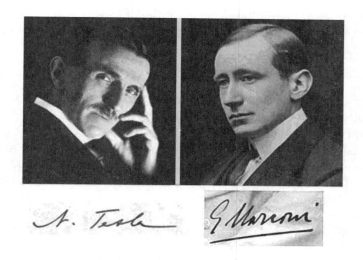

A Silent Tribute...[1]

April 18, 1912
My dear Mr. Tesla:

I attended the Marconi meeting last night, in company with illustrious society... Mr. Marconi gave the history, as he sees it, of wireless up to this date... [He] does not speak any more of Herzian Wave Telegraphy, but accentuates that messages he sends out are conducted along the earth...

Pupin had the floor next, showing that not only was wireless a great achievement, but that it was due entirely to one single man who was too modest to have spoken in the first person...

The only speaker of the evening who understood Mr. Marconi's merit and who did not hesitate to vent his opinion... was Steinmetz. In a brief historical sketch... he maintained that while all elements necessary for the transmission of wireless energy were available, yet it was due to Marconi that intelligence was actually transmitted...

That evening was, without any question, the highest tribute that I ever have heard paid to you in the language of absolute silence as to your name.

Sincerely yours,
Fritz Lowenstein

Marconi's talk, before 1100 members of the New York Electrical Society, was held on the very day the Titanic Sank. Frank Sprague, who gave the eulogy, "visibly affected Mr. Marconi when he credited him with saving the lives of from 700 to 800 persons."[2] Unfortunately, Marconi was unable to save 1500 other individuals including Colonel John Jacob Astor, who went down with the ship after helping his new bride board one of the remaining lifeboats.

LITIGATION

If ever an event epitomized the loss of innocence, the myopic condition of humanity, it was the sinking of the Titanic. Remindful of Tesla's own voyage, this watershed, recapitulated the story of Icarus, the prideful aeronaut who crumbled, because of lack of respect for his limitations. With Tesla's ponderous wish to transmit unlimited energy to the far reaches of the world and bring rain to the deserts, to become, as it were, a master of the universe, it was inevitable that he too would succumb. On the positive side, the tragedy prompted Congress to pass into law an act requiring the use of wireless equipment on all ships carrying 50 or more passengers. On the negative side, the event helped spurn a new wave of litigation causing company after company to elbow themselves into debt and melt away.

Tesla was not the only casualty in the wireless war. Reginald Fessenden's concern "all but ceased functioning in 1912," due to his erratic nature, internal quarrelling with his backers, and "prolonged litigation." And Lee DeForest, who by now had nearly 40 patents in wireless, also went under, when he was convicted with officers of his company in a stock fraud.[3]

Concerning Lowenstein, Tesla's former employee from his days at Colorado Springs, Tesla, for a modest royalty, was backing his protégé's attempts to install equipment on U.S. Navy ships. "He is much abler than the rest of the wireless men," he wrote his secretary, George Scherff, "[so] this has given me great pleasure." [4] The one edge Lowenstein had over Marconi, was the Italian's insistence of an all or nothing deal with the military. Either all ships were hooked into his system or none were. The U.S. Government, however, was loath to relinquish their upper hand, so Marconi had great difficulty integrating his system into the American military arena.

Nevertheless, Marconi was still the only major competitor, so Tesla set out to reestablish his legal right. Conferring with his lawyers, the pioneer began a strategy to sue the pirate in every country he could.

In England, Tesla had allowed an important patent to lapse, so progress there was halted. Oliver Lodge, on the other hand, was able to prevail, and received 1000 £/year for seven years from the Marconi concern there.[5] In the United States, where Tesla was applying for a renewal of his most fundamental patent, he had yet to formalize the suit, but in "the highest court in France" the inventor achieved a resounding success. Sending his written testimony to Judge M. Bonjean in Paris, Tesla explained his work in 1895, where he "erected a large wireless terminal above the building... [and] employed damped and undamped oscillations." He also enclosed two patents from 1897, and specifications for his telautomaton which confirmed that he had displayed it before G.D. Seely, the Examiner in Chief of the U. S. Patent Office in Washington, DC in 1898. Concerning Marconi's June 2, 1896 patent, Tesla testified:

> [This patent is] but a mass of imperfection and error. It describes apparatus shown and experimented with long before and a known method of signaling Hertzian rays, the energy of which is absorbed within a very short distance from the transmitter, and which are wholly unsuited for practical use. This patent has absolutely nothing to do with the successful application of the wireless art and, if anything, it has been the means of misleading many experts and retarding progress in the right direction...It gives no hint as to the length of the transmitting and receiving conductors and the arrangements illustrated preclude the possibility of accurate tuning... [Marconi replaced] the old-fashioned Rhumkorff coil [with] ... the Tesla coil.[6]

Speaking in support of Tesla's cause for Popoff, Ducretet & Rochefort, the French company who had initiated the action, was electrical engineer M.E. Giradeau who stated:

> Indeed, one finds in the American patent extraordinary clearness and precision, surprising even to physicists of today... Tesla as the inventor of the four circuits in resonance... even gives, by way of illustration, the numerical value of the condenser (4/100 of a microfarad) and says that the discharges of the condenser could be effected by a mechanical break... Tesla is also the inventor of the contact detector which is everywhere employed today... and all of this (can be found) in his 1897 patents and articles of 1898 and 1899.
> What a cruel injustice would it be now to try to stifle the pure glory of Tesla in opposing him scornfully.[7]

Judge Bonjean struck down Marconi's patents, and reestablished Tesla's as superseding. Most likely, compensation was also paid to him from the French company who won the case.

THE GERMAN CONNECTION

Marconi's greatest rival in the legal arena may have been Nikola Tesla, but in the battleground of the marketplace, it was Telefunken, the German wireless concern. Although Marconi had patents in Germany, the Telefunken syndicate, had too many important connections on its home front, and easily maintained a monopoly there. Formed through a forced merger under orders of the Kaiser, of the Braun-Siemens-Halske and Arco-Slaby systems, Telefunken had fought the Marconi conglomerate vigorously on every front. It was, without doubt, the number two competitor in the world. Although Marconi had achieved a recent coup in Spain, in America, Telefunken gained an edge when it constructed two enormous transatlantic systems in Tuckertown, New Jersey, and Sayville, New York.

For nationalistic reasons, Tesla had been prohibited from obtaining his rightful royalties in Germany, but Professor Slaby, never hid the fact that he considered Tesla the patriarch of the field. Thus, when Telefunken came to America, Slaby sought out his mentor, not only on moral grounds, but also for gaining a legal foothold against Marconi, and to obtain the inventor's technical expertise.

In 1913, a meeting was held between Tesla and the principles of Telefunken's American holding company, the innocuous sounding Atlantic Cable Company at 111 Broadway, the location of their offices. Present was the director, Dr. Karl George Frank, "one of the best known German [American] electrical experts," and his two managers, Richard Pfund, a frequent visitor to Tesla's lab, who was head of the Sayville plant, and "the monocle," Lieutenant Emil Meyers, head of operations at Tuckertown.

Tesla asked for an advance of $25,000 and royalties of $2,500/month, but settled for $1500/month with a one month advance.[8] Edwin Armstrong also struck a similar deal for his feedback circuit.

Tesla met with Pfund to discuss the turbine deal with the Kaiser, and also fix a transmitter that the German was working on at the Manhattan office. Shortly thereafter, he traveled out to the two transmitting stations with his plan to institute his latest refinements in order to boost their capabilities.[9]

J. P. MORGAN Jr.

Shortly after J. Pierpont Morgan's death in 1913, Tesla began to court his son Jack, trying not to fall into the trap he had laid for himself the first time around. He sent J. P. Morgan Jr. an articulate proposal outlining his plans in the field of "fluid propulsion," i.e., his bladeless turbines, and also his continuing work in wireless.

To reassert his dominance as the preeminent inventor in wireless, the inventor forwarded the entire French litigation proceedings where Marconi's work was overturned in favor of his. If Jack could help fight the legal battle in the U.S., the wireless enterprise, which they contractually shared, could be revived. Jack, however, was not smitten by the same vision of destiny, and graciously declined to become involved in Wardenclyffe. He had not ruled out the turbines, however, and advanced Tesla $20,000 for the

Woolworth Building

endeavor over the next few months. The wizard returned to the Edison Waterside Station where his turbines were located with a new transfusion of 23 Wall Street blood. To reflect the reactivation of the resurrected alliance, the inventor set out to search for more fashionable chambers. Within a few months he took up residence in the loftiest skyscraper in the world, the Woolworth Building, which soared above the city to the dizzying height of 800 feet.

Throughout the latter half of 1913, the inventor prepared a careful marketing plan to exploit his turbines. His plan was to use the profits from this endeavor to finance his triumphant return to Wardenclyffe and prevail at last in the field of wireless. Tesla's best lead in marketing the turbines came from former Edison associate, Sigmund Bergmann who was now one of the leading manufacturers for the Kaiser in Germany.

By the end of the year, however, the inventor was forced to admit that his turbines were not yet perfected, and requested from Morgan an additional $5,000 to help secure the German deal. Morgan agreed to defer interest payments, but decided not to increase the loan. Tesla, however, simply required the funds, and followed up the letter with a testimonial from Excellenz von Tirpitz, Minister of Marine, "who has been requested to the German Emperor relative to the Tesla Turbine who is greatly interested in this invention." Von Tirpitz had "promised his Excellency that the machine will certainly be here on exhibition about the middle of January, so you know what that means." Tesla also informed Morgan that if the deal was consummated, Bergmann would come through with royalties on the turbine of $100,00 /year

Considering Jack's keen antipathy for the Germans, it would seem unlikely that he would reverse his decision. However, unlike his father, the son was able compromise. He graciously changed his mind and forwarded the additional funds.[10]

THE WAR TO END ALL WARS

Within two weeks after World War I started, Germany's transatlantic cable was severed by the British. The only reasonable alternative for communicating with the outside world was through Telefunken's wireless system. Suddenly, the Tuckertown and Sayville plants became of paramount concern. The Germans obviously wanted to maintain the stations to keep the Kaiser abreast of President Woodrow Wilson's intentions, but the British wanted them shut down.

In July of1914, Marconi was decorated by the King of England at Buckingham Palace. Now, the fight against Telefunken would be fought on military as well as commercial grounds as it became clear that the Germans were using their plants to help coordinate submarine and battleship movements. The wireless lines also marked the burgeoning alliance forming between Italy and the English empire.[11]

As a pacifist, Wilson maintained a strict policy of neutrality although the sentiments of the majority of the American population was with England, particularly after Germany stormed through the peaceful kingdom of Belgium. Nevertheless, fully one-tenth of the country was of German stock, and their sentiments were with the other side. To avoid the problem of having the Germans use its American based transmitters for war purposes Wilson ordered the U.S. Navy to appropriate the Tuckertown plant so that the United States could send its own "radio coded messages abroad."

The Kaiser during the Great War

Wilson's presidential decree was published in the papers:

Whereas an order has been issued by me this day, August 5, 1914, declaring that all radio stations within the jurisdiction of the United States of America were prohibited from transmitting or receiving... messages of an unneutral nature and from in any way rendering to any one of the belligerants any unneutral services... It is ordered, by virtue of authority vested in me by the Radio Act... that one of more of the high powered radio stations within the jurisdiction of the United States... shall be taken over by the Government.[12]

THE LEGAL FRONT

Throughout the beginning of the war, Tesla stepped up his legal campaign against Marconi, as he continued to advise and receive compensation from Telefunken. As the country was still officially neutral, (American would dot enter the war for another three years), the arrangement was entirely above board. Nevertheless, few people knew about the German/Tesla link, although the inventor made no secret of it to Jack Morgan.

THE INTER OCEAN SUNDAY MORNING JANUARY 29, 1905.

IVIRELESS ELECTRIC ENERGY THE SOLUTION OF WAR.

BY NIKOLA TESLA.

Tesla must have gotten a kick out of this drawing as it depicts him using Marconi's primitive wireless equipment to blow up ships by remote control. From *Sunday News*, 1905.

February 19, 1915
Dear Mr. Morgan,

I am expecting to embody in their plant at Sayville some features of my own which will make it practicable to communicate with Berlin by wireless telephone and then royalties will be very considerable. We have already drawn papers.[13]

Camouflaged by the smokescreen of the American sounding Atlantic Cable Company, Telefunken swiftly moved to increase the power of its remaining station at Sayville. Located near the town of Patchogue out from the flats of Long Island, just a few miles from Wardenclyffe, the Sayville Complex encompassed 100 acres, and employed many German workers. With its main offices in Manhattan, and its German director, Dr. Karl George Frank, an American citizen, Telefunken was legally covered, as no foreigner could own a license to operate a wireless station in the country. (Thus, Marconi also had an American affiliate.) It was an easy matter for Tesla to confer with Atlantic in the city and also go out to the site of the plant.

Within two months of Tesla's letter to Morgan, through Tesla's promptings to his German counterpart, Jonathan Zenneck, the plant at Sayville tripled its output by erecting two more 500-foot-tall pyramid-shaped transmission towers and vastly increased its ground connection. Resonating accouterments spread out over the land for thousands of more feet. Telefunken's output was thus boosted from 35 kilowatts to over 100. Germany was now perceived as jumping into the number one spot in the wireless race. The *New York Times* reported on their front page, "Few persons outside radio officials knew that Sayville was becoming one of the most powerful transatlantic communicating stations in this part of the world."[14]

TESLA SUES MARCONI ON WIRELESS PATENT
Alleges That Important Apparatus Infringes Prior Rights

A suit brought by Nikola Tesla against the Marconi Wireless Telegraph Company of America, seeking the annulment of one of the chief Marconi patents, will have its first public argument in the United States District Court tomorrow. ...
New York Times, *August 4, 1915.*[15]

The years preceding America's entrance into World War I contained an overwhelming quagmire of litigation involving most countries, and virtually every major inventor in the wireless field. In August of 1914, Lowenstein was sued for violating Marconi's patent 763,772. However, this was only one of a number of legal battles the Marconi Wireless Telegraph Company would have with the Tesla legacy, as Tesla himself sued his arch rival that year, as did the U.S. Navy and also Atlantic Cable Company.

During the following spring, Marconi was subpoenaed by Telefunken. Due to the importance of the case, he sailed off for America on the Lusitania arriving in April of 1915 to testify. "We sighted a German submarine periscope," he told astonished reporters and his friends at dockside.[16] As three merchant ships had already been torpedoed without warning by the German U-Boats the month before, Marconi's inflammatory assertion was not taken lightly.

The *Brooklyn Eagle* reported that this suit brought "some of the world's greatest inventors on hand to testify."[17] Marconi had been declared a victor in a Brooklyn district court by Judge Van Vechter Veeder, during some of the Lownenstein proceedings. Although he certainly had the aura of the press behind him, Signor Marconi was beaten by the Navy in one patent dispute that year.[18] This German case, with all the heavyweights in town, promised to be portentous for establishing, once and for all, the true legal rights.

Aside from Marconi, there was, for the defense, Columbia Professor Michael Pupin, whose testimony was even quoted in papers in California. With braggadocio, Pupin declared, "I invented wireless before Marconi or Tesla, and it was I who gave it unreservedly to those who followed!"[19]

MICHAEL I. PUPIN
ONE WEST SEVENTY-SECOND STREET
NEW YORK CITY

Norfolk, Conn.,
May 29, 1931.

Kenneth M. Swezey,
159 Milton Street,
Brooklyn, N. Y.

My dear Sir:

The friends of Mr. Nikola Tesla are preparing a very graceful compliment for him to be presented to him on the occasion of his seventy Fifth birthday.

I have not seen Mr. Tesla for nearly twenty years. In the beginning of the World War a difference of opinion created a split between Mr. Tesla and myself. Neither he nor I have ever had, since that time, an opportunity to cure that split. In 1915 I offerred through a mutual friend, to forgive and forget, but somehow the offer was not accepted. I regret, therefore, that under the circumstances I could not transmit to Mr. Tesla a letter of greeting or congratulation on his seventy fifth birthday.

Yours very sincerely,

M. I. Pupin.

Michael Pupin rejecting the idea of congratulating Tesla on his 75th birthday

"Nevertheless," Pupin continued, "it was Marconi's genius who gave the idea to the world, and he taught the world how to build a telegraphic practice upon the basis of this idea. [As I did not take out patents on my experiments], in my opinion, the first claim for wireless telegraphy belongs to Mr. Marconi absolutely, and to nobody else."[20]

Watching his fellow Serb upon the stand, Tesla's jaw dropped so hard, it almost cracked upon the floor!

When Tesla took the stand for Atlantic, he came with his attorney, Drury W. Cooper of Kerr, Page & Cooper. Unlike Pupin, who could only state abstractly that he was the original inventor, Tesla proceeded to explain in clear fashion all of his work from the years 1891-1899. He documented his assertions with transcripts from published articles, from the Martin text, and from public lectures such as his well-known wireless demonstration which he had presented to the public in St. Louis in 1893. The inventor also brought along copies of his various requisite patents which he had created while working at his Houston Street Lab during the years 1896-1899.

COURT: *What were the [greatest] distances between the transmitting and receiving stations?*

TESLA: *... From the Houston laboratory to West Point, that is, I think, a distance of about thirty miles.*

COURT: *Was that prior to 1901?*

TESLA: *... Yes, it was prior to 1897 ...*

COURT: *Was there anything hidden about [the equipment], or were they open so that anyone could use them?*

TESLA: *There were thousands of people, distinguished men of all kinds, from kings and greatest artists and scientists in the world down to old chums of mine, mechanics, to whom my laboratory was always open. I showed it to everybody; I talked freely about it.*[21]

JOHN STONE STONE

> *I think we all misunderstood Tesla. We thought he was a... visionary... He did have visions but they were of a real future, not an imaginary one... It has been difficult to make any but unimportant improvements in the art of radio-telegraphy without traveling part of the way, at least, along a trail blazed by this pioneer... The apparatus he devised and constructed... was so far ahead of [its] time that the best of us then mistook him for a dreamer.*[22]

Another jolt to Marconi came from John Stone Stone (his mother's maiden name, by coincidence, was also Stone). Having traveled with his father, a general in the Union Army, throughout Egypt and the Mediterranean as a boy, Stone was educated as a physicist at Columbia University and John Hopkins University, where he graduated in 1890. A research scientist for Bell Labs in Boston for many years, Stone had set up his own wireless concern in 1899.

The following year, he filed for a fundamental patent on tuning, which was allowed by the U.S. Patent Office over a year before Marconi's.[23] Its essential features included his pioneer claim for the idea of "adjustable tuning by means of a variable inductance of the closed circuits of both transmitter and receiver.[24]

Stone, who never considered himself the original inventor of the radio, as President of the Institute of Radio Engineers, and owner of a wireless enterprise, put together a dossier of inventor priorities in "continuous-wave radio frequency apparatus." He wanted to determine for himself the etiology of the invention. Adorned in a formal suit, silk ascot, high starched collar and pince-nez attached by a ribbon around his neck, the worldly aristocrat took the stand:

> Marconi, receiving his inspiration from the experiments of Hertz and Righi...[was] impressed with the electric radiation aspect of the subject...and it was a long time before he seemed to appreciate the real role of the earth in the operation of his system, though he early recognized that the connection of his oscillator the earth was very material value...Tesla's electric earth waves explanation was the more serviceable in that it explained the important and useful function of [potential waves in] the earth, whereby the waves were enabled to travel over and around hills and were not obstructed by the sphericity of the earth's surface, while Marconi's view led many to place an altogether too limited scope to the possible range of transmission... With the removal of the spark gap from the antenna, the development of earthed antenna, and the gradual enlargement of the size of stations as it was realized... greater range could be obtained with larger power used at lower frequencies,... the art returned to the state to which Tesla developed it.

Attributing the opposition, and alas, even himself, to having been afflicted with "intellectual myopia," Stone concluded that although he had been designing wireless equipment and running wireless companies since the turn of the century, it wasn't until he "commenced with this study" that he really understood Tesla's contribution to the development of the field.[25]

THE FDR CONNECTION

Another case which did not receive much publicity, but which became vital to the Supreme Court's ruling in Tesla's favor over Marconi in 1943, was Marconi vs. the United States Navy, which Marconi brought on July 29, 1916. Marconi was suing for infringement of a fundamental wireless patent

(#763,772), which had been allowed in June of 1904. The Italian was seeking $43,000 in damages. Acting Secretary of the Navy, E.F. Sweet, and also Assistant Secretary Franklin D. Roosevelt began a correspondence in September of 1916 to review Tesla's 17 year old file to the Light House Board. Roosevelt wrote:

> *This [Light House Board Tesla File] may be made use of in forthcoming litigation in which the Government is involved... [Tesla's claims provide] suitable proof of priority of certain wireless usages by other than Marconi [and thus] might prove of great aid to the Government.*[26]

1915 dinner of the AIEE in honor of German wireless experts,
Ferdinand Braun and Jonathan Zenneck

Writing from his Experimental Laboratory in Colorado Springs, in 1899 to the Navy, Tesla described seven features of his wireless system which established unequivocally his stature as primary inventor. This discovery, Tesla pointed out, was announced "a few years ago [when] I laid down [these] certain novel principles: (1) an oscillator; (2) a ground and elevated circuit; (3) a transmitter; (4) a resonant receiver; (5) a transformer 'that scientific men have honored me by identifying it with my name' [Tesla coil]; (6) a powerful conduction coil; and (7) a transformer in the receiving apparatus."[27]

The history of Marconi's patent applications to the U.S. Patent Office provided additional ammunition. In 1900, John Seymour, the Commissioner of Patents who had protected Tesla against the demands of Michael Pupin for an SC claim at this same time, disqualified Marconi's first attempts at achieving a patent because of prior claims of Lodge and Braun and particularly Tesla.

"Marconi's pretended ignorance of the nature of a 'Tesla oscillator' [is] little short of absurd," wrote the Commissioner. "Ever since Tesla's famous [1891-1893 lectures]... widely published in all languages, the term 'Tesla oscillator' has become a household word on both continents." The Patent Office also cited quotations from Marconi himself admitting use of a Tesla oscillator.

Two years later, in 1902, Stone was granted a patent on tuning which the government cited as anticipating Marconi, and two years after that, after Seymour retired, Marconi was granted his infamous 1904 patent.[28]

THE FIFTH COLUMN

Due to the dangers that existed on the high seas, and rumors that the Germans were out for Marconi's head, the "Senatore" did not sail back on the Lusitania, but rather returned on the St. Paul in a disguised identity and under an assumed name.

A week after Marconi's departure, in May of 1915, a German submarine torpedoed the great ocean liner, the Lusitania, killing 1,134 individual; only 750 survived. The sinking, in lieu of the alternative procedure of boarding unarmed passenger ships by military vessels, was unheard of. Quite possibly, Marconi could have been a target, however, the Germans used as their reason the cargo of armaments on board headed for Great Britain.

In July of 1915, the Senate chambers in Washington was rocked by a terrorist bomb. The following day, the fanatic who planted it, Dr. Erich Muenter, alias Frank Holt, a teacher of German from Cornell University, and wife murderer, walked into Jack Morgan's Long Island home toting a six guy in each hand. He wanted Morgan to stop the flow of arms to Europe. With his wife and daughter leaping at the assailant, Morgan charged forward. Shot twice in the groin, Morgan and his wife were able to wrestle the guns from the man, and get him arrested. Tesla sent the overnight hero a get well letter,[29] as Holt committed suicide in a Long Island jail cell.

The Fifth Column had emerged. German spies were everywhere. Reports started filtering in that the Germans were creating a secret submarine base around islands off the coast of Maine. It was also alleged that the broadcasting station out at Sayville was not merely sending neutral dispatches to Berlin, but rather coded messages to battleships and submarines. The front pages of the papers were saturated with alarming headlines:

German U-boat sinks Lusitania, 1915

20 OR MORE AMERICANS LOST WHEN GERMANS SINK
[Peaceful Freighter]

NAVY MAY SEIZE SAYVILLE WIRELESS
Plant Under Suspicion
THINK GERMAN STATION MAY SEND
MESSAGES TO SUBMARINES
Evidence Before Congress
Wilson Hears of New Disaster

NEW SUBMARINE ATTACK
However a Surprise in View of Hope from Berlin

BASE FOR GERMAN SUBMARINES HERE?
Von Tirpitz Said to be Launching
New Campaign to Sink Munitions Ships

With Tesla, just a few months earlier, boasting to Morgan that he was working for the Germans, and "Grand Admiral von Tirpitz contemplat[ing] a more vigorous campaign against freight ships... [and planning] a secret base on this side of the Atlantic"[30] it is quite possible that the inventor's name became tainted in some inner circles.

A week later, on Tesla's 59th birthday, the *Times* reported that not only were the Germans dropping bombs over London from Zeppelins, they were also "controlling air torpedoes" by means of radio dynamics. Fired from Zeppelins, the supposed "German aerial torpedo[es] can theoretically remain in the air three hours, and can be controlled from a distance of two miles... Undoubtedly, this is the secret invention of which we have heard so many whispers that the Germans have held in reserve for the British fleet." [31] Although it seemed as if Tesla's telautomatic nightmare prognostication of 1898 had come to be, Tesla himself announced to the press that "the news of these... magic bombs... cannot be accepted as true, [though] they reveal just so many startling possibilities."

"Aghast at the pernicious existing regime of the Germans," Tesla accused Germany of being an "unfeeling automaton, a diabolic contrivance for scientific, pitiless, wholesale destruction of the like of which was not dreamed of before." No doubt, Tesla stopped doing business with von Tirpitz, although he probably continued his relationship with Sayville engineer Jonathan Zenneck who may have been morally opposed to the war although he was later arrested as a potential spy.

Tesla's solution to the war was twofold: (1) a better defense, through an electronic star-wars type shield he was working on, and (2) "the eradication from our hearts of nationalism." If blind patriotism could be replaced with "love of nature and scientific ideal... permanent peace [could] be established." [32]

Spies had infiltrated the Navy Yard in Brooklyn to use the station to send secret coded messages to Berlin; through Richard Pfund, head of the Sayville plant, they had apparently also installed equipment on the roof at 111 Broadway, the building that housed Telefunken's offices. [33] Lieutenant Meyer, "who ran the Tuckertown operation... [was placed] in a Detention Camp in Georgia," suspected of spying. [34] As Tesla had been working with Pfund, Meyer and von Tirpitz, and as he had continued to receive royalty payments from a division of Telefunken right through the first month of 1917, [35] it is possible that he was placed on a list of potential subversives in the mad quest to ferret out all German agents, even though his lengthy condemnation of Germany was published as a major treatise in *THE SUN*.

The job of taking over all wireless stations fell to Josephus Daniels, Secretary of the Navy; his assistant was Franklin Delano Roosevelt. In the summer of 1915, Daniels placed Thomas Edison head of a civilian think-tank called the "Naval Consultant Board." Working with Franklin Roosevelt, Edison appointed numerous inventors to various positions such as Frank Sprague, Elihu Thomson and Michael Pupin, with Tesla's name conspicuously missing from the list.

WIZARD SWAMPED BY DEBTS
Inventor Testifies He Owes the Waldorf
Lives Mostly on Credit Hasn't a Cent in Bank
and "Hocked" Stock in His Company[36]

As 1915 was drawing to a close, Tesla began to find himself in deeper and deeper financial troubles. Although an efficient water fountain which he designed that year, was received favorably,[37] his overhead was still too high. Expenses included outlays for the turbine work at the Edison Station, his office space at the Woolworth Building, salaries to his assistants and secretary, past debts to investors, maintenance costs for Wardenclyffe, legal expenses on the wireless litigation, and his accommodations at the Waldorf-Astoria.

With the publication of his wretched state in the public forum, came also a deep sense of anger and corresponding shame; for now the world had officially branded him a dud. If success is measured in a material way, it became clear that Tesla was the ultimate failure. On the exterior, the inventor kept up appearances, but this event would mark the turning point in his life. He now began a slow, but steady turning away from society. One way or another he rejected the Nobel Prize which was supposedly offered to him at this time, (although it is apparent now from the recently released Swedish Academy files that he was never nominated during that period); and Tesla also dismissed the coveted Edison Medal, although one of his friends, B.A. Behrend, finally convinced him to accept the honor.

THE WALDORF-ASTORIA

Tesla was aghast that Mr. Boldt, manager of the Waldorf-Astoria, had not protected Wardenclyffe adequately as it was valued at a minimum of at least $150,000. Even though he had signed it over to the Hotel, he had done so, according to his understanding, so as to honor his debt "until [his] plans matured." As the property, when completed, would yield $20,000 or $30,000 *a day*, Tesla was simply flabbergasted that Boldt would move to destroy the place. Boldt, and/or "the Hotel management" saw Wardenclyffe now as theirs, free and clear, even though Tesla offered as proof "a chattel mortgage" on the machinery that the inventor had placed at his own expense. The Hotel's insurance was only $5,000, whereas Tesla's covered the machinery valued at $68,000. Why would Tesla independently seek to protect the property if he didn't still have an interest in it? Tesla saw the contract as "a security pledge,"[22] but the paper he signed did not specify any such contingency. According to the Hotel's lawyer, Frank Hutchins of Baldwin & Hutchins, "it was a bill of sale with the deed duly recorded two years earlier.

We fail to see what interest you have" Hutchins callously concluded. Tesla found out that the Smiley Steel Company would be in charge of salvage operations.[38]

Tesla and Marconi wireless stations, early 1900's

THE UNITED STATES NAVY

The wizard decided that the only way to save Wardenclyffe was to extol its virtues as a potential defensive weapon for the protection of the country. Capitalizing upon the excellent Nobel Prize publicity, as it had been announced on the front page of the November 6, 1915 edition of the *NEW YORK TIMES* that he was to share it with Tom Edison, the inventor, once again strained the reader's creditability with another startling vision.

TESLA'S NEW DEVICE LIKE BOLTS OF THOR
He Seeks to Patent Wireless Engine for
Destroying Navies by Pulling a Lever.

TO SHATTER ARMIES ALSO
Nikola Tesla, the inventor, winner of the 1915 Nobel Physics Prize, has filed patent applications on the essential parts of a machine the possibilities of which test a layman's imagination and promise a parallel to Thor's shooting thunderbolts from the sky to punish those who had angered the gods. Dr. Tesla insists there is nothing sensational about it...

"Ten miles or a thousand miles, it will be all the same to the machine," the inventor says. Straight to the point, on land or on sea, it will be able to go with precision, delivering a blow that will paralyze or kill, as it is desired. A man in a tower on Long Island could shield New York against ships or army by working a lever, if the inventor's anticipations become realizations.[39]

FDR as Assistant Secretary of the Navy, circa 1916

In "a serious plight," with nowhere else to turn, the inventor contacted Morgan once again to ask for assistance. He pointed out to Morgan the military was using $10-million of wireless equipment based upon his patents, and that he hoped, some day, to gain compensation for their use. But he had exhausted all other avenues. Morgan provided his last chance to protect their commonly held wireless patents and save the tower. "Words cannot express how much I have deplored the cruel necessity which compelled me to appeal to you again," the inventor explained, but to no avail.[40] He still owed Jack $25,000 plus interest on the turbines; the financier ignored the request, and quietly placed Tesla's account in a bad debt file.[41]

In February of 1917, the United States broke off all relations with Germany, and seized the wireless plant at Sayville. "Thirty German employees... were suddenly forced to leave, and enlisted men of the American Navy have filled their places." Guards were placed around the plant as the high command decided what to do with the remaining broadcasting stations lying along the coast.[42] Articles began springing up like early crocus to announce the potential "existence of [yet another] concealed wireless station [able] to supply information to German submarines regarding the movements of ships."[43]

19 MORE TAKEN AS GERMAN SPIES
Dr. Karl George Frank, Former Head of
Sayville Wireless Among Those Detained[44]

On April 6th, 1917, President Wilson issued a proclamation "seiz[ing] all radio stations. Enforcement of the order was delegated to Secretary Daniels.... It is understood that all plants for which no place can be found in the Navy's wireless system, including amateur apparatus, for which close search will be made, are to be put out of commission immediately."[45] Clearly, an overt decision had to be made about the fate of Wardenclyffe.

Tesla's expertise was well known to Secretary Daniels and Assistant Secretary Franklin Roosevelt, as they were actively using the inventor's scientific legacy as ammunition against Marconi in the patent suit. Coupled with the inventor's astonishing proclamation that his tower could provide an electronic aegis against potential invasions, Wardenclyffe must have been placed in a special category.

However, there were two glaring strikes against it. The first was that Tesla had already turned over the property to Mr. Boldt to cover his debt at the Waldorf; and the second was the transmitter's record of accomplishment: *nonexistent*. What better indication of the folly of Tesla's dream could there be than the tower's own perpetual state of repose. To many, Wardenclyffe was merely a torpid monument to the bombastic prognostications of a not very original mind gone astray. From the Navy's point of view, Tesla may have been the original inventor of the radio, but he was clearly not the one who made the apparatus work.

A HISTORY OF NAVY INVOLVEMENT

In 1899, the U.S. Navy, via Rear-Admiral Francis J. Higginson, requested Tesla to place "a system of wireless telegraphy upon Light-Vessel N. 66 [on] Nantucket Shoals, Massachusetts, which lies 60 miles south of Nantucket Island."[46] Tesla was on his way to Colorado and was unable to comply. Further, the Navy did not want to pay for the equipment, but rather wanted Tesla to outlay the funds himself. Considering the great wealth of the country, Tesla feigned astonishment at the penurious position of John D. Long, Secretary of the Navy, via Captain Perry, who brazenly forwarded the financial disclaimer on U.S. Treasury Department stationary.

Be that as it may, upon Tesla's return to New York in 1900, he wrote again of his interest in placing the equipment aboard their ship. Rear Admiral Higginson, chairman of the Light House Board, wrote back that his committee would meet in October to discuss with the Congress "the estimates of cost."[47] Higginson, who had visited Tesla in his lab in the late 1890's, wanted to help, but he had been placed in the embarrassing position of withdrawing his offer of financial remuneration because of various levels of bureaucratic inanity. Tesla spent the time to go down to Washington to confer face-to-face with the high command, but he was given the proverbial "runaround" and he returned to New York empty handed and disgusted with the way he was treated.

From the point of view of the Navy, this was an entirely new field, and they were unsure what to do. Further, they may have been turned off by Tesla's haughty manner, particularly when it came to being "compared" to Marconi, who Tesla refused to be compared with.

In 1902, the Office of Naval Intelligence called upon Commodore F.M. Barber, who had been in retirement in France, back to the States to be put in charge of the acquisition of wireless apparatus for testing. Although still taking a frugal position, the Navy came up with approximately $12,000 for the purchase of wireless sets from different European companies for testing. Orders were placed with Slaby-Arco and Braun-Siemans-Halske of Germany and Popoff, Ducretet and Rochefort of France. Bids were also requested from DeForest, Fessenden, and Tesla in America and also Lodge-Muirhead in England. Marconi was excluded, because he arrogantly coveted an all or nothing deal.[48]

Fessenden was angry with the Navy for obtaining equipment outside the United States, and so did not submit a bid. Tesla was probably too upset with his treatment from the past, and too involved with Wardenclyffe, which was under active construction at that time, to get involved, and so the Navy purchased additional sets from DeForest and Lodge-Muirhead.

In 1903, a mock battle with the North Atlantic Fleet was held 500 miles off the coast of Cape Cod. The "White Squadron" was commanded by Rear Admiral J.H. Sands and the "Blue Squadron" by Tesla's ally, Rear Admiral Higginson. The use of wireless played a key role in determining the victor. Commander Higginson, who won the maneuver, commented, "To me, the great lesson of the search we ended today is the absolute need of wireless in the ships of the Navy. Do you know we are three years behind the times in the adoption of wireless?"[49]

Based upon comparison testing, it was determined that the Slaby-Arco system out performed all others, and the Navy ordered 20 more sets. Simultaneously, they purchased an 11 year lease on the Marconi patents.[50]

With the onset of World War I, the use of wireless became a necessity for organizing troop movements, surveillance, and intercontinental communication. While the country was still neutral, the Navy was able to continue their use of the German equipment up until sentiments began to shift irreversibly to the British side. Via the British Navy, Marconi had his transmitters positioned in Canada, Bermuda, Jamaica, Columbia, the Falkland Islands, North and South Africa, Ceylon, Australia, Singapore and Hong Kong. This was a mighty operation. In the United States, the American Marconi division, under the directorship of the politically powerful John

Griggs, former governor of New Jersey and Attorney General under President McKinley, had transmitters located in New York, Massachusetts, and Illinois.[51] One key problem, however, was that the Marconi equipment was still using the outmoded spark-gap method.

In April of 1917, when the U.S. Navy took over all wireless stations, this included their allies, the British plants as well. At the same time, Marconi was in the process of purchasing the Alexanderson alternator, which was, in essence, a refinement of the Tesla oscillator. Simultaneously, the Armstrong feedback circuit was becoming an obvious necessity for any wireless instrumentation. However, the Armstrong invention, created a judicial nightmare not only because it used as its core the DeForest audion, but also because DeForest's invention was overturned in the courts in favor of an electronic tube developed by Fessenden—never mind that Tesla, as far back as 1902, had beaten Fessenden in the courts for this development. With the Fessenden patent now under the control of Marconi, the courts would come to rule that no one could use the Armstrong feedback circuit without the permission of the other players.

The most important ruling, concerning the true identity of the inventor of the radio, became neatly sidestepped by the war powers act of President Wilson calling for the suspension of all patent litigation during the time of the war. France had already recognized Tesla's priority by their high court, and Germany recognized him by Slaby's affirmations and Telefunken's decision to pay royalties; but in America, the land of Tesla's home, the government backed off and literally prevented the courts from sustaining a decision. The Marconi syndicate, in touch with kings from two countries, with equipment instituted on six continents, was simply too powerful.

THE FARRAGUT LETTER

With the suspension of all patent litigation and the country in the midst of a world war, Franklin Roosevelt, Assistant Secretary of the Navy, penned the famous "Farragut Letter." This document allowed such major companies as AT&T, General Electric, Westinghouse and American Marconi the right to pool together to produce each other's equipment without concern for compensating the rightful inventors. Further it "assured contractors that the Government would assume liability in infringement suits."[52]

On July 1, 1918, Congress passed a law making the United States financially responsible for any use of "an invention described in and covered by a patent of the United States." By 1921, the United States Government had spent $40 million dollars on wireless equipment, a far cry from Secretary

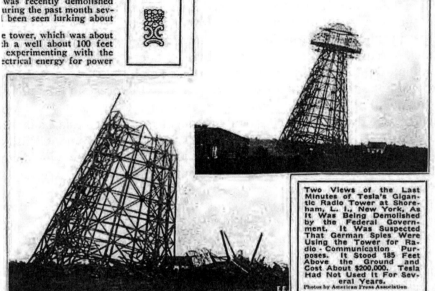

U. S. Blows Up Tesla Radio Tower

G that German spies were
ig wireless tower erected
m, L. I., about twenty
y Nikola Tesla, the Fed-
iment ordered the tower
was recently demolished
uring the past month sev-
l been seen lurking about

e tower, which was about
:h a well about 100 feet
experimenting with the
:ctrical energy for power

gone on record as stating their belief to be
in accordance with Dr. Tesla's. More wonderful still is the fact that this scientist pro-

some of the water into tl
back into the ball by pu
handle, this change in I

Two Views of the Last
Minutes of Tesla's Gigan-
tic Radio Tower at Shore-
ham, L. I., New York, As
It Was Being Demolished
by the Federal Govern-
ment. It Was Suspected
That German Spies Were
Using the Tower for Ra-
dio - Communication Pur-
poses. It Stood 185 Feet
Above the Ground and
Cost About $200,000. Tesla
Had Not Used It For Sev-
eral Years.
Photos by American Press Association

The destruction of Wardenclyffe tower by a salvage company for the Waldorf-Astoria wrongly reported as destroyed by the US government

Long's policy of refusing to pay a few thousand dollars for Tesla's equipment 18 years before. Thus, the Interdepartmental Radio Board met to decide various claims against it. Nearly $3 million in claims were paid out. The big winners were Marconi Wireless, who received $1.2 million for equipment and installations taken over (but not for their patents); International Radio Telegraph received $700,000; AT&T, $600,000, and Edwin Armstrong, $89,000. Tesla received a minuscule compensation through Lowenstein who was awarded $23,000.[53]

In 1921, the Navy published a list of all the inventors in wireless who received compensation from them. The list contained only patents granted after 1902. Inventors included: Blockmen, Braun, Blondel, DeForest, Fuller, Hahnemann, Logwood, Meissner, Randahl, Poulsen, Schiessler, von Arco and Watkins. Note that both Tesla's and Marconi's name is missing.[54] Marconi's could be missing either because his patents had lapsed, or more likely, because they were viewed as invalid from the point of view of the government. In the case of Tesla, all of his 12 key radio patents had "expired and [were] now common property."[55] However, Tesla had renewed one fundamental patent in 1914* and this should have been on the list, as should have Armstrong's feedback patent.

RADIO CORPORATION OF AMERICA

The United States government, through Franklin Roosevelt, knew that Marconi had infringed upon Tesla's fundamental patents. They knew the details of Tesla's rightful claims, through their own files and through the record at the patent office. In point of fact, it was Tesla's proven declaration which was the basis and central argument that the government had against Marconi when Marconi sued in the first place, and this same claim, and the same Navy Lighthouse Board files that would eventually be used by the U.S. Supreme Court to vindicate Tesla three months after he died, nearly 25 years later in 1943.

Rather than deal with the truth, and with a difficult genius whose present work appeared to be in a realm above and beyond the operation of simple radio telephones and wireless transmitters, Roosevelt, Daniels, President Wilson and the U.S. Navy took no interest in protecting the Tesla Tower.

It was during the height of the world conflagration, when the Smiley Steel Company's explosives expert approached the gargantuan transmitter soaring above as he circled it to place a charge around each major strut. With the Associated après recording the event, and military personnel apparently present, the magnifying transmitter was leveled, the explosion alarming many of the Shoreham residents. Tesla became essentially a non-person the day his magnifying transmitter was leveled.

With the death of the Tesla World Telegraphy Center came the birth of the Radio Broadcasting Corporation, a unique conglomerate of private concerns under the auspices of the U.S. government. Meetings were held behind closed doors in Washington between President Wilson, who wanted America to gain "radio supremacy,"[56] Navy Secretary Daniels, his assistant Franklin Roosevelt, and representatives from General Electric, American Marconi, AT&T and the Westinghouse Corporation.

---

* *Patent 1,119,732, Apparatus for transmitting electrical energy, was applied for January 16, 1902. The application was renewed May 4, 1907, and granted December 1, 1914. This patent, in essence, contains all of Tesla's key ideas behind the construction of Wardenclyffe. Armstrong's invention, although a necessity, was still tied up in court because of DeForest.*

RADIO SECTION

HARRISBURG, PA., SATURDAY EVENING, MARCH 22, 1924.

Radioed Light, Heat and Power Perfected by Tesla

INVENTOR ANNOUNCES FINAL SUCCESS OF EXPERIMENTS BEGUN THIRTY YEARS AGO

Tesla's full vision of distributing information & even power by means of wireless

With J.P. Morgan & Company on the Board of Directors, and the Marconi patents as the backbone of the organization, RCA was formed. It would combine resources from the mega-corporations of General Electric, American Marconi, AT&T, and Westinghouse, all who had cross-licensing agreements with each other* and all who co-owned the company.[57]

Here was another entente cordial reminiscent of the AC polyphase days, which was not so for the originator of the invention. It was a second major time Tesla would be carved from his creation.** A secret deal was probably concocted which absolved the government for paying any licensing fee to Marconi in lieu of them burying their Tesla archives. David Sarnoff, as managing director, would soon take over the reins of the entire operation.

The *New York Sun* inaccurately reported:

U.S. BLOWS UP TESLA RADIO TOWER

Suspecting that German spies were using the big wireless tower erected at Shoreham, L.I., about twenty years ago by Nikola Tesla, the Federal Government ordered the tower destroyed and it was recently demolished with dynamite. During the past month several strangers had been seen lurking about the place.[58]

The destruction of Nikola Tesla's famous tower... shows forcibly the great precautions being taken at this time to prevent any news of military importance of getting to the enemy.[59]

"One of the first actions of the Board of Directors was to invite President Wilson to nominate a naval officer of the rank of captain or above... to present the Government's views and interests concerning matters pertaining to radio communication." Simultaneous with the end of the war, President Wilson also returned all confiscated radio stations to their rightful owners. American Marconi, now RCA, of course, was the big beneficiary.[60]

In 1920, The Westinghouse Corporation, was granted the right to "manufacture, use and sell apparatus covered by the [Marconi] patents."[61]

---

* *Cross-licensing agreements also existed with the government who also owned some wireless patents.*

** *Tesla would also be cut out of a secret agreement between GE and Westinghouse to hold back production of efficient fluorescent lighting equipment as they did not want to undermine the highly profitable sale of normal Edison light bulbs, not "cut too drastically the demand for current" (Gilfillan, S.C. Invention & the Patent System. Washington, DC: U.S. Printing Office, 1964, p. 100)*

Westinghouse also formed an independent radio station which became as prominent at RCA. At the end of the year, Tesla wrote a letter to E.M. Herr, president of the company, offering his wireless expertise and equipment. Westinghouse replied:

> Dear Mr. Tesla,
>   I regret that under the present circumstances we cannot proceed further with any developments of your activities.[62]

A few months later, Westinghouse requested that Tesla "speak to our 'invisible audience' some Thursday night in the near future [over our...] radio telephone broadcasting station."[63]

> November 30, 1921
> Gentlemen,
>   Twenty-one years ago I promised a friend, the late J. Pierpont Morgan, that my world-system, then under construction... would enable the voice of a telephone subscriber to be transmitted to any point of the globe...
>   I prefer to wait until my project is completed before addressing an invisible audience and beg you to excuse me.
>   Very truly yours,
>   N. Tesla[64]

END NOTES

1.   Fritz Lowenstein/Nikola Tesla 4/18/1912 [Swezey Papers, Smithsonian].

2.   Marconi lecture before NY Electrical Soc. *Electrical World*, 4/20/1912, p.835.

3.   Sobel, Robert. *RCA*. New York, NY: Stein & Day, 1986, pp. 19-20; Harding, Robert. *George H. Clark Radiona Collection*. Smithsonian Institute, 1990.

4.   Nikola Tesla/George Scherff 1/18/1913 [Library of Congress Microfilm].

5.   Jolly, W. *Marconi.* New York, NY: Stein & Day, 1972, p. 190.

6.   Nikola Tesla/JP Morgan Jr. 3/19/1914 [Archives, J. Pierpont Morgan Library].

7.   Nikola Tesla/JP Morgan Jr. 7/23/1913 re: Girandeau, M.B. testimony [Archives, J. Pierpont Morgan Library].

8.   1917 royalty payment from *Hochfrequenz Maschiemen Aktievgesell* [no doubt, a division of Telefunken] for $1567 in 1917 [Swezey Collection, Smithsonian Institute].

9.   Nikola Tesla/JP Morgan Jr. 2/19/1915 [Archives, JPM Library]; Nikola Tesla and Nikola Tesla/Pfund corresp., circa 1912-1922 [Nikola Tesla Museum]; 19 More Taken as German Spies, *New York Times*, I, 1:3; Find Radio Outfit in Manhattan Tower, *New York Times*, 3/5/1918, 4:4.

10.   Nikola Tesla/JP Morgan Jr. 1/6/1914 [Morgan Library].

11.   Jolly, 1972.

12.   Nation to Take Over Tuckertown plant. *New York Times*. 9/6/1914, II, p.14:1.

13.   Nikola Tesla/JP Morgan Jr. 2/19/1915 [Morgan Library].

14.   Germans Treble Wireless Plant. *New York Times*, 4/23/1915, 1:6.

15.   Tesla Sues Marconi. *New York Times*, 8/4/1915, 8:1.

16.   Jolly, 1972, p. 225.

17.   R & A, p.100.

18.   Marconi Loses Navy Suit, *New York Sun*, 10/3/1914 [Nikola Tesla/JP Morgan Jr. corresp., Morgan Library.].

19.   Prof. Pupin Now Claims Wireless His Invention. *Los Angeles Examiner*, May 13, 1915; Ratzlaff, J & Anderson, L., *Dr. Nikola Tesla Bibliography 1884-1978*. Palo Alto, CA: Ragusen Press, 1979, p. 100 [literary license on quote].

20.   When Powerful High-frequency Electrical Generators Replace the Spark-gap. *New York Times*, 10/6/1912, VI, 4:1.

21.   Marconi Wireless vs. Atlantic Communications Co., 1915 [Archives, L. Anderson].

22.   Anderson, Leland (Ed.). John Stone Stone on Nikola Tesla's Priority in Radio and Continuous-Wave Radiofrequency Apparatus. *The Antique Wireless Review*, Vol 1, 1986.

23.   Dunlap, Orin. *Radio's 100 Men of Science*. New York, NY: Harper & Brothers, 1944.

24.   Marconi Wireless vs. United States. Cases Adjudged in the Supreme Court. 10/1942, v. 320, p. 17. This feature was obviously also part of Tesla's design, although the court eventually ruled Stone as the originator.

25.   Anderson, L. *John S. Stone on Nikola Tesla*, 1986, pp. 37 to 40.

26.   E.F. Sweet and FDR correspondence re: Tesla, 9/14/1916; 9/16/1916; 9/26/1916 [National arch.]

27.   Nikola Tesla/Light House Board corresp., 8/11/1899 [Nat. Arch.].

28.   Anderson, Leland. Priority in the Invention of the Radio: Tesla vs. Marconi. *The Tesla Journal*, vol. 2/3, 1982/83, pp. 17-20.

29.   Nikola Tesla/JP Morgan Jr. 7.1915 [Library of Congress Microfilm].

30.   Germany to Sink the Armenian. Navy May Seize Sayville Wireless. *New York Times*, 7/1/1915, 1:4-7.

31.   Wireless Controls German Air Torpedo. *New York Times*, 7/10/1915, 3:6, 7.

32.   Nikola Tesla. Science and Discovery are the Great Forces Which Will Lead to the Consummation of the War. *New York Sun*, 12/20/1914, in Nikola Tesla, 1956, pp. A-162-171.

33.   Federal Agents Raid Offices Once Occupied by Telefunken. Former Employee Richard Pfun Charged; No Arrests Made. *New York Times*, 3/5/1918, 4:4.

34.   Nikola Tesla/George Scherff 12/25/1917 [Library of Congress Microfilm].

35.   Royalty Check to Nikola Tesla for $1567 from *Hochfrequenz Maschienen Aktievgesell Schaft for drachlose Telegraphic*, 1917 [Swezey Collection, Smithsonian Institute]. Tuckertown was still owned by the Germans, although seized by the U.S. Navy, and Tuckertown, with full knowledge of the "Director of Naval Communications," had agreed to pay Tesla royalties, see Nikola Tesla/George Scherff 10/12/1717 [Library of Congress Microfilm].

36.   Tesla No Money; Wizard Swamped by Debts. *New York World*, 3/16?/1916.

37.   Nikola Tesla's Fountain. *Scientific American*, 1915.

38.   Lester S. Holmes was represented for the Hotel as owner of said Tesla property. Baldwin & Hutchins/Nikola Tesla corresp., 7/13/1917, from: Wardenclyffe property foreclosure proceedings, *NY Supreme Ct*, circa 1923 [L. Anderson files].

39.   Tesla's New Device Like Bolts of Thor. *New York Times*, 12/8/1915, 8:3.

40.   Nikola Tesla/JP Morgan Jr. 4/8/1916 [L. Anderson files].

41.   Nikola Tesla/JP Morgan Jr. 2/19/1915 [Morgan Library.].

42.   Reason for Seizing Wireless. *New York Times*, 2/9/1917, 6:5.

43.   Spies on Ship Movements. *New York Times*, 2/17/1917, 8:2.

44.   19 More Taken as German Spies. *New York Times*, 4/8/1917, 1:3.

45.   Navy to Take Over All RadioS. *New York Times*, 4/7/1917, 2:2.

46.   F. Higginson/Nikola Tesla 5/111899 [Nat. Arch.].

47.   F. Higginson/Nikola Tesla 8/8/1900 [Nat. Arch.]. For the full correspondence of this event, see *Wizard: The Life & Times of Nikola Tesla* by Marc Seifer.

48.   Howeth, L.S. *History of Communications-Electronics in U.S. Navy.* Washington, DC: U.S. Government Printing Office, 1963, pp. 518-519; Hezlet, A. Electronics & Sea Power. New York, NY: Stein & Day, 1975, p. 41.

49.   Howeth, 1963, p. 64.

50.   Hezlet, A.R., *Electronics & Sea Power*. New York: Stein & Day, 1975, pp.41-42.

51.   Sobel, 1986, p.43; Hezlet, 1975, p. 77.

52.   Howeth, 1963, p. 256.

53.   Howeth, 1963, pp. 375-376; George Scherff/Nikola Tesla corresp. [Library of Congress Microfilm].

54.   Howeth, 1963, pp. 577-580.

55.   Nikola Tesla, Electric drive for battleships. *New York Herald*, 2/25/1917; in Nikola Tesla, 1956, p. A-185.

56.   Howeth, 1963, p. 354.

57.   The breakdown was as follows: GE 30%, Westinghouse 20%, AT&T 10%, United Fruit 4%, others 34%. Sobel, 1986, pp. 32-35.

58.   U.S. Blows Up Tesla Radio Tower. *Electrical Experimenter*, 9/1917, p. 293.

59.   Destruction of Tesla's Tower at Shoreham, LI Hints of Spies. *New York Sun*, 8/5/1917.

60.   Howeth, 1963, pp. 359-360.

61.   Howeth, 1963, p. 361.

62.   E.M. Herr/Nikola Tesla 11/16/1920 [Library of Congress Microfilm].

63.   GW Corp/Nikola Tesla 11/28/1921 [Library of Congress Microfilm].

64.   Nikola Tesla/GW Co. 11/30/1921 [Library of Congress Microfilm].

DR. MARC J. SEIFER is one of the world's leading Tesla experts. A graduate of Saybrook Institute and the University of Chicago, he has lectured at the United Nations, Brandeis University, West Point Military Academy, in Zagreb Croatia, before the Serbian Academy of Sciences in Belgrade, at the University of Vancouver, in Jerusalem, Israel, at Cambridge University and Oxford University in England and at numerous Tesla conferences throughout the United States. Featured in *The New York Times, New Scientist, Rhode Island Monthly, The Washington Post* and on the back cover of Uri Geller's book, *Mind Medicine*, his publications include articles in *The Historian, Civilization, Lawyer's Weekly* and *Wired*. His works include in fiction: ***Rasputin's Nephew, Doppelgänger, Crystal Night*** and ***Fate Line***, and in non-fiction: ***Where Does Mind End?, Transcending the Speed of Light*** and ***FRAMED!***, and the definitive Tesla biography ***WIZARD: The Life & Times of Nikola Tesla*** (Citadel Press/Kensington). Called "Revelatory" by *Publisher's Weekly* and "Highly Recommended" by the American Association for the Advancement of Science, ***WIZARD*** has also been performed as a screenplay reading at Producer's Club Theater in New York, screenplay co-written with Tim Eaton, visual effects editor at Sony ImageWorks and Industrial Light & Magic.

7   A Striking Tesla Manifesto

Nikola Tesla, <u>Electrical World & Engineer</u> February 6th, 1904, P. 256

     We reproduce herewith in slightly reduced facsimile the first page of a four-page circular which has been issued this week by Mr. Nikola Tesla in a large square envelope bearing a large red wax seal with the initials, "N.T." At the back of the page, which we reproduce, is given a list of 93 patents issued in this country to Mr. Tesla. The fourth page is blank. The third page has a little vignette of Niagara Falls and is devoted to quotations from various utterances of Mr. Tesla. The first of these is from his lecture delivered in 1893 before the Franklin Institute and the National Electric Light Association, as to transmission of intelligible signals and power to any distance without the use of wires. The second quotation is from his article on the problem of increasing human energy, which appeared in the <u>Century Magazine in June, 1900</u>, dealing with virtually the same subject. The third item quotes from his patents, Nos. <u>645,576</u> and <u>649,621</u>, dealing with the transmission of electrical energy in any quantity to any distance, with transmitting and receiving apparatus movable as in ships or balloons. The circular is an extremely interesting one. It is most sumptuously got up on vellum paper and altogether constitutes a manifesto worthy of the original genius issuing it. It is to be gathered from the circular that Mr. Tesla proposes to enter the field of consulting engineership, in which he already has enjoyed an extensive connection here and abroad.

One page from the Tesla Manifesto.

New York, January 1, 1904

I wish to announce that in connection with the commercial introduction of my inventions I shall render professional services in the general capacity of consulting electrician and engineer. The near future, I expect with confidence, will be a witness of revolutionary departures in the production, transformation and transmission of energy, transportation, lighting, manufacture of chemical compounds, telegraphy, telephony and other arts and industries.

In my opinion, these advances are certain to follow from the universal adoption of high-potential and high frequency currents and novel regenerative processes of refrigeration to very low temperatures.

Much of the old apparatus will have to be improved, and much of the new developed, and I believe that while furthering my own inventions, I shall be more helpful in this evolution by placing at the disposal of others the knowledge and experience I have gained. Special attention will be given by me to the solution of problems requiring both expert information and inventive resource – work coming within the sphere of my constant trailing and predilection.

I shall undertake the experimental investigation and perfection of ideas, methods and appliances, the devising of useful expedients and in particular, the design and construction of machinery for the attainment of desired results. Any task submitted to and accepted by me, will be carried out thoroughly and conscientiously.

Nikola Tesla

8   The True Wireless

Nikola Tesla, The Electrical Experimenter, May, 1919, p. 28

In this remarkable and complete story of his discovery of the "True Wireless" and the principles upon which transmission and reception, even in the present day systems, are based, Dr. Nikola Tesla shows us that he is indeed the "Father of the Wireless." To him the Hertz wave theory is a delusion; it looks sound from certain angles, but the facts tend to prove that it is hollow and empty. He convinces us that the real Hertz waves are blotted out after they have traveled but a short distance from the sender. It follows, therefore, that the measured antenna current is no indication of the effect, because only a small part of it is effective at a distance. The limited activity of pure Hertz wave transmission and reception is here clearly explained, besides showing definitely that in spite of themselves, the radio engineers of today are employing the original Tesla tuned oscillatory system. He shows by examples with different forms of aërials that the signals picked up by the instruments must actually be induced by earth currents — not etheric space waves. Tesla also disproves the "Heaviside layer" theory from his personal observations and tests.

- EDITOR.

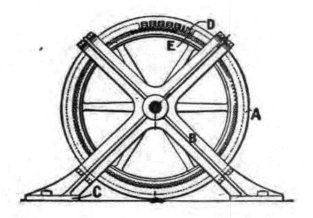

Fig. 1. — Alternator of 10.000 Cycles p.s., Capacity 10 K.W., Which Was Employed by Tesla in His First Demonstrations of High Frequency Phenomena Before the American Institute of Electrical Engineers at Columbia College, May 20, 1891.

Ever since the announcement of Maxwell's electro-magnetic theory scientific investigators all the world over had been bent on its experimental verification. They were convinced that it would be done and lived in an atmosphere of eager expectancy, unusually favorable to the reception of any evidence to this end. No wonder then that the publication of Dr. Heinrich Hertz's results caused a thrill as had scarcely ever been experienced before. At that time I was in the midst of pressing work in connection with the commercial introduction of my system of power transmission, but, nevertheless, caught the fire of enthusiasm and fairly burned with desire to behold the miracle with my own eyes. Accordingly, as soon as I had freed myself of these imperative duties and resumed research work in my laboratory on Grand Street, New York, I began, parallel with high frequency alternators, the construction of several forms of apparatus with the object of exploring the field opened up by Dr. Hertz. Recognizing the limitations of the devices he had employed, I concentrated my attention on the production of a powerful induction coil but made no notable progress until a happy inspiration led me to the invention of the oscillation transformer. In the latter part of 1891 I was already so far advanced in the development of this new principle that I had at my disposal means vastly superior to those of the German physicist. All my previous efforts with Rhumkorf coils had left me unconvinced, and in order to settle my doubts I went over the whole ground once more, very carefully, with these improved appliances. Similar phenomena were noted, greatly magnified in intensity, but they were susceptible of a different and more plausible explanation. I considered this so important that in 1892 I went to Bonn, Germany, to confer with Dr. Hertz in regard to my observations. He seemed disappointed to such a degree that I regretted my trip and parted from him sorrowfully. During the succeeding years I made numerous experiments with the same object, but the results were uniformly negative. In 1900, however, after I had evolved a wireless transmitter which enabled me to obtain electro-magnetic activities of many millions of horse-power, I made a last desperate attempt to prove that the disturbances emanating from the oscillator were ether vibrations akin to those of light, but met again with utter failure. For more than eighteen years I have been reading treatises, reports of scientific transactions, and articles on Hertz-wave telegraphy, to keep myself informed, but they have always imprest me like works of fiction.

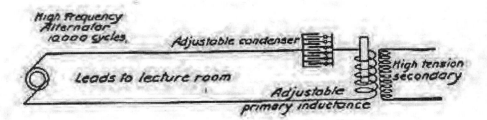

Fig. 2. — Diagram Illustrating the Circuit Connections and Tuning Devices Employed by Tesla In His Experimental Demonstrations Before the American Institute of Electrical Engineers With the High Frequency Alternator Shown in Fig. 1.

The history of science shows that theories are perishable. With every new truth that is revealed we get a better understanding of Nature and our conceptions and views are modified. Dr. Hertz did not discover a new principle. He merely gave material support to a hypothesis which had been long ago formulated. It was a perfectly well-established fact that a circuit, traversed by a periodic current, emitted some kind of space waves, but we were in ignorance as to their character. He apparently gave an experimental proof that they were transversal vibrations in the ether. Most people look upon this as his great accomplishment. To my mind it seems that his immortal merit was not so much in this as in the focusing of the investigators' attention on the processes taking place in the ambient medium. The Hertz-wave theory, by its fascinating hold on the imagination, has stifled creative effort in the wireless art and retarded it for twenty-five years. But, on the other hand, it is impossible to over-estimate the beneficial effects of the powerful stimulus it has given in many directions.

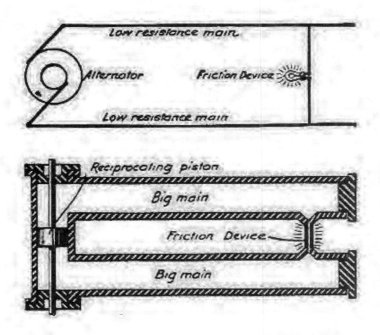

Fig. 3. — Electric Transmission Thru Two Wires and Hydraulic Analog.

As regards signaling without wires, the application of these radiations for the purpose was quite obvious. When Dr. Hertz was asked whether such a system would be of practical value, he did not think so, and he was correct in his forecast. The best that might have been expected was a method of communication similar to the heliographic and subject to the same or even greater limitations.

In the spring of 1891 I gave my demonstrations with a high frequency machine before the American Institute of Electrical Engineers at Columbia College, which laid the foundation to a new and far more promising departure. Altho the laws of electrical resonance were well known at that time and my lamented friend, Dr. John Hopkinson, had even indicated their specific application to an alternator in the Proceedings of the Institute of Electrical Engineers, London, Nov. 13, 1889, nothing had been done towards the practical use of this knowledge and it is probable that those experiments of mine were the first public exhibition with resonant circuits, more particularly of high frequency. While the spontaneous success of my lecture was due to spectacular features, its chief import was in showing that all kinds of devices could be operated thru a single wire without return. This was the initial step in the evolution of my wireless system. The idea presented itself to me that it might be possible, under observance of proper conditions of resonance, to transmit electric energy thru the earth, thus dispensing with all artificial conductors. Anyone who might wish to examine impartially the merit of that early suggestion must not view it in the light of present day science. I only need to say that as late as 1893, when I had prepared an elaborate chapter on my wireless system, dwelling on its various instrumentalities and future prospects, Mr. Joseph Wetzler and other friends of mine emphatically protested against its publication on the ground that such idle and far-fetched speculations would injure me in the opinion of conservative business men. So it came that only a small part of what I had intended to say was embodied in my address of that year before the Franklin Institute and National Electric Light Association under the chapter "On Electrical Resonance." This little salvage from the wreck has earned me the title of "Father of the Wireless" from many well-disposed fellow workers, rather than the invention of scores of appliances which have brought wireless transmission within the reach of every young amateur and which, in a time not distant, will lead to undertakings overshadowing in magnitude and importance all past achievements of the engineer.

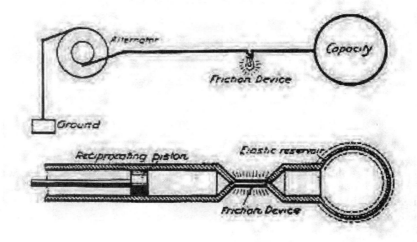

Fig. 4. — Electric Transmission Thru a Single Wire Hydraulic Analog.

The popular impression is that my wireless work was begun in 1893, but as a matter of fact I spent the two preceding years in investigations, employing forms of apparatus, some of which were almost like those of today. It was clear to me from the very start that the successful consummation could only be brought about by a number of radical improvements. Suitable high frequency generators and electrical oscillators had first to be produced. The energy of these had to be transformed in effective transmitters and collected at a distance in proper receivers. Such a system would be manifestly circumscribed in its usefulness if all extraneous interference were not prevented and exclusiveness secured. In time, however, I recognized that devices of this kind, to be most effective and efficient, should be designed with due regard to the physical properties of this planet and the electrical conditions obtaining on the same. I will briefly touch upon the salient advances as they were made in the gradual development of the system.

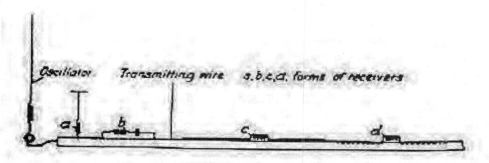

Fig. 5. — Illustrating Typical Arrangements for Collecting Energy In a System of Transmission Thru a Single Wire.

The high frequency alternator employed in my first demonstrations is illustrated in Fig. 1. It comprised a field ring, with 384 pole projections and a disc armature with coils wound in one single layer which were connected in various ways according to requirements. It was an excellent machine for experimental purposes, furnishing sinusoidal currents of from 10,000 to 20,000 cycles per second. The output was comparatively large, due to the fact that as much as 30 amperes per square millimeter could be past thru the coils without injury.

The diagram in Fig. 2 shows the circuit arrangements as used in my lecture. Resonant conditions were maintained by means of a condenser subdivided into small sections, the finer adjustments being effected by a movable iron core within an inductance coil. Loosely linked with the latter was a high tension secondary which was tuned to the primary.

The operation of devices thru a single wire without return was puzzling at first because of its novelty, but can be readily explained by suitable analogs. For this purpose reference is made to Figs. 3 and 4.

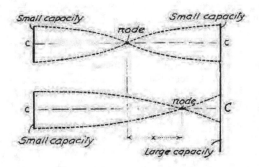

Fig. 6. — Diagram Elucidating Effect of Large Capacity on One End.

In the former the low resistance electrical conductors are represented by pipes of large section, the alternator by an oscillating piston and the filament of an incandescent lamp by a minute channel connecting the pipes. It will be clear from a glance at the diagram that very slight excursions of the piston would cause the fluid to rush with high velocity thru the small channel and that virtually all the energy of movement would be transformed into heat by friction, similarly to that of the electric current in the lamp filament. The second diagram will now be self-explanatory. Corresponding to the terminal capacity of the electric system an elastic reservoir is employed which dispenses with the necessity of a return pipe. As the piston oscillates the bag expands and contracts, and the fluid is made to surge thru the restricted passage with great speed, this resulting in the generation of heat as in the incandescent lamp. Theoretically considered, the efficiency of conversion of energy should be the same in both cases.

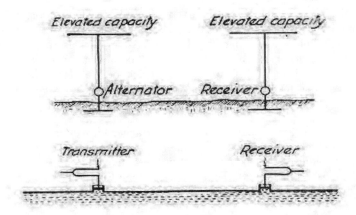

Fig. 7. — Transmission of Electrical Energy Thru the Earth as Illustrated in Tesla's Lectures Before the Franklin Institute and Electric Light Association in February and March, 1893, and Mechanical Analog of the Same.

Granted, then, that an economic system of power transmission thru a single wire is practicable, the question arises how to collect the energy in the receivers. With this object attention is called to Fig. 5, in which a conductor is shown excited by an oscillator joined to it at one end. Evidently, as the periodic impulses pass thru the wire, differences of potential will be created along the same as well as at right angles to it in the surrounding medium and either of these may be usefully applied. Thus at *a*, a circuit comprising an inductance and capacity is resonantly excited in the transverse, and at *b*, in the longitudinal sense. At *c*, energy is collected in a circuit parallel to the conductor but not in contact with it, and again at *d*, in a circuit which is partly sunk into the conductor and may be, or not, electrically connected to the same. It is important to keep these typical dispositions in mind, for however the distant actions of the oscillator might be modified thru the immense extent of the globe the principles involved are the same.

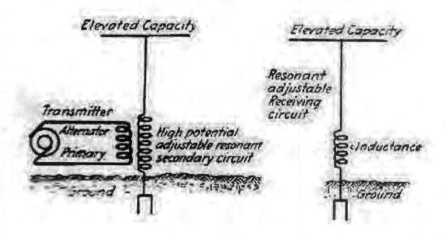

Fig. 8. — Tesla's System of Wireless Transmission Thru the Earth as Actually Exposed In His Lectures Before the Franklin Institute and Electric Light Association in February and March, 1893.

Consider now the effect of such a conductor of vast dimensions on a circuit exciting it. The upper diagram of Fig. 6 illustrates a familiar oscillating system comprising a straight rod of self-inductance $2L$ with small terminal capacities cc and a node in the center. In the lower diagram of the figure a large capacity C is attached to the rod at one end with the result of shifting the node to the right, thru a distance corresponding to self-inductance X. As both parts of the system on either side of the node vibrate at the same rate, we have evidently, $(L + X) c = (L - X) C$ from which $X = L (C - c / C + c)$. When the capacity C becomes commensurate to that of the earth, X approximates L, in other words, the node is close to the ground connection. *The exact determination of its position is very important in the calculation of certain terrestrial electrical and geodetic data* and I have devised special means with this purpose in view.

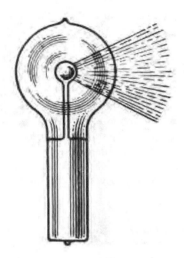

Fig. 9. — The Forerunner of the Audion — the Most Sensitive Wireless Detector Known, as Described by Tesla In His Lecture Before the Institution of Electrical Engineers, London, February, 1892.

My original plan of transmitting energy without wires is shown in the upper diagram of Fig. 7, while the lower one illustrates its mechanical analog, first published in my article in the *Century Magazine* of June, 1900. An alternator, preferably of high tension, has one of its terminals connected to the ground and the other to an elevated capacity and impresses its oscillations upon the earth. At a distant point a receiving circuit, likewise connected to ground and to an elevated capacity, collects some of the energy and actuates a suitable device. I suggested a multiplication of such units in order to intensify the effects, an idea which may yet prove valuable. In the analog two tuning forks are provided, one at the sending and the other at the receiving station, each having attached to its lower prong a piston fitting in a cylinder. The two cylinders communicate with a large elastic reservoir filled with an incompressible fluid. The vibrations transmitted to either of the tuning forks excite them by resonance and, thru electrical contacts or otherwise, bring about the desired result. This, I may say, was not a mere mechanical illustration, but a simple representation of my apparatus for submarine signaling, perfected by me in 1892, but not appreciated at that time, altho more efficient than the instruments now in use.

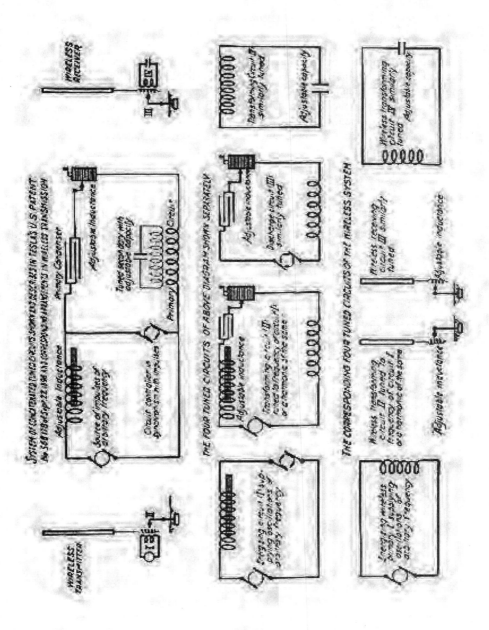

Fig. 10. — Tesla's System of Concatenated Tuned Circuits Shown and Described In U. S. Patent No. 568,178 of September 22, 1896, and Corresponding Arrangements in Wireless Transmission.

The electric diagram in Fig. 7, which was reproduced from my lecture, was meant only for the exposition of the principle. The arrangement, as I described it in detail, is shown in Fig. 8. In this case an alternator energizes the primary of a transformer, the high tension secondary of which is connected to the ground and an elevated capacity and tuned to the imprest oscillations. The receiving circuit consists of an inductance connected to the ground and to an elevated terminal without break and is resonantly responsive to the transmitted oscillations. A specific form of receiving device was not mentioned, but I had in mind to transform the received currents and thus make their volume and tension suitable for any purpose. This, in substance, is the system of today and I am not aware of a single authenticated instance of successful transmission at considerable distance by different instrumentalities. It might, perhaps, not be clear to

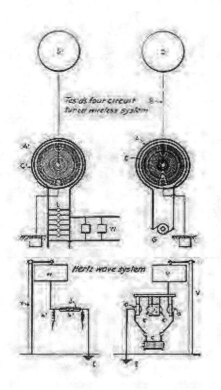

those who have perused my first description of these improvements that, besides making known new and efficient types of apparatus, I gave to the world a wireless system of potentialities far beyond anything before conceived. I made explicit and repeated statements that I contemplated transmission, absolutely unlimited as to terrestrial distance and amount of energy. But, altho I have overcome all obstacles which seemed in the beginning unsurmountable and found elegant solutions of all the problems which confronted me, yet, even at this very day, the majority of experts are still blind to the possibilities which are within easy attainment.

Fig. 11. — Tesla's Four Circuit Tuned System Contrasted With the Contemporaneous Hertz-wave System.

My confidence that a signal could be easily flashed around the globe was strengthened thru the discovery of the "rotating brush," a wonderful phenomenon which I have fully described in my address before the Institution of Electrical Engineers, London, in 1892, and which is illustrated in Fig. 9. This is undoubtedly the most delicate wireless detector known, but for a long time it was hard to produce and to maintain in the sensitive state. These difficulties do not exist now and I am looking to valuable applications of this device, particularly in connection with the high-speed photographic method, which I suggested, in wireless, as well as in wire, transmission.

Possibly the most important advances during the following three or four years were my system of concatenated tuned circuits and methods of regulation, now universally adopted. The intimate bearing of these inventions on the development of the wireless art will appear from Fig. 10, which illustrates an arrangement described in my U. S. Patent No. 568178 of September 22, 1896, and corresponding dispositions of wireless apparatus. The captions of the individual diagrams are thought sufficiently explicit to dispense with further comment. I will merely remark that in this early record, in addition to indicating how any number of resonant circuits may be linked and regulated, I have shown the advantage of the proper timing of primary impulses and use of harmonics. In a farcical wireless suit in London, some engineers, reckless of their reputation, have claimed that my circuits were not at all attuned; in fact they asserted that I had looked upon resonance as a sort of wild and untamable beast!

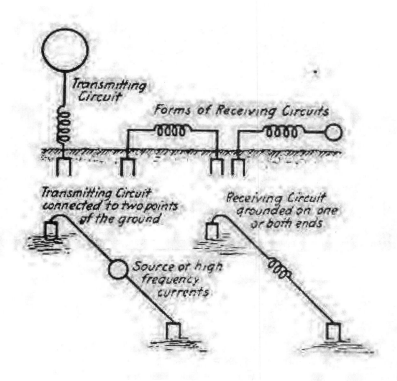

Fig. 12. — Arrangements of Directive Circuits Described In Tesla's U. S. Patent No. 613,809 of November 8. 1898, on "Method of and Apparatus for Controlling Mechanism of Moving Vessels or Vehicles."

It will be of interest to compare my system as first described in a Belgian patent of 1897 with the Hertz-wave system of that period. The significant differences between them will be observed at a glance. The first enables us to transmit economically energy to any distance and is of inestimable value; the latter is capable of a radius of only a few miles and is worthless. In the first there are no spark-gaps and the actions are enormously magnified by resonance. In both transmitter and receiver the currents are transformed and rendered more effective and suitable for the operation of any desired device. Properly constructed, my system is safe against static and other interference and the amount of energy which may be transmitted is *billions of times greater* than with the Hertzian which has none of these virtues, has never been used successfully and of which no trace can be found at present.

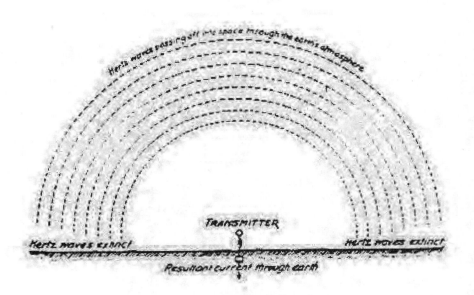

Fig. 13. — Diagram Exposing the Fallacy of the Gilding Wave Theory as Propounded In Wireless Text Books.

A well-advertised expert gave out a statement in 1899 that my apparatus did not work and that it would take 200 years before a message would be flashed across the Atlantic and even accepted stolidly my congratulations on a supposed great feat. But subsequent examination of the records showed that my devices were secretly used all the time and ever since I learned of this I have treated these Borgia-Medici methods with the contempt in which they are held by all fair-minded men. The wholesale appropriation of my inventions was, however, not always without a diverting side. As an example to the point I may mention my oscillation transformer operating with an air gap.

This was in turn replaced by a carbon arc, quenched gap, an atmosphere of hydrogen, argon or helium, by a mechanical break with oppositely rotating members, a mercury interrupter or some kind of a vacuum bulb and by such *tours de force* as many new "systems" have been produced. I refer to this of course, without the slightest ill-feeling, let us advance by all means. But I cannot help thinking how much better it would have been if the ingenious men, who have originated these "systems," had invented something of their own instead of depending on me altogether.

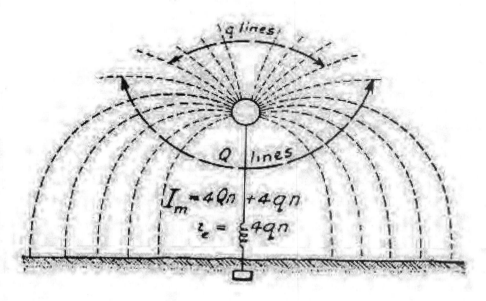

Fig. 14. — Diagram Explaining the Relation Between the Effective and the Measured Current In the Antenna.

Before 1900 two most valuable improvements were made. One of these was my individualized system with transmitters emitting a wave-complex and receivers comprising separate tuned elements cooperatively associated. The underlying principle can be explained in a few words. Suppose that there are n simple vibrations suitable for use in wireless transmission, the probability that any one tune will be struck by an extraneous disturbance is $1/n$. There will then remain $n-1$ vibrations and the chance that one of these will be excited is $1/n-1$ hence the probability that two tunes would be struck at the same time is $1/n(n-1)$. Similarly, for a combination of three the chance will be $1/n(n-1)(n-2)$ and so on. It will be readily seen that in this manner any desired degree of safety against the statics or other kind of disturbance can be attained provided the receiving apparatus is so designed that is operation is possible only thru the joint action of all the tuned elements. This was a difficult problem which I have successfully solved so that now *any desired number of simultaneous messages is practicable in the transmission thru the earth as well as thru artificial conductors.*

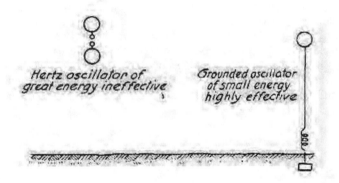

Fig. 15. — Illustrating One of the General Evidences Against the Space Wave Transmission.

The other invention, of still greater importance, is a peculiar oscillator enabling the transmission of energy without wires in any quantity that may ever be required for industrial use, to any distance, and with very high economy. It was the outcome of years of systematic study and investigation and wonders will be achieved by its means.

The prevailing misconception of the mechanism involved in the wireless transmission has been responsible for various unwarranted announcements which have misled the public and worked harm. By keeping steadily in mind that the transmission thru the earth is in every respect identical to that thru a straight wire, one will gain a clear understanding of the phenomena and will be able to judge correctly the merits of a new scheme. Without wishing to detract from the value of any plan that has been put forward I may say that they are devoid of novelty. So for instance in Fig. 12 arrangements of transmitting and receiving circuits are illustrated, which I have described in my U. S. Patent No. 613809 of November 8, 1898 on a Method of and Apparatus for Controlling Mechanism of Moving Vessels or Vehicles, and which have been recently dished up as original discoveries. In other patents and technical publications I have suggested conductors in the ground as one of the obvious modifications indicated in Fig. 5.

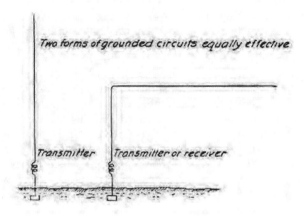

Fig. 16. — Showing Unimportance of Relative Position of Transmitting and Receiving Antennae In Disproval of the Hertz-wave Theory.

For the same reason the statics are still the bane of the wireless. There is about as much virtue in the remedies recently proposed as in hair restorers. *A small and compact apparatus has been produced which does away entirely with this trouble*, at least in plants suitably remodeled.

Nothing is more important in the present phase of development of the wireless art than to dispose of the dominating erroneous ideas. With this object I shall advance a few arguments based on my own observations *which prove that Hertz waves have little to do with the results obtained even at small distances.*

In Fig. 13 a transmitter is shown radiating space waves of considerable frequency. It is generally believed that these waves pass along the earth's surface and thus affect the receivers. I can hardly think of anything more improbable than this "gliding wave" theory and the conception of the "guided wireless" which are contrary to all laws of action and reaction. Why should these disturbances cling to a conductor where they are counteracted by induced currents, when they can propagate in all other directions unimpeded? The fact is that the radiations of the transmitter passing along the earth's surface are soon extinguished, the height of the inactive zone indicated in the diagram, being some function of the wave length, the bulk of the waves traversing freely the atmosphere. Terrestrial phenomena which I have noted conclusively show that there is no *Heaviside layer*, or if it exists, it is of no effect. It certainly would be unfortunate if the human race were thus imprisoned and forever without power to reach out into the depths of space.

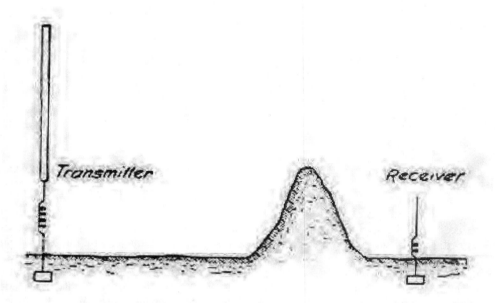

Fig. 17. — Illustrating Influence of Obstacle In the Path of Transmission as Evidence Against the Hertz-wave Theory.

The actions at a distance cannot be proportionate to the height of the antenna and the current in the same. I shall endeavor to make this clear by reference to diagram in Fig. 14. The elevated terminal charged to a high potential induces an equal and opposite charge in the earth and there are thus Q lines giving an average current $I = 4Qn$ which circulates locally and is useless except that it adds to the momentum. A relatively small number of lines q however, go off to great distance and to these corresponds a mean current of $ie = 4qn$ to which is due the action at a distance. The total average current in the antenna is thus $Im = 4Qn + 4qn$ and its intensity is no criterion for the performance. The electric efficiency of the antenna is $q / Q+q$ and this is often a very small fraction.

Dr. L. W. Austin and Mr. J. L. Hogan have made quantitative measurements which are valuable, but far from supporting the Hertz wave theory they are evidences in disproval of the same, as will be easily perceived by taking the above facts into consideration. Dr. Austin's researches are especially useful and instructive and I regret that I cannot agree with him on this subject. I do not think that if his receiver was affected by Hertz waves he could ever establish such relations as he has found, but he would be likely to reach these results if the Hertz waves were in a large part eliminated. At great distance the space waves and the current waves are of equal energy, the former being merely an accompanying manifestation of the latter in accordance with the fundamental teachings of Maxwell.

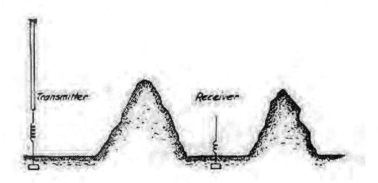

Fig. 18. — Showing Effect of Two Hills as Further Proof Against the Hertz-wave Theory.

It occurs to me here to ask the question — why have the Hertz waves been reduced from the original frequencies to those I have advocated for my system, when in so doing the activity of the transmitting apparatus has been reduced a billion fold? I can invite any expert to perform an experiment such as is illustrated in Fig. 15, which shows the classical Hertz oscillator and my grounded transmitting circuit. It is a fact which I have demonstrated that, altho we may have in the Hertz oscillator an activity thousands of times greater, the effect on the receiver is not to be compared to that of the grounded circuit.

This shows that *in the transmission from an airplane we are merely working thru a condenser,* the capacity of which is a function of a logarithmic ratio between the length of the conductor and the distance from the ground. The receiver is affected in exactly the same manner as from an ordinary transmitter, the only difference being that there is a certain modification of the action which can be predetermined from the electrical constants. It is not at all difficult to maintain communication between an airplane and a station on the ground, on the contrary, the feat is very easy.

To mention another experiment in support of my view, I may refer to Fig. 16 in which two grounded circuits are shown excited by oscillations of the Hertzian order. It will be found that the antennas can be put out of parallelism without noticeable change in the action on the receiver, this proving that it is due to currents propagated thru the ground and not to space waves.

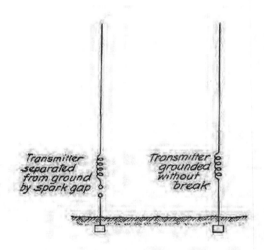

Fig. 19. — Comparing the Actions of Two Forms of Transmitter as Bearing Out the Fallacy of the Hertz-wave Theory.

Particularly significant are the results obtained in cases illustrated in Figures 17 and 18. In the former an obstacle is shown in the path of the waves but unless the receiver is within the effective *electrostatic* influence of the mountain range, the signals are not appreciably weakened by the presence of the latter, because the currents pass under it and excite the circuit in the same way as if it were attached to an energized wire. If, as in Fig. 18, a second range happens to be beyond the receiver, it could only strengthen the Hertz wave effect by reflection, but as a matter of fact it detracts greatly from the intensity of the received impulses because the electric niveau between the mountains is raised, as I have explained with my lightning protector in the Experimenter of February.

Again in Fig. 19 two transmitting circuits, one grounded directly and the other thru an air gap, are shown. It is a common observation that the former is far more effective, which could not be the case with Hertz radiations. In a like manner if two grounded circuits are observed from day to day the effect is found to increase greatly with the dampness of the ground, and for the same reason also the transmission thru sea-water is more efficient.

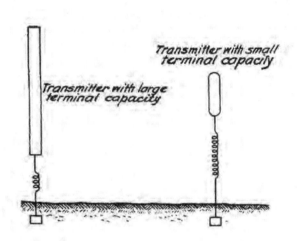

Fig. 20. — Disproving the Hertz-wave Theory by Two Transmitters, One of Great and the Other of Small Energy.

An illuminating experiment is indicated in Fig. 20 in which two grounded transmitters are shown, one with a large and the other with a small terminal capacity. Suppose that the latter be 1/10 of the former but that it is charged to 10 times the potential and let the frequency of the two circuits and therefore the currents in both antennas be exactly the same. The circuit with the smaller capacity will then have 10 times the energy of the other but the effects on the receiver will be in no wise proportionate.

The same conclusions will be reached by transmitting and receiving circuits with wires buried underground. In each case the actions carefully investigated will be found to be due to *earth currents*. Numerous other proofs might be cited which can be easily verified. So for example *oscillations of low frequency* are ever so much more effective in the transmission which is inconsistent with the prevailing idea. My observations in 1900 and the recent transmissions of signals to very great distances are another emphatic disproval.

The Hertz wave theory of wireless transmission may be kept up for a while, but I do not hesitate to say that in a short time it will be recognized as one of the most remarkable and inexplicable aberrations of the scientific mind which has ever been recorded in history.

9   Nikola Tesla and the Wireless Transmission of Energy

A. S. Marincic
Department of Electrical Engineering University of Belgrade
Belgrade, Yugoslavia

ABSTRACT

Nikola Tesla, the inventor of the polyphase-current system, is best known for his contribution regarding induction and other types of alternating-current machines. His patents and his published and unpublished notes about wireless transmission of energy are less known and, if known to some extent, they are usually wrongly interpreted. For many years the author studied Tesla's works on wireless transmission of energy and that what is given here is a review of relevant documents, unpublished notes and letters from the archives of the Nikola Tesla Museum in Belgrade. An attempt is made to explain Tesla's physical model on the basis of which he concluded that the wireless transmission of energy on a global scale is possible. His model is critically examined in view of the present day knowledge of extremely low frequency propagation phenomena.

INTRODUCTION

Nikola Tesla, a prolific inventor, well known for his contributions to the present day alternating current system, and less known for his important discoveries in the high frequency field, devoted nearly 50 years to developing his ideas of wireless transmission of energy. When he began to think about it, some time in 1890-91, only a few facts were known about electromagnetic waves. Following the publication of Hertz's research in 1888, which provided confirmation of J. Clark Maxwell's dynamic theory of the electromagnetic field of 1865, scientists became more and more convinced that electromagnetic waves behaved like light waves, propagating in straight lines. This led to pessimistic conclusions about the possible range of radio stations, which were soon refuted by experiments using the antenna-earth system designed by Tesla in 1893. Tesla invented "open resonant circuit" with one condenser plate in the earth and the other elevated above the earth surface. Such open resonant circuits were used at the transmitting and the receiving sides. Tesla thought that the electric energy could be transmitted through the globe just as through a good conductor. For many years the author searched through Tesla's written and unpublished notes in order to obtain a complete picture of the Tesla wireless transmission model. That which was published about it at the beginning of this century did more harm than good to Tesla. It was only recently that Tesla's name reappeared in papers dealing with the propagation of extremely low frequency (ELF) radio waves and resonance of the earth.[1,2,3] In spite of the fact that Tesla's model of energy transmission is oversimplified and lacking a mathematical rigor, it is fascinating in that it provides some insight to the very recent development in the ELF field where electromagnetic waves with low attenuation, as predicted by Tesla, are possible.

SOME EARLY EXPERIMENTS ON WIRELESS TRANSMISSION OF ENERGY

In 1881 Nikola Tesla invented the high frequency, high voltage transformer known since that time as "The Tesla Coil." This transformer contained little or no iron and still could easily, with high efficiency, produce extremely high voltages. With this kind of transformer, connected as a part of a coupled oscillatory circuit with a spark gap driven by the low frequency AC or DC source, Tesla produced strong electric field in his laboratory which could produce light in passing through a rarefied gas tube. He was impressed by the efficiency of this kind of light generation and invented several types of gas lamps which produced light that is "soft and agreeable to the eye, closely resembling daylight." The first record of the Tesla coil is to be found in Patent No. 454 622 of June 23, 1891, under the title "System of electric lighting."

The idea that the electric energy could be transmitted through one wire, or no wires, seems to appear for the first time in connection with the Tesla system of electric lighting. Subsequently, Tesla presented much new information about the discharge oscillator and his further research on high frequency currents in the lecture he gave to the IEE in London, February 1892 which he repeated in London and then in Paris.[4] It was there that he gave once again open demonstration of his "single wire" lamps. He noted that HF currents readily passed through slightly rarefied gas and suggested that this might be used for driving motors and lamps at considerable distance from the source, the high frequency resonant transformer being an important component in such a system.3

In February 1893 Tesla held a third lecture on high frequency currents before the Franklin Institute in Philadelphia, and repeated it in March before the National Electric Light Association in St. Louis.4 The most significant part of this lecture is that which refers to a system for "transmitting intelligence or perhaps power, to any distance through the earth or environing medium." What Tesla described here is recognized to be the foundation of some important principles and ideas of radio engineering, viz.: "the idea of inductive coupling between the driving and the working circuits; the importance of tuning both circuits, that is, the idea of an "oscillation transformer," and the idea of a capacitance loaded open secondary circuit."[5]

To Tesla, transmitting intelligence in 1893 seemed to be an established fact as can be judged from his words: "I no longer look upon this plan of energy or intelligence transmission as a mere theoretical possibility, but as a serious problem in electrical engineering, which must be carried out some day." From his numerous experiments he concluded that the transmission of HF currents through a single conductor was possible. The next step in his reasoning was to replace the single conductor with the earth. To him the earth behaved as a smooth perfectly conducting ball. In 1893 he wanted to find out the capacity of the earth, and what was its charge if electrified. He thought of disturbing the electric charge of the earth and detecting the period of the charge oscillation. It is fascinating that at that moment of radio development he thought of the earth resonance! It was about five years after Hertz's experiments, and at the moment when the radio waves could be detected within the laboratory distance. Tesla, also in 1893, speculated about "the upper strata of the air that are conducting" and their possible role in the transmission process.

In the period of 1893 to 1897 Tesla invented and conducted many types of HF spark-gap oscillators and improved the design of his HF transformer. In 1897 he was granted a patent entitled "Electrical Transformer" (No. 593 138, Nov. 2, 1897) in which he thoroughly explained how to design his HF transformer. The transformer had no iron core and the amount of copper was small. The primary coil had few turns of stranded copper wire wound around the secondary coil. Tesla tried cylindrical, conical and spiral coil shapes. For the best results he recommended that the length of the secondary coil of his transformer should approximately be "one quarter of the wave length of the electrical disturbance in the secondary circuit." This was somehow an optimized version of the Tesla coil from 1891, designed to produce a maximum secondary voltage. In the above mentioned patent Tesla described a single wire power transmission system with step up and step down resonant transformers. In fact, it was a single phase transmission system with the ground replacing one wire.

Developing further his "single wire" transmission system, Tesla concluded that the remaining copper conductor, connecting the two high tension terminals, could be removed, providing each terminal was connected to a body of large surface. He thought that by using extremely high voltages applied to the body of large surface (antenna) would render higher air strata conductive. Since the limiting pressure at which the gas becomes a good conductor is higher the higher the voltage, he maintained that it would not be necessary to elevate a metal conductor to an altitude of some 15 miles above sea level, but that layers of the atmosphere which could be good conductors could be reached by an antenna at much lower altitudes. Claims like this did not seem much convincing in 1897, so Tesla had to perform an experimental demonstration of power transmission through rarefied gas before an official of the Patent Office. From an original Tesla slide and the Patent No. 645 576, March 20, 1900 (applied Sept. 2, 1897), it was found that Tesla connected the two high tension terminals through an air column under pressure between 120 and 150 mm Hg isolated from the surrounding air by a glass tube. At this air pressure, and with the transformers tuned to resonance, he claimed that efficient power transfer was achieved with a voltage of 2-4 million volts on the transmitter antenna. In the application, Tesla also claims patent rights to another, similar method of transmission, also using the earth as one conductor, and conductive high layers of the atmosphere as the other.

COLORADO SPRINGS LABORATORY

In the Houston Street laboratory in New York, Tesla made several high voltage Tesla coils but the space was limited there, and in 1898 he began looking for a site for a new laboratory. In mid-1899 he finally decided on Colorado Springs, a plateau about 2000 m above sea level, where he erected a shed large enough to house a high-frequency transformer with a coil diameter of 15 meters!3 In his article "The transmission of electric energy without wires," written in 1904 in The Electrical World and Engineer (reprinted in 1), Tesla writes that he came to Colorado Springs with the following goals:

1. To develop a transmitter of great power.
2. To perfect means for individualizing and isolating the energy transmitted.
3. To ascertain the law of propagation of currents through the earth and the atmosphere.

During his eight months stay in Colorado Springs Tesla wrote notes with a detailed day-by-day description of his research. Unlike many other records in the archives of the Nikola Tesla Museum in Belgrade, the Colorado Springs diary is continuous and orderly. Most likely, it was not intended for publication but he kept it as a record of his research. Tesla had an unhappy remembrance of the fire which in 1895 destroyed his laboratory in New York. In Colorado Springs he performed dangerous experiments and he could easily burn his own laboratory. So the notes could also have been a safety measure.

In Colorado Springs Tesla certainly thought a lot about his wireless transmission of energy but did not write much about it in his diary. He devoted most of his time (56%) to the development, measurements and tests of his huge Tesla coil, about 21% to developing receivers for small signals, about 16% to measuring the capacity of the vertical antenna, and about 6% to miscellaneous other research. Among the latter were some interesting remarks on the fire balls which he claimed to have produced. He described some wireless energy transmission but all one could see in the diary points to transmission over short distances.

One day (July 4, 1899) Tesla recorded unusual behavior of his thunderstorm recording instrument, a rotating coherer device. He noticed periodic recording of the instrument when the storm approached and receded from his laboratory. He concluded that it showed clearly the existence of stationary waves. He speculated about the point of their reflection but, what is most important, he was convinced that the stationary waves could be produced with his oscillator. In his idealized earth model he expected to notice the reflected waves from the opposite side of the earth. However, it was in vain. In Colorado Springs Tesla did not use sufficiently low frequencies so that he could not detect expected-peculiar behavior of the electromagnetic waves radiated from his mighty Tesla coil. It will be shown later that he made the necessary correction to the choice of operating frequency, but, unfortunately, he did not perform an experiment that would prove what he claimed.

LONG ISLAND LABORATORY

After returning to New York from Colorado Springs laboratory, Tesla took energetic steps to get backing for the implementation of a system of "World Telegraphy." With the significant financial help of J. Pierpont Morgan and "unselfish and valuable assistance" of his friend Stanford White, architect, Tesla built a laboratory and nearly finished the transmitting tower (capacitively loaded monopole antenna). The work on this plant began in 1900 and was very intensive in the following three to four years.

In a letter to his sponsor J. Pierpont Morgan, early in 1902, Tesla explained in details his research aims in which he envisaged three "distinct steps to be made: 1) the transmission of minute amounts of energy and the production of feeble effects, barely perceptible by sensitive devices; 2) the transmission of notable amounts of energy dispensing with the necessity of sensitive devices and enabling the positive operation of any kind of apparatus requiring a small amount of power; and 3) the transmission of power in amounts of industrial significance. With the completion of my present undertaking the first step will be made." For the experiments with transmission of large power he envisaged the construction of a plant at Niagara to generate about 100 million volts. In 1904 he wrote to Morgan that "The Canadian Niagara Co. will agree in writing to furnish me 10,000 H.P. for 20 years without charge, if I put up a plant there to transmit this power without wires to other parts of the world."[6] As far as is known nothing came out of this plan. At that moment Tesla was "financially in a dreadful fix," as he explained his situation to Morgan after he refused to advance more money to Tesla in order to finish his telegraphy plant at Long Island.

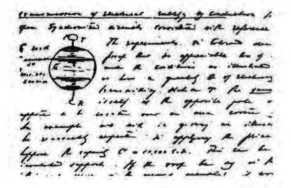

In searching through the archives of the Nikola Tesla Museum in Belgrade, many notes have been found relative to his Long Island research. The notes are scarce in 1901, more numerous in 1901 to 1903 and again rare in 1904, and hardly any in 1905 and 1906. In 1900 he prepared to work in the new laboratory and had to complete several patent applications. Among these patents is the one submitted on May 16, 1900 which is closely related to his wireless transmission plans. It was renewed June 17, 1902, and issued on April 18, 1905, under the title "Art of transmitting electrical energy through the natural mediums" (No. 787 412). Many of Tesla's complicated explanations in this patent become clearer and even more acceptable, if his Long Island notes are studied. For example, when he states that "the rate of radiation of energy in the form of Hertzian or electromagnetic waves is very small," one may think that this is nonsense. But when one reads in his notes that he tried to select the operating frequency in such a way that the di-pole formed by the globe and the elevated ball (antenna) radiates little according to Hertz and Maxwell's formulas, one has to reconsider the judgment and think in terms of ELF propagation phenomena.[7] Tesla made calculations for operating frequencies 4, 6, 60 Hz, and at somewhat higher frequencies of 15 and 20 kHz. Particularly at the lower frequencies he expected the resonance of the earth to occur. It is enchanting that he calculated the lowest resonant frequency of the earth to be 6 Hz, while a very elaborate theory of today predicts 10.5 Hz (losses neglected), and the actual measurements gave resonant peak at about 8 Hz. Tesla assumed that the earth behaves like a conducting ball through which the current passes in Figure 1. Tesla also predicted that at the antinodes R, the effects will be pronounced as the current density is the highest there. He also predicted that the standing waves could be set around the globe, and that, in contrast to classical electromagnetic wave propagation; the field at some points may increase with the distance from the transmitter! Just to illustrate that this is possible at ELF a plot is given of the principal fields in the Earth-ionosphere waveguide radiated by a vertical dipole from a recent book[8] on this subject (Figure 2). As is clearly seen, at the operating frequency of 75 Hz, the fields, from about 4 Mm show the characteristic standing wave pattern, most pronounced around the antinodes at 20 mm, and pronounced around the antinode at 20 mm, and do not decay inversely with distance which is the characteristic of the radiation field.

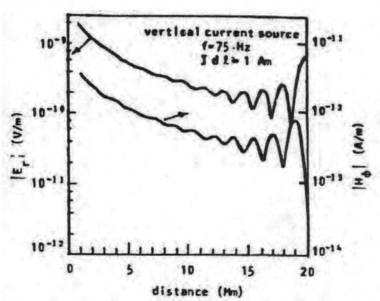

Unfortunately, the radiating systems at ELF are very inefficient. Tesla thought about that, too. As the radiating antenna he suggested the use of a vertical monopole (15 miles long or so) or two plates buried some distance apart. To illustrate this point some of his drawing, probably prepared for a patent application, is shown in Figure 3.

The similarity with some present day ELF antennas cannot be overlooked. However, one knows that the efficiency of ELF antennas is, in general, very small. As an example let be quoted some relevant data for a typical ELF antenna operating at 45 Hz. For a grid antenna of total length of 106m, the efficiency of the antenna is only 0.026 percent! Even a very low attenuation of the elector-magnetic wave at this frequency, which is only 0.8 db/Mm, cannot help much. If another similar antenna is used at the receiving side, the efficiency of the whole system would be, roughly, (0.0026)2 percent.* As the operating frequency is increased, the efficiency of the transmission system increases, but then the attenuation of the wave in the propagation rapidly increases and the overall efficiency is again very low. From various experimental data, approximate values of attenuation at several frequencies of interest are given in the following table. [7]

Carrier frequency Hz	10	50	100	300	1000
Attenuation dB/Mm	0.25	0.80	1.00	3.00	20.00

If the antenna size is specified, it would be possible to calculate the overall efficiency of ELF system as a function of distance and frequency. Such a calculation is hardly necessary if a large distance system is required. The antenna efficiency of reasonable size is high enough at about 20 kHz, where the attenuation of the wave in propagation is too high. For example, an antenna that is made in the form of two umbrella shaped top loaded monopoles, each of 2 km in diameter and some 250 m high, with the grounding system of some 3000 km of copper wire, has 86% efficiency at 20 kHz and only 0.1% at 100 Hz.8 Hence, it is possible that the efficiency of the system may vary with frequency and distance but it will be always very low. Thus, at present, ELF systems are used in communication with objects under the sea, where one can make use of the property of ELF waves to penetrate to deeply submerged receiving antennas.

CONCLUSIONS

For many years Nikola Tesla tried to realize wireless transmission of energy on a large scale. From the initial steps in 1891, when he produced light in single wire or no wire gas tubes within laboratory distance, he gradually improved his wireless energy transmission system. In Colorado Springs he lighted a few lamps some distance apart from the huge transmitter that produced open circuit voltage of about ten million volts. He was convinced that at sufficiently low frequencies, low attenuation and standing waves characterize the behavior of his transmission system. He also predicted that at very low frequency the earth resonances could be set. His concept of energy flow (through the earth!) was peculiar and to some extent misleading. Tesla did not manage to turn attention of the world to his ideas of wireless energy transmission, but some of his statements and visions were proved much later when the research into ELF propagation phenomena began about a half century later. Today is known that Tesla was right in that at low frequencies certain waves can propagate with little attenuation, that the standing waves can be set and that the earth resonates. Regarding energy transmission is known that the overall efficiency of an ELF system is very low for antennas of reasonable sizes irrespective of the fact that the attenuation of electromagnetic waves at very low frequencies is very small. Thus, at present, ELF transmission systems seem to be of use only in some special communication systems with buried receiving antennas.

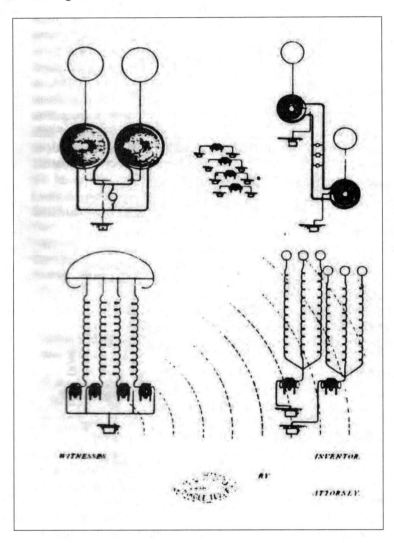

REFERENCES

1. J.R. Wait, "Historical background and introduction to the special issue on extremely low frequency (ELF) propagation," IEEE Transactions on

Communications, vol. COM-22, pp. 353-354, April 1974.

2. J.D. Jackson, Classical Electrodynamics. New York: John Wiley, 1975.

3. A.S. Marincic, "Introduction and Commentaries" in Nikola Tesla, Colorado Springs Notes 1899-1900, Belgrade: Nolit, 1976.

4. N. Tesla, Lectures, Patents, Articles. Belgrade: Museum Nikola Tesla, 1976.

5. L.P. Wheeler, "Tesla's contribution to high frequency," Electr. Engineering, New York, August 1943.

6. Letters of Nikola Tesla to J. Pierpont Morgan: Jan. 9, 1902 and Jan. 13, 1904. Belgrade: Archives of Nilola Tesla Museum.

7. J.R. Wait, "Propagation of ELF electromagnetic waves and project Sanguine/Seafarer," IEE Journal of Oceanic Engineering, vol. OE-2, pp. 161-172. April 1977.

8. M.L. Burrows, ELF Communication Antennas. Stevenage Peter Peregrinus, 1978.

10 Wireless Energy Transmission: Nikola Tesla "UNPLUGGED"

Nikolaos Simos, Ph.D., P.E.

Senior Scientist
Brookhaven National Laboratory

Proceedings of the Seventh Conference on Future Energy
Albuquerque NM, July 29-August 1, 2015

OBJECTIVE - SUMMARY

By staying within the framework of the accepted governing physics laws:

- Attempt to arrive to an understanding of how Tesla conceived and planned
- WIRELESS Energy Transmission at his Shoreham Experimental Station
- In particular try to understand the Principles & his Methods
- The objective is NOT to prove or disprove his ideas BUT rather to understand what he was after
- Aim is to steer clear from convoluted concepts which, after all have hurt rather than helped N. Tesla's scientific contributions and reputation

Presented are connections of Tesla's experiments to Maxwell's equations in an effort to understand what he was after and the physics behind the ideas. Effort was put forward to study the fields that Tesla claimed he created at Colorado Springs and the fields he intended to create at the Shoreham, NY laboratory. Central to his approach was the existence and/or excitation of longitudinal-type waves. Given that there exists a debate regarding these types of waves or wave-like disturbances of the electric and magnetic fields a discussion on the subject is presented based on the solution of Maxwell's equations under the presence of sources which in turn were part of Tesla's work on the subject of energy transmission.

Finally, Tesla's envisioned approaches of wireless energy transmission based on atmospheric conduction (closed circuit) and resonance (open circuit) are discussed on the basis of their feasibility.

Resonance.......and Nikola Tesla (Cornerstone of his work)

Heraclitus: "τα παντα ρει ..."
    Translated as "everything is in motion"

Tesla understood that "everything oscillates"
-        earth
-        Fields (electric, mechanical, etc.)

Finally Nikola Tesla is in the company of pioneers and immortal scientists!

Figure 1.2: Immortal scientists of electromagnetic theory. From left to right: Jean-Baptiste Biot (1774–1862), French physicist, astronomer, and mathematician. Heinrich Rudolf Hertz (1857–1894), German physicist. Hendrik Antoon Lorentz (1853–1928), Dutch physicist. Nikola Tesla (1856–1943), Serbian inventor, mechanical engineer, and electrical engineer.

Figure 1.1: The pioneers of electromagnetic theory. From left to right: André Marie Ampère (1775–1836), French physicist. Michael Faraday (1791–1867), English chemist and physicist. James Clerk Maxwell (1831–1879), Scottish physicist and mathematician.

After Gerhard Kristensson 2012, Lund, January 30, 2012

SOME BACKGROUND

Maxwell Equations .... and TESLA

While these form the basis of our understanding of electromagnetism we also accept that inconsistencies (both mathematical and physical) MAY EXIST in them. Stemming from abrupt changes or DISCONTINUITIES (not everything is smooth)

In an effort to understand what Tesla had in mind, resorting to some mathematical formulations and Maxwell Equation solutions will unavoidable (Kept to a minimum but necessary as a matter of record)

In all, we will NOT DEVIATE from the Maxwell Equations Domain (thus, no new physics)

We will try to explain everything starting every time from the universally accepted Maxwell governing relations

Maxwell Equations......?   What are they?

Maxwell's Equations: Ingenious way of explaining the connection between electric and magnetic fields (electric field E and magnetic field B)

$$\nabla \times E = -\mu \frac{\partial B}{\partial t} \, (Faraday's - Law)$$

$$\nabla \times B = J + \varepsilon \frac{\partial E}{\partial t} \, (Ampere - Law)$$

$$\nabla \bullet E = \frac{\rho_v}{\varepsilon} \, (Gauss - Law)$$

$$\nabla \bullet B = 0 \, (Gauss - Law - magnetism)$$

ρ_v is volume electric charge density,
J is the electric current density (ε is the permittivity
and
µ is the permeability

All is smooth and easy UNTIL one gets close to abrupt electric and magnetic sources

Classical Maxwell equations are supposed to be satisfied point-wise at any instant of time.

Therefore, discontinuous or diverging solutions are not allowed.

However, when electromagnetic fields vary with time, we can re-arrange and decouple these relations and express them in forms that describe propagation of waves

$$\Delta E(r,t) - \varepsilon\mu\frac{\partial^2 E(r,t)}{\partial t^2} = \frac{1}{\varepsilon}\nabla\rho(r,t) + \mu\frac{\partial J}{\partial t}(r,t)$$

$$\Delta B(r,t) - \varepsilon\mu\frac{\partial^2 B(r,t)}{\partial t^2} = -\mu\nabla \times J(r,t)$$

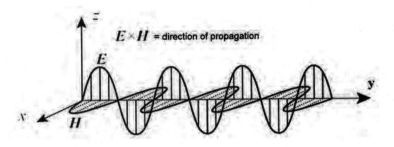

Transverse waves (**Hertz type**). An antenna radiates high frequency **transverse electromagnetic waves** as shown

But is it all Hertz-type? .... **Not necessarily!**
Can other types of waves be generated and propagated?

How about longitudinal-like waves?
Tesla bet on it ➔ will revisit the point

To understand how Tesla, while remaining true to the Maxwell governing principles, conceptualized and pursued his goals, we need to go back and start at *Colorado Springs*

TESLA AT COLORADO SPRINGS

- Pre-cursor to Wardenclyffe
- What did he do there?
- What did he observe?
- How relevant to his work in Wardenclyffe (Shoreham)?

What did he do there?

A. Wirelessly transmitted electric pulses to large distances
B. Observed the undiminished return of an electric pulse (which he speculated traveled to the antipode and back)

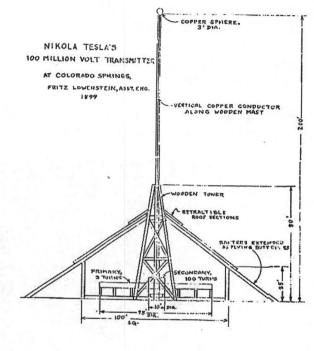

NIKOLA TESLA'S
100 MILLION VOLT TRANSMITTER
AT COLORADO SPRINGS,
FRITZ LOWENSTEIN, ASS'T. ENG.
1899

There has been speculation that what Tesla did he excited was resonances of the Schumann Cavity, but did he?

IN HIS WORDS:

"This mode of conveying electrical energy to a distance is not 'wireless' in the popular sense, but a transmission through a conductor, and one which is incomparably more perfect than any artificial one. All impediments of conduction arise from confinement of the electric and magnetic fluxes to narrow channels. The globe is free of such cramping and impediments........."

Some Background on the Cavity

Earth and Ionosphere (both are conducting, capacitor-like or a cavity that can resonate)

Plenty is happening in the space between: Electric field between this spherical capacitor exhibits characteristic modes when excited

Electrical discharges can excite them (lightning is a broadband or white noise excitation exciting all the modes in the cavity)

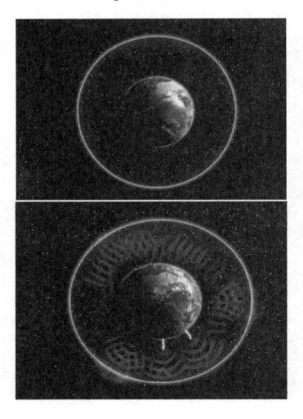

BUT why is it important?

Most importantly, it proves that certain electromagnetic types EXIST: such as longitudinal Extremely Low Frequency (ELF) and Very Low Frequency (VLF) waves.

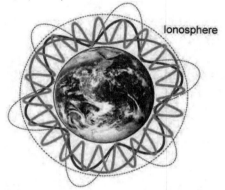

Graphical representation of modes in the cavity

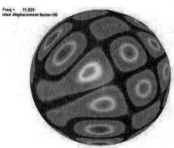

Simulation of mechanical equivalent of cavity modes

Schumann Resonances – "Earth Breathing"

STANDING waves (modes-like) are formed in the cavity which functions like a wave guide

Solving the CLASSICAL Maxwell Equations one can arrive at the conclusion that both:

Longitudinal and Transverse Modes (or standing waves) are present in the cavity, namely

- Longitudinal ~ Cavity mean radius
- Transverse ~ Cavity height

Two (2) orders of magnitude difference between eigenfrequencies

Specifically,

ELF (0-100 Hz) and VLF (0-10 kHz) waves or resonances develop.

These are considered *longitudinal* modes

Lightning (or man-made) discharges can EXCITE such modes and especially those considered as Extremely Low Frequency (ELF) – **including the 7.83 Hz mode**

Prof. O. Schumann around 1952 mathematically estimated the 7.83 Hz mode that was soon after (1954) confirmed with measurements

But wait a second ….

According to Wikipedia:

The first documented observations of global electromagnetic resonance were made by **Nikola Tesla at his Colorado Springs laboratory in 1899.** The observations led to certain conclusions about the electrical properties of the Earth, and which made the basis for his idea of wireless energy transmission.

Tesla researched ways to transfer power wirelessly over long distances (via transverse and longitudinal waves) transmitting ELF through the earth and as well as between the Earth's surface and the Kennelly-Heaviside layer (standing waves).

Making calculations based on the experiments, Tesla discovered that the resonant frequency of the Earth was ~8 Hz. In the 1950's researchers confirmed that the resonant frequency of the Earth's ionospheric cavity was in that range **(later named the Schumann resonance)**

Wikipedia Quote:

"Physicist Nikola Tesla back in 1890's was FIRST to experiment with the CAVITY, powerful discharges emulating lightning and exciting ELF waves based on which he "discovered" the resonance frequency of the earth at 8 Hz. Unfortunately Tesla was before his time and his discoveries were not taken seriously"

Wireless Transmission based on "Schumann Resonances"?

Not quite *FEASIBLE*, here is why:

- Spherical waves in Cavity or Modes

- Small Q-factor (ratio electric field energy stored in cavity per cycle/average power input)

While Tesla understood the Cavity effect and assessed its fundamental resonance, he did not ride these resonances to wirelessly transfer electric pulses.

So, what did Tesla excite at Colorado Springs?

He disturbed the earth's electric field (or blanket) with extreme electric discharges

- Electrically excite receivers at great distances
- Observed the undiminished electric pulses return from antipode

Determined a DIFFERENT frequency (not the 8 Hz Schumann fundamental resonance) which will guide his wireless transmission concepts

Determined that time required for transmitted pulse to travel to the antipode and back is **0.08484** seconds

This corresponds to a fundamental earth resonance frequency of **11.786892** Hz.

Electromagnetic waves at these frequencies!!! ELF & VLF shown in Schumann Cavity

Tesla quoted:
Power would be transmitted by creating "standing waves" in the earth by charging the earth with a giant electrical oscillator that would make the earth vibrate electrically in the same way a bell vibrates mechanically when it is struck with a hammer. . . ."

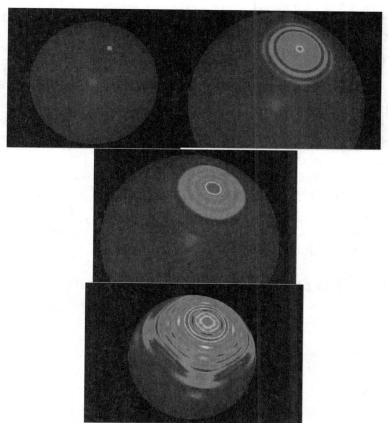

Figure set represents mechanical equivalent of earth response to impulse and wave generation, propagation to antipode and back

Tesla at Shoreham
Wireless power transmission
From Colorado Springs to Shoreham

Fundamental principles underlying Tesla's wireless energy transmission: (what he counted on)

1. Low frequency alternating current can be transmitted through the earth with low loss (net resistance between earth's antipodes <1 ohm)
2. Low frequency, high voltage alternating current via electric displacements by electrostatic induction, b) electrical conduction, or a combination of the two
3. Earth's naturally existing electrostatic potential (electrostatic field) 400,000 V potential with ionosphere. Downward directed E-field of about 100 V/m near surface ability to create disturbances in this charge as annular distortion of the background electric field

AIMED AT TWO (2) Methods, or Circuit Types, namely:

OPEN CIRCUIT   ➜ Earth Resonance

CLOSED Circuit   ➜ Atmospheric Conduction

How do his two concepts work?

CENTRAL to Tesla's concepts is the

Capacitor Principle

He is envisioning however a capacitor of enormous proportions.

In a typical capacitor (energy indeed transfers wirelessly across the two plates through the dielectric) AC current jumps the gap that is between 1/6-1/2 wavelength.

Because AC frequencies are high, wavelengths are small and so GAP must be small

In Tesla's model the two distant elevated terminals are electrically coupled together in a manner similar to the transfer of electrical energy between two closely spaced capacitor plates but at distances greatly exceeding 1/6 - 1/2 wavelength

His concept/model, at first glance, considered to be inconsistent with mainstream physics UNLESS different types of waves between the plates can be excited!

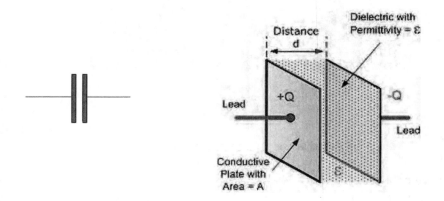

Needs to have receiving capacitor plate to resonate with its counterpart

TESLA Coil and System: What is it?

A capacitor and some other stuff!

- The wireless system is comprised of two sympathetically tuned electrical oscillators
- RF power supply connected to a grounded top-loaded helical resonator
- Distance apart = several wavelengths (unlike capacitor!!!)

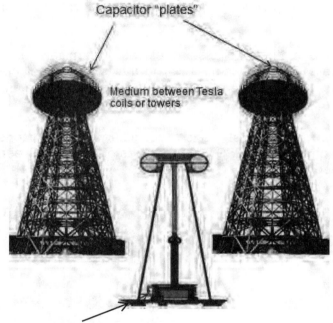

Capacitor "plates"

Medium between Tesla coils or towers

Primary (power in) and secondary coils

- Primary coil connected to power source
- Secondary (pancake) coil and connector to capacitor plate: *high voltage & **selected frequency (harmonic of 11.78 Hz)***
- Ionization of air between capacitor plates:
- *excitation of **LF longitudinal modes or waves***

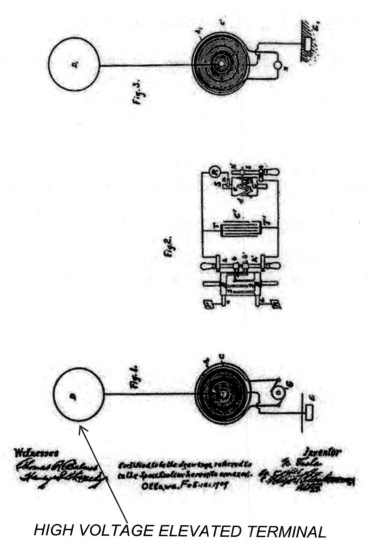

HIGH VOLTAGE ELEVATED TERMINAL

Tesla's Earth RESONANCE Method

Open Circuit working principle – Earth Resonance Method

- A Tesla coil earth resonance transmitter creates a local disturbance in the earth's charge

- He is now zooming-in to not simply excite the earth's electric field with a broadband source BUT with a source that operates at some harmonic of the earth field-resonance (**~11.78Hz**)

- Such source is the basis of Tesla's earth resonance method.

- The receiver (from the two Tesla coils) is Passive. Earth's electric field resonance will provide the synchronization of the receiver

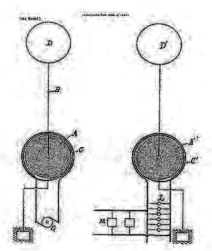

Recall that Tesla estimated that the fundamental earth resonance frequency is of the order **11.786892 Hz**.

That came from his Colorado Springs work!

So, there is the connection!

Governing Principles of Resonance Method

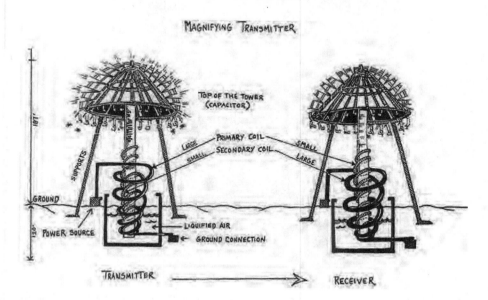

What must be the medium between similar coils (transmitter and receiver)? Types of waves it can support (VLF, ELF) Longitudinal waves? (Not Hertzian)That is the debate that is discussed later in the document in more detail.

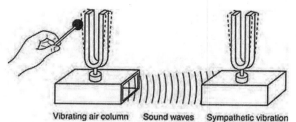

Vibrating air column   Sound waves   Sympathetic vibration

Medium between tuning forks (transmitter and receiver) is **AIR**

Sound (compression) waves propagate in the medium   Energy transfers between transmitter and receiver

If the characteristic tone of fork B is different than A, then NOTHING happens!

How will Tesla's Resonance System look like?

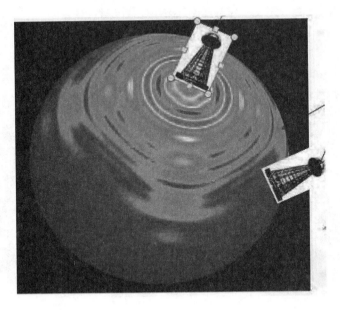

Receiver will be tuned to transmitter and will be excited by the field distortion, itself excited by the transmitter at the same frequency (much like tuning fork below)

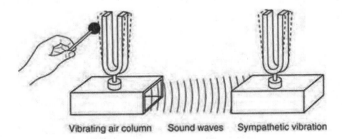

Vibrating air column   Sound waves   Sympathetic vibration

SAME THING, Only ELECTRICALLY!

OPEN CIRCUIT WORKING PRINCIPLE—
EARTHLY RESONANCE METHOD

System MUST operate at some harmonic of the earth-resonance frequency (**~11.78Hz**)

Receiver waiting to be SYMPATHETICALLY activated, once it does, NEEDS to draw load (transfer of energy).

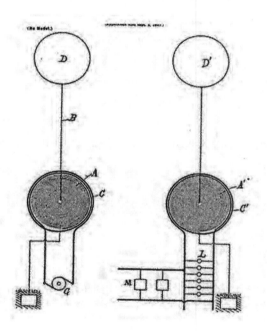

Success of the method rides on the ability of the transmitter to excite and induce waves in the medium between them **OTHER than of the Hertz-type**

Let's go back to the Maxwell's Equations and take a closer look.

Longitudinal and Transverse Waves

According to J. Nitsch, et al. in the study of dynamical and non-dynamical components of the electromagnetic field, *any vector field which is finite, uniform, and continuous and square integratable may be split into a longitudinal and a transverse part.*

$$F = F_\perp + F_{||} \qquad \nabla \times F_{||} = 0$$

$$\nabla \bullet F_\perp = 0$$

$$\nabla \bullet D_{||} = \rho(r,t)$$

$D_{||}$ = longitudinal electric excitation
$E_{||}$ = longitudinal electric field

Maxwell equations for the long part of electric displacement field:

Equation 1

$$D_{||}(r,t) = \frac{1}{4\pi} \int \rho(r',t) \frac{r-r'}{|r-r'|^3} d^3r'$$

$$E_{||}(r,t) = \frac{1}{4\pi\varepsilon} \int \rho(r',t) \frac{r-r'}{|r-r'|^3} d^3r'$$

Which, after some mathematical manipulations leads to:

Equation 2

$$E_{||}(r,t) = \frac{1}{4\pi\varepsilon} \int \rho(r',t) \frac{r-r'}{|r-r'|^3} d^3r'$$

Above is an interesting finding which says that the longitudinal electric displacement $D_{||}$ and electric strength $E_{||}$ are fully coupled to **the instantaneous Coulomb charge!**

$$E_{||}(r,t) \overset{\Longleftrightarrow}{} \rho(r',t)$$

Or whatever the SOURCE is doing at (r',t) the field is responding at (r, t). In other words the longitudinal component of the electromagnetic field is determined from the instantaneous Coulomb field generated by the electric charge.[1] Therefore the degrees of freedom of longitudinal components are tied to those of the electric charge.

This assertion or finding is very important!

If the source is oscillating with a frequency *f* so will any point on the domain receiver will oscillate sympathetically at the same frequency (which is made to be identical to that of the transmitter).

Let's attempt to see if the types of waves Tesla was counting on can be generated:

$$\Delta E(r,t) - \varepsilon\mu\frac{\partial^2 E(r,t)}{\partial t^2} = \frac{1}{\varepsilon}\nabla\rho(r,t) + \mu\frac{\partial J}{\partial t}(r,t)$$

$$\Delta B(r,t) - \varepsilon\mu\frac{\partial^2 B(r,t)}{\partial t^2} = -\mu\nabla \times J(r,t)$$

From the above equations interesting solutions due to charge/current singularities can be deduced.

---

[1] **Ed. Note**: see Jackson, *Classical Electrodynamics*, and other physics textbooks. This is the accepted definition of the coulomb or scalar potential and wave created from the appearance of an electric charge. The propagation speed is not limited by the speed of light and therefore, according to the Standard Model, is instantaneous in its appearance at a distance from the source, which explains what the author is referring to with (r,t) and (r',t) coupling at a distance.

Consider one of the "wave-like" equations

$$c^2 \Delta E(r,t) - \frac{\partial^2 E(r,t)}{\partial t^2} = \delta(\xi - x, \eta - y, \zeta - z, \tau - t)$$

Laplace transform on τ leading to

$$(r\overline{E})_{rr} - (s/c)^2 (r\overline{E}) = (r/c^2)\delta(\xi - x; \eta - y; \zeta - z)$$

$$\overline{E} = -e^{-[t+(r/c)s]} / 4\pi \cdot c^2 \cdot r$$

In 3-D space the solution of the field E takes the form

$$E = -\delta[\tau - t - (r/c)] / 4\pi \cdot c^2 \cdot r$$

While the same solution in 2-D space is of the form

$$E = \{ \begin{array}{cc} 0 & r > c(\tau - t) \\ \dfrac{1}{2\pi c \sqrt{c^2 (\tau - t)^2 - r^2}} & r < c(t - t) \end{array}$$

"Near-field" and "wave-zone" effects

Let's look closer into the area of serious controversy as to existence of longitudinal or longitudinal-like waves that can be generated or propagated.

Dr. G. W. Bruhn in "Can Longitudinal Electromagnetic Waves Exist?" in *Journal of Scientific Exploration, Vol. 16, No. 3, pp. 359–362, 2002* argues that such waves (i.e. plane and spherical longitudinal) do not exist and bases the thesis on the solution of Maxwell's Equations. While the mathematical solutions of the equations presented are correct in his argument, however the solutions are based on the following considerations (exactly as stated): "We consider a homogeneous medium of constant dielectricity ϵ and constant permeability μ that is *free of electric sources* and currents. Then all electromagnetic processes are governed by the **homogeneous** Maxwell equations that are source-free." With an *inhomogeneous* solution however and point-wise defined field with a source, there is no argument except for the fact that ELF and VLF waves of the longitudinal type exist and can be excited within the ionospheric cavity. One must keep in mind that the Hertzian-type electromagnetic waves argued to only exist MUST have been generated somewhere and by some source. The proof that it is only that type of waves (Hertzian) that exist is for a zone far from any source generating them.

Now let's take a look closer to the source producing Hertzian waves and look at the field due to an oscillating dipole (for harmonic oscillations this is known as Hertzian Oscillator). Using the notation and the basic derivations of Maxwell field equations presented in M. Mason and W. Weaver "The Electromagnetic Field," Dover Pub., 1929, we analyze the fields generated around a dipole consisting of an stationary positive charge +e at point P accompanied by an oscillating negative charge −e vibrating about point P forming a dipole with vector moment p as a function of time. Following elimination of higher order terms and retaining only leading terms of the ratio *u/c*, where *u* is the velocity of the negative charge (<<c), the magnetic and electric fields (**B** and **E** respectively) can be deduced

$$B = \frac{1}{4\pi c R^2} f[\dot{p}, R_1] + \frac{f[R_1, \ddot{p}]}{4 \cdot \pi \cdot c^2 \cdot R}$$

$$E = \frac{e}{4\pi R^2} (R_1 - r_1) + \frac{f[R_1, \ddot{p}]}{4 \cdot \pi \cdot c^2 \cdot R}$$

Near-Field Wave-zone

We observe two distinct regions as expressed by the two equations, a near-field in the neighborhood of the dipole where the term $1/R^2$ predominates over the $1/R$ term, and the far region where the term $1/R^2$ will become negligible as compared to the $1/R$ which in turn will dominate the solution. The region where $1/R$ is dominant is referred to as "wave-zone".

This brings a very important point that an oscillating dipole which is the source of Hertzian (or transverse waves) induces a near field of disturbance that decays according to $1/R^2$ or a unique wave type if the dipole oscillation is harmonic. So in the dipole neighborhood the field is described by

$$B = \frac{1}{4\pi c R^2}\{[\dot{p}, R_1]\} \qquad E = \frac{e}{4\pi R^2}(R_1 - r_1)$$

From these relations one can deduce that the steady-state B values generated by the motion of the charge propagate outward with velocity c. Similarly, the E vector represents the electrostatic vector resulting from instantaneous moment possessed at the retarded time $t-R/c$ and so the electrostatic values E that characterize the instantaneous state of the dipole also propagate outward with constant velocity c.

The portion of the total E and B fields that varies as 1/R and dominates in the wave zone is not due to the velocity of the charge (as the near field is) but the acceleration $\ddot{p}$ normal to the line that links the far-field point and the oscillator[2]. This represents the electromagnetic field.

We now turn our attention to the special case where the oscillation of the charge forming the dipole is harmonic,

$$p = p_o \cos(\omega t)$$

So the E and B fields become simple harmonic functions of time and

$$\lambda = \frac{2\pi c}{\omega}$$

If one forms the ratio of the 1/R term in either B or E fields with that of $1/R^2$ then will arrive at an expression such as

---

[2] **Ed. Note:** $\ddot{p}$ is the usual shorthand notation in physics for the second derivative with respect to time or acceleration.

$$\frac{wave-zone}{near-field} = 2\pi\frac{R}{\lambda}$$

So the wave-zone is the region where R is large compared to λ.

This relation, however, brings about an interesting observation which relates to the applicability of other-than Hertzian electromagnetic waves that are in terms related to the near-field. The region of dominance is dependent on the value of λ.

What if the wavelength is very large (as in the case of the cavity resonances and the harmonic of ~11 Hz that Tesla was after)? This means that the zone of influence of the near field effects would extend to a distance R of the order of λ. This immediately implies large distances. Therefore, we may infer that Tesla, following his study at Colorado Springs where he manage to excite the cavity resonances, had realized that he can created these fields that are non-Hertzian at these low frequencies and based his Tesla Coil on a driving frequency of **11.78Hz**.

Following these near-field and far-field observations about sources we revisit the relationships developed to capture the discharge-induced fields of Tesla at Colorado Springs, called a "**Singular Discharge**."

$$E = -\delta[\tau - t - (r/c)]/4\pi \cdot c^2 \cdot r$$

And compare with the solution generated due to an *Oscillating Dipole* (charge) or a continuous discharge.

$$E = \frac{e}{4\pi R^2}(R_1 - r_1) + \frac{f[R_1, \ddot{p}]}{4 \cdot \pi \cdot c^2 \cdot R}$$

From the *Singularity Discharge* the solution indicates a radially (longitudinal) expanding thin spherical shell

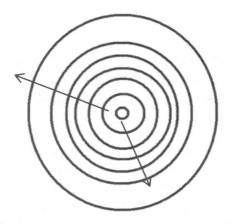

From *Oscillating Dipole* in the far field (or wave zone),

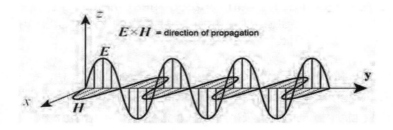

the transverse wave depends on the disturbance, or is a "source-type."

From a Singularity Discharge, can the receiver respond instantly to a charge? It can only happen if the Transmitter EXCITES a mode of the field (standing wave). ONLY then will the field be disturbed instantly and Tesla AIMED at exactly that:

Hence, RIDE the fundamental mode of the earth electric charge field OR a multiple harmonic of it.

Tesla's ATMOSPHERIC CONDUCTION Method

Closed Circuit Method

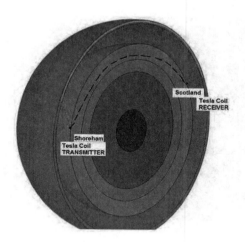

What will the atmospheric conduction system look like?

- Tesla coil transmitter
- Ionized path connecting the transmitter upper atmosphere
- Ionosphere
- Ionized path connecting the upper atmosphere back down to receiving location
- Tesla Coil (receiver)
- Closed circuit is completed with current back to the transmitter <u>through the earth</u>

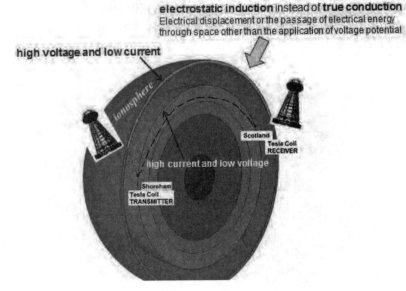

DETAILS of the Atmospheric Conduction Method

An oscillating electric field is induced by High Potential RF current applied to the HELICAL Resonator of a Tesla coil:

oscillating magnetic field ➔ oscillating electric field

A plasma state is formed in the High Voltage Terminal (capacitor plate) making it conductive

Current is pushed in the 10,000,000 to 12,000,000 volts line created to the upper strata (high voltage will minimize loss due to plasma transmission-line resistance)

An ionizing beam of ultraviolet radiation will also form a high-voltage plasma transmission line

If a plasma state is induced in the upper troposphere (part of the scheme) then that becomes a conductor

End result is a flow of true conduction currents between transmitter and receiver

The process relies on **plasma waves** developing in the ionized region between the two terminals:

Electrostatic waves or
Magneto-Hydrodynamic (MHD) waves

Plasma wave

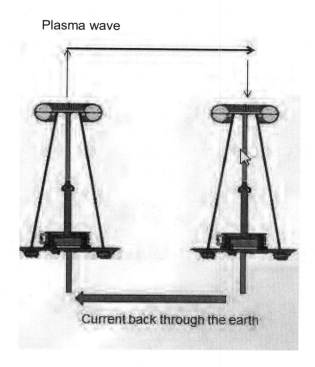

One can see a **purely conduction method** and a **hybrid one (part conductive, part inductive)**

Longitudinal Waves & Maxwell Equations

Electrostatic or magneto-hydrodynamic plasma wave model
Electrostatic Waves (K. T. McDonald, *An Electrostatic Wave*)

Assertion: A time-varying purely ELECTRIC field E(r,t) can exist
and propagate as a longitudinal wave WITHOUT coupling to a time-
varying magnetic field

$$\nabla \times E = 0 \qquad \frac{\partial B}{\partial t} = 0$$

But under the Condition that Electric Displacement D is zero in a
plasma medium

These **Longitudinal Electric Waves** can coexist with background
electrostatic and magneto-static fields.

Solution of Maxwell Equations

$$E = E_x \hat{x} e^{i(kx - \omega t)}$$

$$\nabla \times E = -\frac{1}{c}\frac{\partial B}{\partial t} = 0$$

$$E = -\nabla V$$

$$\nabla \bullet E = 4\pi\rho$$

$$V = i\frac{E_x}{k} e^{i(kx - \omega t)}$$

$$P = -\frac{E}{4\pi}$$

$$\rho = -\nabla P$$

$$Condition:\ D = E + 4\pi P = 0$$

P = volume density of electric moments
D = Electric Displacement of Long. Wave
ρ = charge density

Further analysis reveals that the longitudinal waves can only have
the plasma frequencies!

We observe the same in the separation of transverse and longitudinal waves of Maxwell's equations.

So, what happens with a COUPLED Transmitter and Receiver? The Transmitter "feels" the load in the receiver. Below is what happens at the receiver:

ENERGY SUCKING ANTENNAS and TESLA

Parallelism with Atoms (1 Am) which are strongly interacting with LIGHT (5000 Am wavelength)

Flood the atmosphere with standing waves (ionosphere keeps most of this EM energy from escaping into space) then *a small resonator[3]* can grab significant wattage right out of the air (effective disk in figure below). A **small resonator** can produce an extensive and intense AC field of its own, and act as an **"EM funnel"** (simple desktop experiments demonstrated this principle).

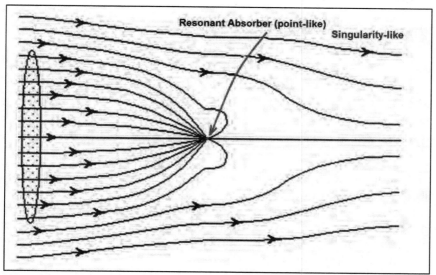

This is "circuitry", where wavelength is huge, and circuits are small.

---

[3] **Ed. Note**: Here the author offers an intriguing suggestion of a receiver that is not a duplicate of the transmitter but smaller.

How Feasible or Workable are Tesla's Concepts?

Resonance or Open Circuit Method:

Tesla's model involves two very distant, electrically coupled elevated terminals (capacitor plates) in a manner similar to the transfer of electrical energy between two closely spaced capacitor plates in a typical AC circuit, **but at distances greatly exceeding 1/6 - 1/2 wavelength**

This model considered to be inconsistent with a basic tenet of mainstream physics (related to the scalar derivatives of the electromagnetic potentials)

So we thought, BUT "Witricity" and Qualcomm are proving to be otherwise. MIT tests along with Resonant Inductive Coupling distances > ¼ wavelength may not be the distances Tesla envisioned BUT much greater than what was thought as the threshold of a capacitor.

Atmospheric Conduction or Closed Circuit Method:

This Model has no inconsistencies. It also involves two very distant, electrically coupled elevated terminals (capacitor plates) that are both active and rely on in electrostatic induction alone. It also involves the ionization of the space between the two "capacitor" plates which in turn (being in a state of plasma) will permit ELF and VLF waves. While consistent, it may be impractical for its size.

Maybe!

References

www.teslasciencecenter.org – TESLA Science Center at Wardenclyffe

www.teslasociety.org – Tesla Society of USA and Canada

www.teslaradio.com

J. Nitsch, F. Gronwald, G. Wollenberg, "Equivalent Circuit Method", J. Wiley & Sons

ELECTROMAGNETISM AND THE STRUCTURE OF MATTER, World Scientific Publishing Co. Pte. Ltd., "Something is wrong with classical electromagnetism"

Z. Turakulov, "The Dipole Radiation - Retarded Potentials and Maxwell Equations, Ulugh Beg Astronomical Institute, Uzbek Academy of Sciences

11 Tesla's Self-Sustaining Electrical Generator

Oliver Nichelson
Adapted from *Proceedings of the Tesla Centennial Symposium*, 1984
Reprinted from *Harnessing the Wheelwork of Nature: Tesla's Science of Energy*

Abstract

Before the discovery of the electron, the principle theory used to describe the electrical activity was that of the ether. At the turn of this century, the ether theory in use by science was a remnant of the concept common in western thought for several centuries. This situation favored the rise of atomic theory. This change in scientific paradigm requires a translation from 19th century terminology into 20th century language in order to understand Tesla's later research. Of particular interest is his magnifying transformer which claimed to produce resistanceless current.

The Historical Ether

Though science aims at giving accurate descriptions of the workings of nature, these descriptions change from historical period to historical period. In the same way that an object in one European country is called by a different name in an adjoining European country, so do the descriptions of nature change during different periods of man's history.

In the 19th century western science the broadest view of the physical world was that all objects were somehow each connected to one another through pre-material ether. Solid bodies were believed to be made from condensation of this ether. In this worldview, atoms and electrons did not exist as scientific realities.

Toward the end of the last century the atomic picture of the world emerged in steps. Solid bodies were explained by minute vortices in the ether – small whirlpools – forming lumps of matter. Lord Kelvin, the virtual spokesman of Victorian science, developed an ether vortex model of the electron in an effort to explain some of the properties of electricity. The electron as a discrete particle did not become a fact of science until Thompson discovered it in 1897.

The view of nature as a single entity formed out of the ether changed to the modern one of matter being made of collections of individual particles in 1905 [1]. In that year, Einstein presented his paper on Brownian motion explaining the movement of pollen particles on the surface of water in terms of discrete units of matter. From then until today, the atomic view has prevailed.

This difference between the 19th century description of nature and our presented description makes it difficult to have a complete picture of the work of the early electrical researchers. Today, Faraday, Maxwell and Tesla are recognized as valued contributors of the understanding of electricity, but their work was carried out before the electron- the fundamental carrier of electrical charge – was discovered. All of these scientist held a belief in an physical ether. Though Faraday's laws of induction are still accepted, and Maxwell's equations from electromagnetism are still used routinely, and Tesla's generators are still powering our lights, the 19th century physics that they learned and out of which their physics came, has been judged scientifically wrong.

The curious situation in which Faraday, Maxwell and Tesla can be seen be both right in their results but wrong in their beliefs about physics comes from an inability to translate the concepts of their historical periods into the language of our period. This lack of chronological translation, in contract to the spatial translation between European languages, is also an obstacle to understanding the physics of self-sustaining electrical ("free energy") generators based on the 19th century views.

In the last half of the 19th century, when researchers had to deal with the ether in practical engineering terms in order to guild their electrical devices, the concept of the ether, then, several centuries old, was a watered down theory. At that time, the ether was considered something like a thin gas that could be found everywhere. However, that was not a historically correct view of the ether.

The ether had been pictured traditionally as a non-material substance capable of condensing into ponderable matter. Gas, no matter how thin, is still ponderable matter; and because of that, could not qualify as the ether.

To find out what was meant historically by the concept of ether, an early writer on the subject can be cited. Robert Fludd, in 1659 described the "Ethericall...influences" as "far subtler condition than is the vehicle of visible light...so thin, so mobile, so penetrating, so lively, that they are able, and also do continually penetrate, and that without manifest obstacles or resistance, even unto the center or inward bosom of the earth where they generate metals of sundry kinds."[2]

Fludd quotes an even older source on the nature of ether, the writings of Plotinus (3rd Century AD) where the ether is described as being so fine "that it doth penetrate all bodies and... it maketh them not a jot bigger for all that because this inward spirit doth nourish and preserve all bodies." [3]

From these older descriptions of the ether, the following attributes can be seen missing from the late 19th century concept. First, the ether was held to be truly non-material – it does not make bodies "a jot bigger". If the ether were a gas, its addition to anything would be measurable. Second, the ether is a substance less material than "the vehicle of visible light", that is, something less than what today is known as a photon. Third, the ether was credited with generating metals and nourishing all bodies, clearly a distinct property not belonging to gases.

Whether or not the reality of the ether as put forth by these authors is accepted, it is historical fact that the tether Michelson and Moraly did not find in their experiments and that the modern atomists ridiculed so strongly when they came to scientific power in the early 20th century was never claimed to exist by people who first used the term. Taking a longer view of science, modern theorists fought a battle against an issue that never existed.

If, on the other hand, the ether is looked at in the earlier description of its properties, something can be learned about the operation of a least one type of self sustaining electrical generator. To do this, the ether concept has to be translated into an artifact of contemporary science.

The Modern Ether

The properties of having less mass than a massless photon, being able to interpenetrate a body but not add to it, and generating material bodies are encompassed in the modern view of the quantum wave nature of matter. In quantum theory, an object can be viewed as either made of particles or waves. It is not an idea everyone is comfortable with even now but one that is widely accepted and known to be verifiable by experiment. Transistors, tunnel diodes and even digital watches are a few of the real world objects operating on physical principles that are explained best by the quantum wave nature of matter.

If an object can be both a quantum wave and a particle, then it its wave state, it can be said to interpenetrate an object without making it "a jot bigger". Also, being a wave equivalent to a particle, the wave would not have the mass of a particle. It has amplitude instead. The quantum wave is also responsible for the generation of solid bodies. Present theory has it that a particle exists in its quantum wave state until a measurement is made, when the wave is then said, to collapse to form an object. The collapse of the quantum wave defines the state of the object, that is, it generates the particle.

The quantum wave state of nature very much resembles the 17th century picture of the ether.

With this conceptual parallel in mind, it is possible to understand better the work of Nikola Tesla, who held the ether theory as a scientific concept, who, also no the basis of this theory, build working electrical machines, and who is associated with the idea of an electrical generator which could maintain a current without an external prime mover.

Schooled during the 1860's, Nikola Tesla's understanding of physics was pre-atomic. In his biographical articles Tesla does not comment on the theoretical aspects of his education, but in his technical writings, he uses the term "the ether" in a positive sense and only in his later writings are found grudging references to atomic particles and electrons.

Tesla Magnifying Transformer

Tesla's most famous device was what he called a Magnifying Transformer, the principal tests of which were carried out in Colorado Springs during 1899. The device is described in his U.S. Patent as an "Apparatus for Transmitting Electrical Energy" [4] and claims some unusual characteristics among which were the propagation of waves faster than the speed of light, the transmission of signals, not around the earth, but through the earth, and doing this by eliminating as much as possible electromagnetic waves - the only electrically related waves known today capable of transmitting signals.

Tesla did this using a coil with $10,000 - 11,000$ feet of cable [5], with what he claimed to be little or no resistance. This last fact, giving rise to the belief that in addition to tits other unusual characteristics, the device had the property of maintaining its current for a measurable period of time after disconnection from an outside power source.

Taking these ideas together – that the ether is equivalent to quantum wave energy, that Tesla held a belief in a physical ether, and that Tesla build a device capable of maintaining an electrical current without an external prime mover, a conclusion that can be reached, is that the quantum wave theory can be used to understand the dynamics of Tesla's magnifying transformer. This follows from the work of Dr. Andrija Puharich who, in a 1976 paper, put forth the idea that the magnifying transformer could not be explained by the laws of classical electrodynamics, but, rather in terms of high energy particle transformations [6].

The wave theory of matter gained its present popularity in 1923 through the efforts of deBroglie. When experiments showed that light could be considered both a particle and a wave he reasoned that an electron, clearly a particle, could behave like a wave. He deduced the wavelength of the electron from the equation $E=hf$ which equates the energy of a particle to the product of Planck's constant times the frequency. (Lambda works out to be 2.4×10^{-12} meters, which is the classical wavelength for the electron.)

In analyzing the Tesla magnifying transformer, this mathematical relationship can be used to determine the quantum energy of a wave in the transformer's operating frequency (here we use the pulse repetition rate of 7.5 Hz, following Corum [10] instead of the author's originally suggested kilohertz oscillation frequency – Ed. note) and putting that value into the equation gives:

$$E = hf = (6.63 \times 10^{-34} \text{ Js}) (7.5 \text{ Hz}) \qquad (1)$$

$$E = 4.97 \times 10^{-33} \text{ J/e}$$

which would be the radiated energy <u>per accelerated charge carrier</u> (electron) in the conductor.

If the magnifying transmitter were operating at a current $I = 100$ amperes, the total charge can be found. Current is charge per time $(I = q / t)$ and by definition, 1 Ampere = 1 Coulomb / second. This relationship can be used in turn to determine the number of charge carriers per second in the conductor for a 100 A current:

$$I = \frac{100 \quad \text{C/s}}{1.6 \times 10^{-19} \text{ C/e}} \qquad (2)$$

$$I = 6.25 \times 10^{20} \text{ electrons per second} \qquad (3)$$

The total number of charge carriers times the emitted energy per charge carrier would equal the quantum energy of the wave at a given frequency (7.5 Hz in this case):

$$E_Q = E I = (6.25 \times 10^{20} \text{ e/s}) (4.97 \times 10^{-33} \text{ J/e}) \qquad (4)$$

$$= 3.1 \times 10^{-12} \text{ J/s} = 3.1 \text{ picowatts}$$

If the highest reported current that Tesla used, 1000 amperes, is put into the calculation, the energy range would be 3.1×10^{-12} J/s to 31×10^{-12} J/s.

Converting to a more commonly used system of measures, the energy of a quantum wave at 7.5 Hz would be:

$$e = (6.2 \times 10^{12} \text{ Mev / J })(3.1 \times 10^{-12} \text{ J/s}) \qquad (5)$$

$$= 19 \text{ Mev}$$

If the highest current of 1000 amperes is put into the calculation, the energy of a quantum wave would be 190 Mev.[1]

In order to generate a wave of this energy, an electron would have to undergo a potential difference in the range of 19 to 190 million volts.

Tesla's magnifying transformer was reported to operate in the range of tens of millions of volts. At 20 million volts there would be more than sufficient electrical force to create a

---

[1] Compare to Corum [10] who calculates about 225 coulombs in a volume of 10,000 cubic meters of glow discharge. Using 2.5 eV per molecule of air, the amount of power Tesla used for a pulse repetition rate of 7.5 Hz is found to be only 6.5 hp, consistent with what Tesla reported. For reasons explained in the article, the Corums find that Tesla generated <u>10 MeV</u> electrons at 1000 amperes.

vacuum wave for the amount of charge in motion at 7.5 Hz. At 200 million volts there would be enough force to produce such a wave for a current of 1000 amperes at that frequency.

The generation of a quantum wave by the magnifying transformer goes a long way in explaining some of the properties Tesla claimed for the device. For one, he said that electromagnetic waves were reduced to a minimum and, indeed, it would seem hard to propagate any e.m. radiation with the blunt topped tower used in his transmission experiments. If, however, the waves that were being emitted were quantum waves, or waves of the ether, his claims for radiating energy from one point to another without the use of electromagnetism becomes clear.

Also, Tesla's statement that electromagnetic radiations were similar to the waves transmitted by an ordinary whistle through the air [7] makes sense. According to his view, e.m. waves would be nothing but undulations in the atmospheric gases, while his transmissions were taking place in a wholly different medium, that of the ether.

Tesla's claim for instantaneous transmission of energy has a basis in modern theory too, for a quantum wave is non-local in nature. That is, its effect is not limited to one particular point, but, through a physical process still not completely agreed upon, the effect can be measured at great distances from the point of origin at the moment of origin.

The Superconducting State

As to maintaining a current in the transformer without an external power source, the only condition known today for achieving this, is the state of superconduction, which seemed to be ruled out in the case of Tesla's device which operated far above the almost zero temperatures needed for superconduction. However, what is understood as the superconducting state in today's science is in fact a description of the conductor. If a material has a certain type of atomic configuration and is cooled to a certain temperature, a superconducting condition exists in which a perpetual current can be maintained. The superconducting state, though, can exist without there being a current in the conductor. The state is a characteristic of the conductor.

Tesla may have discovered that superconductivity can be a property not of the conductor but of the current itself.

To examine how this might be the case, a specific model of electrical activity will be used. Instead of picturing an electric current composed of billiard ball particles of of little satellites of nuclear suns, or as an electron gas, or as electron plasma, it can be imagined as an electron liquid. At this point the make up of the liquid is not as important as is its fluid nature and that the fluid is electrical.

The model of a liquid is useful because it provides an easy example of how a substance can remain the same and yet become radically different under certain conditions. With water, when heat is removed from it, a phase change takes place which transforms it into solid ice. When thermal energy is added to water, it undergoes a different phase change and becomes a gas. The substance remains the same, but it exists in three difference states.

One of the extreme states that a fluid can achieve is superfluidity during which a liquid will move up the walls of its container. This, of course, is a property of the liquid, not of the container.

Perhaps the same phase change phenomenon takes place in the electron liquid. Under certain conditions, high voltage and or high current, the electron liquid will remain the same substance but will take on radically different properties, similar to the state of superfluidity. This condition would be a state change in the current, not in whatever material is serving as the conductor.

A state of superfluidity in an electron liquid would explain how Tesla was able to send a current through the earth. When in its commonly known state a current does not travel far through the earth's resistance, but if the current has undergone the proper phase change, it could easily travel with no resistance.

Likewise, a phase changed current would travel through a generator coil with no resistance. Having undergone the change it would become a super-current in a non-superconducting conductor. Such a condition would allow a generator to maintain a current without an external power source.

This particular solution, which of course has to be tested, of Tesla's self-sustaining generators, is not an explanation of all the other similar devices such as the Figuera, Hubbard, and Herdershot devices [8]. There are probably as many engineering solutions to such generators as there are inventors of them.

One characteristic all the other devices have in common in contrast to Tesla's magnifying transformer, is that they did not require the high voltage and currents Tesla used. They do not, though, represent an engineering advancement over Tesla's engineering methods.

Tesla put his main efforts into high energy devices as a matter of mere practicality in marketing a product. A year after his Colorado Springs experiments, he wrote in his Century magazine article, 1900, that he had spent a great deal of time on a smaller generator but realized that negative market pressures would not allow such a machine to see the commercial light of day [9]. And he was right; it is not possible yet to by a Hubbard or a Hendershot generator to light our homes.

Tesla believed he had a greater chance for introducing a new electrical technology if it made use of the generators then being sold, but which used their output in novel ways – which is why he concentrated on the wireless power transmission project, though even that idea proved too much for his time.

A careful study of his later writings shows that many of his more advanced concepts were based on earlier work with lower voltage versions of generators capable of maintaining a super-current. These designs appear to be based on intricate configurations of coil geometries. The peak of this line of research might have been just before the fire of his New York City laboratory in which, many of his prototypes and papers were lost. The task of uncovering the precise nature of these designs becomes very complex, because after the fire, Tesla spoke of his more advanced work only obliquely and never in detail.

Recovering these earlier designs would bring about the second stage of electrical technology – one that Nikola Tesla started, here, a century ago.

References

1. There have been several such paradigm changes in western ideas about nature. Theories alternate between a one substance universe out of which everything is made and a many substance universe in which the constituent particles are separated by a vacuum.

2. Robert Fludd, *Mosaical Philosophy*. London, Humphrey Moseley, 1659, p. 221.

3. Fludd, p. 221

4. U.S. Patent #1,119,732 of December 1, 1914; application filed January 18,1902.

5. Nikola Tesla, *Colorado Springs Notes*, 1899-1900. Beograd: Nolit. 1978, p.43.

6. Andrija Puharich, "The Physics of the Tesla Magnifying Transmitter and the Transmission of Electrical Power without Wires". Planetary Association of Clean Energy, (Ottawa, Ontario, 1976).

7. Nikola Tesla, quoted in the *NY Herald Tribune*, Sept 22, 1929, pg. 21.

8. C. Bird and O. Nichelson, "Nikola Tesla, Great Scientist, Forgotten Genius", *New Age*, Feb. 1977. p. 41.

9. Nikola Tesla, "On the Problems of Increasing Human Energy", in Nikola Tesla, Lectures, Patents, Articles, Biograd, Nikola Tesla Museum, 1956, p. A-143. (Also in *Century*, June, 1900 – Ed. note)

10. Corum, James & Kenneth, "Critical Speculations Concerning Tesla's Invention and Applications of Single Electrode X-Ray Directed Discharges for Power Processing, Terrestrial Resonances and Particle Beam Weapons," *Proc. Inter. Tesla Symposium*, 1986, p.7-21

Oliver Nichelson, 670 W 980 N, Provo, UT 84604

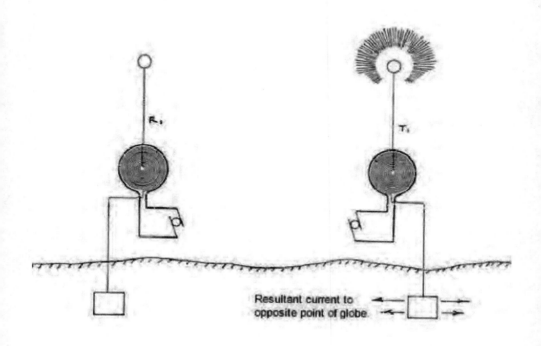

12 Nikola TESLA's Wireless Systems

André Waser*
www.andre-waser.ch

Issued (German): 29.07.2000
Last changes (English): 05.08.2000

After his inventions about the polyphase powering systems Nikola TESLA has focused himself more to experiments with high voltages, high currents and high frequencies. One of his goals was to transmit electrical energy without a power network directly from a central plant to the different consumers. In New York TESLA has done his first trials for this new technology. Then at the change of the century 1899-1900 TESLA moved to the high lands of Colorado Springs. There he has executed so many experiments, which has not been repeated in all its details and specialties until these days. Now, exactly one hundred years after a review about this impressive and important experiments may be of a particularly interest.

Introduction

It is surprising how little information can be found in literary about the work of the famous Serbian experimenter Nikola TESLA. In the contrary his antagonist Thomas EDISON, which mainly promoted the direct current systems, is mentioned where ever one looks. But it was Nikola TESLA who invented the today used polyphase power system in all its part of generation, transmission and consumption. It was Tesla, not EDISON, who has made the world-wide use of electricity even become possible.

And today almost all publications about TESLA's work are looking at his high frequency and high voltage transformers, known under the summary term „Tesla-Coil". From time to time some papers has been published about this specific topic; for example for a repetition of some experiments[3],[7],[16], about applications of this transformers[1],[9],[17], about the measurement on such devices[49] or about some theoretical considerations[2],[4].

Very special arrangements of the TESLA coils are the power transmitting and receiving devices of Tesla. Konstantin MEYL has recently published many papers about this topic. MEYL[10]-a,[11]-a has used the same speculative explanation hypothesis as the author[44] has used at an earlier time and which are – as suggested now – not necessary anymore.

It is typical for an experimental explorer that he discovers unexpected results and finds new facts only because he makes some leading experiments on the basis of speculative models. And because of this TESLA was far ahead of the theoretical knowledge of that time with his experimental practices. Therefore a communication with the established science was not always easy for him, what could be a reason (beneath of commercial interests) that Tesla has more or less stopped his publications in scientific journals after the year 1899 and since then only published in popular daily or weekly newsletters.

Wireless transmission of electrical energy

In the years 1884-1889 TESLA got different patents for his alternating polyphase technology, which has been a substantial breakthrough at that time against the direct current technology. But leading economists and companies in Europe didn't understand TESLA's visions and he was forced to emigrant in the USA. Together with George WESTINGHOUSE TESLA made it possible to build the first alternating power station of a large scale at the Niagara Falls in 1893. But the first patent[19], which reveals the landmark thoughts TESLA's, was filed in the year 1891 and is a fully description of a high frequency lighting system. The specific feature of this system is the use of only one supplying single wire to the particularly build and patented single terminal carbon lamps without a return wire. (The patent has been granted in the record time of only two months.)

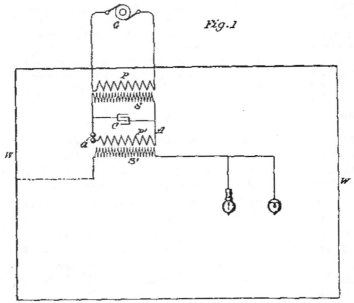

Figure 1:   US-Patent 454,622 „System of Electric Lighting" issued on June 23rd, 1891[19]

With the first transformer P-S the alternating voltage of generator G (about 5 kHz) is transformed to high voltage. The resonance circuit S-C is then vastly discharged along the spark gap. As a result there are high current peaks in the primary winding P' of the second transformer. With this second transformation the high frequency part of this current peaks is again transformed upwards and feeds the load circuit. One end of the second secondary S' is connected to a long wire or wire grid W positioned along the room walls. The other end is connected to TESLA's invented single terminal lamps. In opposite to the lamps used today this lamps have only one connector. And this connection leads to an electrode – mostly made from carbon – inside the fully or partly evacuated glass bulb. On different occasions TESLA[18],[20],[22] has demonstrated, that this high frequency currents and voltages do not cause immediate injury to the experimenter (himself) or the audience.

This patent shows all characteristics of the high frequency circuits with high voltage and high currents as used by TESLA. In the following steps TESLA optimized the technology of generation and utilization of high frequency and high voltage apparatus, which he mostly applied to lighting systems with different kinds of bulbs. In the year 1897 he applied for three patents about the transmission of electrical energy. The first patent[23] he registered on March 20th about a high frequency transformer with high power capabilities. Besides a common ground connection this transmission method needs only one transmission wire.

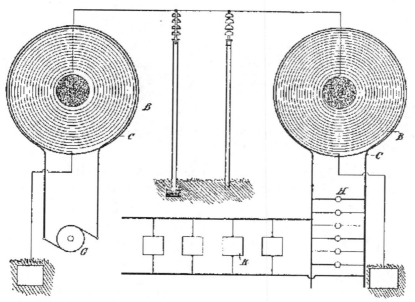

Figure 2: US-Patent 593,138 „Electrical Transformer" issued on November 2nd, 1897

The generator G supplies the primary of the flat coil C. This simplified diagram does not come very close to the real experimental setup[27]. Then as previously shown with the patent about the lighting system an intermediate step-up transformation with a spark gap and a high voltage transformer is necessary to achieve a resonant frequency of some million cycles per second. With some advantages it is also possible to use this step-up transformation after the flat secondary coil B. This flat coil TESLA[21] has extra patented because of its excellent performance with high voltage and high frequency signals. On one end the secondary B is connected to ground and on the other end to the transmission wire which is connected to a receiving device with a flat coil B' of a symmetrical form. With a step-down transformation with the coil C' the electrical energy is finally transmitted from the generator G to the load L with only one conducting wire.

Some months later TESLA[27] has shown that the transmission wire can be dropped completely and can be replaced by a glass tube filled with air of low pressure.

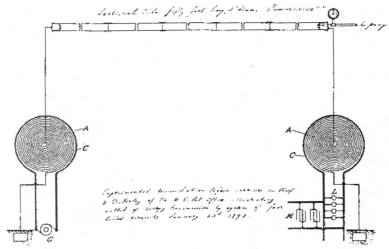

Figure 3:   Slide of Nikola TESLA[27] about the energy transmission through a partly evacuated glass tube; dated of January 23rd 1898.

In figure 3 the arrangement of figure 2 can be found again. With this discovery of the good electrical conductivity of air of low pressure the path was free for further developments.

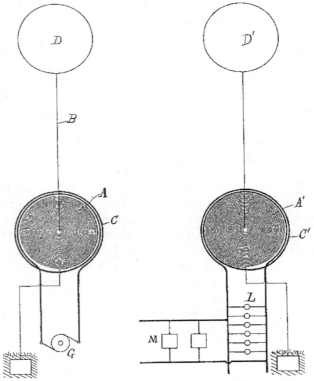

Figure 4:   US-Patent 645,576 „System of Transmission of Electrical Energy" filed on September 2nd 1897, issued on March 20th, 1900

Then on September 1897 TESLA[24],[25] has filed two other patents for the transmission of electrical energy (figure 4). But the granting of this patent has been made dependent of the experimental success as a corresponding part in the patent shows (Pat. 645'576, p. 3, col. 2).

In this patents TESLA writes of a grounded high frequency emitter with a highly elevated ball electrode which was in resonant connection with a symmetrical, grounded resonant circuit (receiver) to enable the energy transmission through the upper atmosphere, which in great heights becomes more and more conductive for electrical currents.

The *Electrical Review*[24] of London published on May 1899 a summary of articles about the work of Nikola TESLA previously published by their New York colleges. Here TESLA stated that the air will have a sufficient conductivity for his experiments, if the ball electrodes are placed in a height of four miles (~6.5 km). This could probably be done by balloons, TESLA suggested.

Trained with many experiments TESLA left New York on May 11th 1899 to the highlands of Colorado Springs (2000 m about sea level) where he experimented[27] with several systems for the transmission for electrical energy until the turn of the century on January 11th 1900. One of the goals was to prove by experiment the feasibility of his patent applications of 1897. As a result of his experiments he got his second patent[24] on March 20th 1900 and his third patent[25] on May 15th 1900. And only one day after he got this third patent he filed an other, very important patent[32]. In this patent he describes for the first time in detail the energy transmission through the earth and gives more information about signal detection (figure 5).

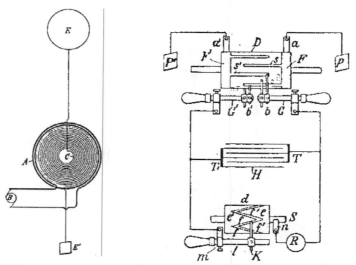

Figure 5: US-Patent 787,412 - „Art of Transmitting Electrical Energy Through the Natural Mediums" filed on May 16th 1900, issued on April 18th 1905.

Obviously he was only able to file this patent after the other two patents from 1987 has been granted. And this is because the older patents and this new one does contradict each other in the description of the method of the energy transmission in essential points! The results of the Colorado Springs experiments has motivated TESLA to replace his previous patents – based on his New York experiments – with a newer and accurate one.

Almost during his work in Colorado Springs TESLA[28] to [31] filed continuously some patents which report his experimental progress in detail but which are mainly focused on the receiving devices only and not on the full system of transmitter and receiver.

The topic of signal transmission through the earth has engaged Tesla further and two months later he again filed a patent[39] which shows some different methods for signaling with and without the use of transmitting wires.

All his efforts culminated in a project for the transmission of electrical power of 10MW in Wardencliffe[1], USA, which has never been completed probably because of low fundings. The basic arrangement for the large scale power transmission was published in his last patent[40] file of this kind.

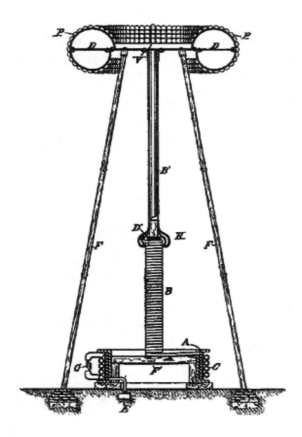

Figure 6: US-Patent 1,119,732 - „Apparatus for Transmitting Electrical Energy" filed on January 18th 1902, issued on December 1st 1914.

This enormous work TESLA's, which has not – or only perfunctory – been published in the scientific publications of that time, is worth to be reconsidered at least partly on the basis of today's knowledge and theories.

On the first glance to the series of figure 2 to 4 one supposes that the energy transmission finally occurs through the air by the means of an increasing electric conductivity of the upper atmosphere. Actually TESLA[24],[25] has written in his first patents that this is the case. But with a closer look at his drawings there can be recognized that all his circuits has – beneath the high voltage transformers – an other common thing: the ground connection. In later publications TESLA[32],[36],[32] has mentioned explicitly that the really conductor of the power transmission is the Earth itself. The Earth acts like a giant reservoir for electrical charges which can be set into oscillation by his powerful equipment. Is now a very sensitive resonant circuit (receiver) placed on an other place on Earth, which is tuned to the transmitter's frequency, then the receiver couples to this oscillations and gains its signals due to resonance.

This electrical excitement of the Earth TESLA[27]-S.61, [36] has discovered in a stormy night from July 03rd to 04th 1899 in Colorado Springs. To his great surprise he detected standing waves on the Earth surface after heavy lightning. With his sensible equipment he was able to record that the signals first diminished when the storm passes away but then again increased and later on diminished again and so fourth. Of a special interest for Tesla was the fact that the different maximum readings almost increased the more the storm was moved away form the receiver, and this to an estimated distance of about 200 miles.

The receiver must be constructed according to figure 4 by enabling powerful oscillations between Earth and the elevated charge terminal D', if it is used for energy transmission. If only signals are to be detected, then it is sufficient to have a receiving device according to figure 5, which only detects and demodulates electrical signals on Earth's surface.

After his discovery on July 3rd 1899 TESLA obviously has done further measurements, which he has not published in great detail, but on which he has made some insinuations[32] after his time in Colorado Springs. Beneath some distortions due to lightning and other influences due to sun eruptions and aurora borealis he also discovered a week periodic signal. He was only able to speculate about the origin of this signal, which he recorded with his very sensitive devices.

Today we can assume with great certain that TESLA has detected radio signals from pulsars, from which he erroneously thought[35] they are signals from intelligence of civilizations on other planets. Since 1967 the radio signals of pulsars has not been detected again by science. The team of Antony HEWISH[6] in the Cavendish-Labor of Cambridge has rediscovered this signals, for which HEWISH received 1974 the Nobel price of physics, which really had been admitted to TESLA. It is typical for the awarding of this Nobel price, that Jocelyn BELL-BURNELL, who has worked in HEWISH's team and who first has noticed the absolutely unknown and curious peaks of a period of $1^1/_3$ seconds on the recorded signals, also not has been nominated .

TESLA intended to transmit huge amount of electricity through the atmosphere and discovered with his experiments in Colorado Springs[27] the surprising fact of Earth's electric conductivity. By using the whole planet as a receiving device Tesla had the biggest radio telescope ever used on Earth to detect signals from outer space. He was not able to determine the exact direction of the incoming signals but the sensitiveness of his receiving equipment was so extremely high (for that time), that he was able to detect this signals from which we know today that they come from pulsars and magnetars. This is – beneath the discovery of the X-rays (later named by Roentgen) – his second missed Nobel price.

Analysis

The force of an oscillating HERTZ dipole on a stationary charge is well known. This can also be described as a sum of forces between relatively resting, moving and accelerating charges as the author[45] has shown for the case of large distances to the HERTZ dipole. A transmission of electrical energy from one point to an other is certainly possible with a HERTZ dipole, too. But with increasing distance r form the transmitter the energy density diminishes rapidly. This law of distance can be undergone when instead of air under normal pressure a conducting medium (electrical wire) is used. An almost frictionless transmission of electrical energy between two points on Earth without wires only can be done by using some sort of a 'connecting wire', a voltage or current source and a load. This connecting wire is the Earth. The voltage or current source is the transmitter and the receiver is the load.

The elevated terminals D and D' function as a charge reservoir (electric capacitor), but they do not act as the transmitting terminal itself, whereas the energy is given off to the air. If no transmission through air is planned, it is preferred to insulate this terminals so that no charges are lost to the atmosphere. This has been sometime described by TESLA. For a simpler construction TESLA could have placed the capacitor terminals D and D' beneath the transformers A-C and A'-C' respectively. But obviously the specific arrangement of the terminals D and the supply wire B as shown in all patent drawings is very important for the correct function of the apparatus.

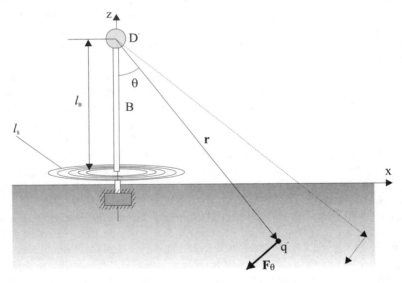

Figure 7: Accelerating dependent forces of a TESLA transmitter on negative charges in the Earth.

TESLA has operated the Earth as a ball capacitor. The transmitter "pumps" with a frequency between 20...250kHz[24],[32] electrons between the Earth and the elevated terminal back and fourth. To minimize the HERTZ radiation losses this frequency has to be as low as possible, as TESLA has mentioned explicitly. To achieve an optimal effect it is necessary to use high voltages. Tesla has tuned the whole conductor length consisting of the secondary coil A and of the conductor B to the wave length of the resonant frequency of the secondary.

With this tuning the voltage between ground and terminal rised up to more than four Million Volts. To produce such high voltages a resonant circuit with high efficiency (low damping) is requested, as TESLA mentioned many times. For that he sometimes used his flat spiral coil. The main goal is to move as many charges as possible in a short time from the terminal down to the usually bad conducting ground and back to the terminal again.

If we look at the transmitter oscillating with the resonant frequency ω then the equation of the force between resting, moving and accelerating charges [45],[46] can be applied. At the considered time the terminal D should be fully charged with electrons.

The conductor part l_B is much shorter than the conductor l_S used in the coil S as arranged in the TESLA experiments. So we can assume with high accuracy that the current is not a function of the direction z. Then, for example, on a distance r » l_B the acceleration dependent force[45] acts on charges in the vicinity of the transmitter proportional to 1/r:

$$\frac{\mathbf{F}}{q} = \frac{I l_B}{4\pi\varepsilon_0} \frac{i\omega}{c^2 r} sin\theta \, \mathbf{r}_\theta^0 \tag{0.1}$$

That means, the electrons previously sitting in D are not only locally pressed into ground but in addition there acts a force $\mathbf{F}_\theta$ on every ‚free' charge in the Earth (and atmosphere), which is inverse proportional to the distance to the transmitter. This force pushes (or pulls) the negative charges in the Earth down to deeper layers (or up again to the transmitter). Additionally there acts also a force proportional $1/r^2$ to on every charge in the ground around the transmitter. Only the simple "injection" of electrons into ground has a much smaller effect than the forces of the moving and accelerating charges in the wire element B.

With this explanation it is clear why Tesla used such high voltages or why he always intended to use as much charges as possible in his circuits. The effects of the moving and accelerating charges in the wire B depends directly on the number of involved electrons and of the frequency of the apparatus. The acceleration can not be made higher in ordinary conductors but the number of electrons can be increased with higher voltages. And the increasing of the voltage was always TESLA's intention.

The energy of the transmitter is used for the acceleration of the free charges in Earth, which in turn again accelerate more distant charges in the ground. The result is a longitudinal wave of oscillating electrons across the Earth's diameter. And exactly this is what TESLA[32] always has claimed to do. If the Earth would be a body of unlimited size, the impressed wave would be dissipated as well as the involved energy. But because of the finite size of the Earth the longitudinal wave soon approaches the borderline to the atmosphere where it will be reflected similar to sound waves. The really astonishing fact is, that the longitudinal wave through the Earth is close to the speed of light in vacuum as can be calculated form TESLA's patent information.

If the Earth is electrically struck – for example by lightning – there will always be at least two basically different resonances. The main resonance between Earth and atmosphere is known as SCHUMANN resonance[14],[15] and has a frequency of about 7.9 Hz, whereas the TESLA resonance is 11.8 Hz. Both different resonances are again presented in figure 8.

The Earth behaves like a perfect electrical conductor: „...the planet behaves like a per-fectly smooth or polished conductor of inappreciable resistance with capacity and self-induction uniformly distributed along the axis of symmetry of wave propagation and trans-mitting slow electrical oscillations without sensible distortion and attenuation." This wave is concentrated and reflected exactly at the opposite pole of the planet as Charles YOST[49] and HARTHUN et. al.[5] has shown. TESLA describes in one[32] his patents the velocity of the surface wave along the Earth's circumference form pole to pole in words as to be $v_O = 471'0240$ km/s. This means, the wave velocity through the Earth along the diameter $2r_E$ is close to the speed of light in vacuum, then it is:

$$\frac{v_O}{c} = \frac{l_O}{2r_E} = \frac{\pi}{2} \quad \rightarrow \quad v_O = c\frac{\pi}{2} \tag{0.2}$$

So the speed of the longitudinal wave through the Earth is close to c. The main longitudinal resonance is 11.79 Hz. With this longitudinal wave of free electrons in the ground the whole Earth is set in resonance. The Earth diameter must be an odd multiple of a quarter wavelength of the transmitter. Then, to produce a forward and backward wave front, the signal must be applied at least for 0.085 seconds to achieve a standing wave. And exactly all this numbers are given in TESLA's patent[32].

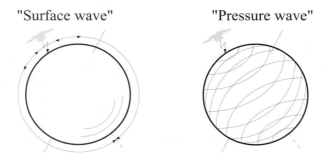

Figure 8: Difference of SCHUMANN (left) and TESLA (right) resonance

Once there are built such standing waves it is possible to produce on different places on the globe wave knots, where the excitation is a maximum and places where no oscillation can be measured. Preferably on the places of maximum oscillations a receiver is placed. This receiver is built symmetrically to the transmitter. Because of its low OHM'ic losses the receiver gains its amplitude due to resonance. Then the receiver becomes a transmitter, too. The receiver also builds a standing wave in strong synchrony with the transmitter and as a result the energy transmission can be started, if a load is placed on the receiver as shown in figure 4. Principally this energy transmission is possible in both directions.

First the transmitter must supply the energy to build up the standing wave in Earth and the to build up the receiver's oscillations. This does not require a high energy throughput. Then, as Tesla states, if this standing waves are established in perfect synchrony an energy transmission can be done without heavy losses. The energy consumed by the receiver (and the losses) out of the standing waves must be supplied by the transmitter to hold the oscillating system through the Earth alive. According to TESLA[2] the requested energy transmission can be made with an efficiency of 99.5%.

The assumption[10]-b, that TESLA has received more energy than transmitted – and therefore gave his system the name „Magnifying Transmitter" – can neither be definitively confirmed nor rejected with the presented analysis. There exists a text passage[4] that may support this assumption, but in most other original publications TESLA[2],[3],[3] always claims of an efficiency of about 99%.

If only the receiving of signals is requested, the receiver can be built much cheaper, because it has not to induce also a standing wave into Earth. Also the transmitter does not necessarily have to produce a standing wave in Earth so that every desired frequency can be used. According to figure 5 (right side) the receiver can detect the potential difference between two distant points on the Earth surface. It is possible to receive signals around the globe as well as under water with this method.

TESLA has used high voltages and high currents, as he often said. The noise of his experiments in Colorado Springs was detectable many miles. Despite the fact, that he doesn't involve such large amounts of energy as assumed to be released in thunderstorms, the analogy to lightning is allowed. So the Earth's longitudinal resonance should be detectable if lightning strikes the Earth surface. Actually it can be seen by eyes that the lightning brightness appears to flicker. It is known[8], that with ground lightning the flash strikes two to four times the same location within a time duration between each stroke between 40...80 milliseconds. This corresponds to the propagation time of a forward and backward wave through Earth close to the speed of light in vacuum.

Up to this point the transmitter and receiving devices are described in its basic functionality. It is desirable that this particular TESLA devices would be reconstructed in fully detail as done hundred years ago in Colorado Springs to get an even better understanding what has happened.

TESLA has made much more progress after his experiments with the transmission of electrical energy as mentioned above. Over 30 years he has made much more discoveries, which he has published only partly or even nothing. But in his later years he always mentioned a new energy source he already has found in the years, where he worked with the wireless systems. About this part of TESLA's work an other paper[48] will be published.

References

[1]   ABRAMYAN E. A., "Transformer Type Accelerators for Intense Electron Beams", *IEEE Transactions on Nuclear Science* **NS-18** (1971) 447-455

[2]   BARRETT Terence W., "TESLA's Nonlinear Oscillator-Shuttle-Circuit (OSC) Theory", *Annales de la Fondation Louis de Broglie* **16** No.1 (1991) 23-41

[3]   BRUNS Donald G., "A solid-state low-voltage Tesla coil demonstrator", *American Journal of Physics* **60** No.9 (September 1992) 797-803

[4]   HEISE Werner, "Tesla Transformatoren", *Elektrotechnische Zeitschrift A* **85** /1 (10 Jan. 1964) 1-8

[5]   HARTHUN Norbert und Axel BERNHARDT, "Tesla Transmitter", *Mensch & Technik*, Verlüßmoor, Vollersode, Deutschland (1984)

[6]   HEWISH Antony, Jocelyn BELL, J.D.H. PILKINGTON, P.F. SCOTT and R.A. COLLINS, „Observation of a rapidly pulsating radio source", *Nature* **217** (24 February 1968) 709-713

[7]   KELLY James B. and Lee DUNBAR, "The Tesla Coil", *Am. Journal of Physics* **20** (1952) 32-35

[8]   KRINDER Philip E., „Physics of Lightning", The Earth's Electrical Environment, *CPSMA* ISBN 0-309-03680-1 (1986)

[9]   LAURITSEN Charles C. and Richard CRANE, "A Combined Tesla Coil and Vacuum Tube", *Review of Scientific Instruments* **4** (September 1933) 497-500

[10]   MEYL Konstantin, „Elektromagnetische Umweltverträglichkeit", *Indel Verlag, Villingen-Schwenningen* Teil **1** ISBN 3-9802542-8-3 (Dezember 1996) a: 207, b: 205, c: 157

[11]   MEYL Konstantin, „Elektromagnetische Umweltverträglichkeit", *Indel Verlag, Villingen-Schwenningen* Teil **2** ISBN 3-9802542-9-1 (1999) a: 133

[12]   RATZLAFF John, „Tesla Said", *Tesla Book Company, Chula Vista*, ISBN 0-914119-00-1

[13]   RATZLAFF John T., „Reference Articles for Solutions to Tesla's Secrets", *Tesla Book Company, Chula Vista, CA-91912*, ISBN 0-9603536-3-1, Part II (1981)

[14]   SCHUMANN W. O., „Über die strahlungslosen Eigenschwingungen einer leitenden Kugel, die von einer Luftschicht und einer Ionosphärenhülle umgeben ist", *Zeitschrift für Naturforschung* **7a** (1952) 149-154

[15] SCHUMANN W. O., „Über die Dämpfung der elektromagnetischen Eigenschwingnugen des Systems Erde – Luft – Ionosphäre", *Zeitschrift für Naturforschung* **7a** (1952) 250-252

[16] SKELDON Kenneth D., Alstair I. GRANT, Gillan MACLELLAN and Christine MCARTHUR, "Development of a portable Tesla coil apparatus", *European Journal of Physics* **21** (2000) 125-143

[17] SLOAN David H., „A Radiofrequency High-Voltage Generator", *Physical Review* **47** (01 January 1935) 62-71

[18] TESLA Nikola, „Experiments with Alternate Currents of very high Frequency and their Application for methods of artificial lighting", *Lecture before the American Institute of Electrical Engineers, Columbia College* (20 May 1891), Deutsche Fassung in: „TESLA's Verschollene Erfindungen", *VAP Verlag* ISBN 3-922-367-93-3 (1994)

[19] TESLA Nikola, "System of Electric Lighting", *US Patent* **454'622** (Application filed: 25. April 1891; Patented: 23 June 1891)

[20] TESLA Nikola, "Experiments with Alternate Currents of High Potential and High Frequency", *Lecture before the Institution of Electrical Engineers, London* (03, 04 February 1892), Reprint in: *Lindsay Publications* ISBN 0-917914-39-2 (1986)

[21] TESLA Nikola, "Coil for Electro Magnets", *US Patent* **512'340** (09 January 1894)

[22] TESLA Nikola, "The Streams of LENARD and ROENTGEN and Novel Apparatus for Their Production", *Lecture before the New York Academy of Sciences* (06 April 1897), Reprint in: *Twenty First Century Books* ISBN 0-9636012-7-X (1994)

[23] TESLA Nikola, "Electrical Transformer", *US Patent* **593'138** (Application filed: 20 March 1897; Patented: 02 November 1897)

[24] TESLA Nikola, "System of Transmission of Electrical Energy", *US Patent* **645'576** (Application filed on 02 September 1897, Patented on 20 March 1900)

[25] TESLA Nikola, "Apparatus for Transmission of Electrical Energy", *US Patent* **649'621** (Application filed on 02 September 1897, Patented on 15 May 1900)

[26] TESLA Nikola, "Tesla's High Potential and High Frequency Work", *Electrical Review (London)* **44** /1,119 (05 May 1899) 730-733

[27] TESLA Nikola, „Colorado Spring Notes", *Nikola TESLA Museum, Belgrad, Yugoslavien* Edited by Aleksandar MARINCIC (1899-1900); http://www.etf.bg.ac.yu/Prez/MuzejTesla/index.htm

[28] TESLA Nikola, "Method of Intensifying and Utilizing Effects Transmitted Through Natural Media", *US Patent* **685'953** (Application filed: 24 June 1899; Patented: 05 Nov. 1901)

[29] TESLA Nikola, "Method of Utilizing Effects Transmitted Through Natural Media", *US Patent* **685'954** (Application filed: 01 August 1899; Patented: 29 May 1901)

[30] TESLA Nikola, "Apparatus for Utilizing Effects Transmitted from a Distance to a Receiving Device Through Natural Media", *US Pat.***685'955** (filed: 08 September 1899; Pat: 05 Nov. 1901)

[31] TESLA Nikola, "Apparatus for Utilizing Effects Transmitted Through Natural Media", *US Patent* **685'956** (Application filed: 02 November 1899; Patented: 05 Nov. 1901)

[32] TESLA Nikola, "Art of Transmitting Electrical Energy Through the Natural Mediums", *US Patent* **787'412** (Application filed: 16. May 1900; Renewed: 17 June 1902; Patented 18 April 1905)

[33] TESLA Nikola, "System of Signaling", *US Patent* **725'605** (Application filed: 16 July 1900; Patented: 14 April 1903)

[1] TESLA Nikola, "Tesla's New Discovery – Capacity of Electrical Conductors is Variable", *New York Sun* (30 January 1901); reprinted in [12] 57-58

[2] TESLA Nikola, "Talking with the Planets", *Collier's Weekly*, (09 February 1901) 4-5; reprinted in [12] 61-65

[36] TESLA Nikola, "Method of Signaling", *US Patent* **723'188** (Application filed: 14 June 1901; Patented: 17 March 1903)

[37] TESLA Nikola, "The Transmission of Electrical Energy Without Wires", *Electrical World and Engineer* (05 March 1904)

[3] TESLA Nikola, "The Transmission of Electrical Energy Without Wires As a Means for Furthering Peace", *Electrical World and Engineer* (07 January 1905) 21-24; Reprinted in [12] 78

[39]   TESLA Nikola, "Method of Signaling", *US-Patent* **723'188** (Application filed: 16 July 1900, divided on 14 June 1901, Patented: 17 March 1903)

[40]   TESLA Nikola, "Apparatus for Transmitting Electrical Energy", *US Patent* **1'119'732** (Application filed: 18 January 1902; renewed: 04 May 1907; Patented: 01 December 1914)

[4]   TESLA Nikola, "The Peoples Forum", *New York World* (19 May 1907); reptrinted in [12]

[5]   TESLA Nikola, "Famous Scientific Illusions", *Electrical Experimenter* (February 1919); reprinted in [12] 192-199

[6]   TESLA Nikola, "The Magnifying Transmitter", *Electrical Experimenter* (June 1919) 112-113, 148, 173, 176-178; reprinted in [13] 69-75

[44]   WASER André, „The Puzzling Nature – Die rätselhafte Natur", *AWVerlag* (May 1996) 126

[45]   WASER André, „On Electrodynamics of uniformly moving charges ", www.andre-waser.ch (28 June 2000)

[46]   WASER André, „Force of a HERTZ dipole on a stationary charge", www.andre-waser.ch (28 June 2000)

[47]   WASER André, „ Elektrische Skalarwellen: Review zum Experiment von Prof. Dr. Konstantin Meyl", *raum&zeit* **107** (August 2000) 46-54

[48]   WASER André, „Nikola Tesla's Radiations and the Cosmic Rays", AW-Verlag, www.andre-waser.ch (29 July 2000)

[49]   YOST Charles A., "The Tesla Experiment – Lightning & Earth Electrical Resonance", *Tesla Book Company, Chula Vista, CA-91912* (1983)

Figure 9: TESLA in his New York laboratory. Has for example be published 1897 in *Electrical Review* (New York and London).

Figure 10: TESLA's Colorado Springs experiment in 1899. The metal ball (~75cm diameter) can be moved in height up to 50 meter about ground.

Figure 11: The skillet of the first TESLA plant (constructed ~1901-1903) for the transmission of energy and broadcasting signals in Wardencliffe, Long Islands, New York. It was never completed.

"Tesla III Millennium" 5th International Tesla Conference, Oct. 15-19, 1996, Belgrade, Yugoslavia

Caption from previous page poster (in Serbian and English):

"My project was retarded by laws of nature. The world was not prepared for it. It was too ahead of its time. But the same laws will prevail in the end and make it a triumphal success."

- Nikola Tesla, about his World Wireless project on Long Island, 1919

Papers from the conference – www.teslasociety.com

Nikola Tesla at Colorado Springs laboratory, 1899-1900

Nikola Tesla, *Time,* July 20, 1931   "All the world's his power house"

13 Worldwide Wireless Power Prospects

Kurt Van Voorhies, Ph.D., P.E.
Proceedings of IECEC, 1991, Reprinted from *Harnessing the Wheelwork of Nature*

ABSTRACT

Worldwide wireless power began as a concept with the pioneering work of Nikola Tesla about 100 years ago. His principal approach is summarized. The viability of such a system must still be demonstrated and many questions remain. Potentially, a wireless system can transfer power more efficiently and flexibly, especially to and from remote regions. The principal elements of worldwide wireless power transfer include: 1) the source: an oscillator/transmitter, 2) the path: the cavity bounded by the earth and the ionosphere and 3) the receiver: a means of extracting power from the path. The system transfers and stores energy via the resonance modes of the cavity. The key challenges facing demonstration of technical feasibility are in finding an efficient means of coupling power into and out of the earth-ionosphere cavity, and in devising a feasible receiver that is both small and efficient. Along with demonstrating technical feasibility, new research must consider safety, environmental impact, susceptibility to weather, and effects on weather.

INTRODUCTION

Nikola Tesla pioneered the concept of worldwide wireless power transfer about 100 years ago, beginning with work on high voltage, high frequency single electrode lighting systems, and following with development of the Tesla Coil, The Magnifying Transmitter, and the single electrode x-ray tube. The Tesla Wireless system and concepts leading thereto are documented in Tesla's notes [1,2] patents [3,4], lectures [4-8] and published articles [4, 5, 9-11] and described by Tesla's biographers [12,13] and others [14, 15]. Following the death of Tesla in 1943, the concept lay dormant until referenced by Wait in 1974 [16,17] in conjunction with extremely low frequency communications, followed by Marincic's illuminating review in 1982 [18] and subsequent technical analysis by Corum and Corum [19-24], Golka [25,26] replicated the oscillator used in Tesla's Colorado Springs experiments for studying ball lightning and plasma containment for nuclear fusion. Corum and Corum [27-31] have also replicated Tesla's ball lightning experiments but with smaller scale equipment. However, Tesla's worldwide wireless power concept remains unverified.

PRINCIPLES OF WORLDWIDE WIRELESS POWER TRANSFER

Consider the earth as a large spherical capacitor or cavity resonator, comprising the *terra firma* as the inner conductor, the lower atmosphere as the insulating dielectric, and the upper atmosphere (electrosphere) and ionosphere as the outer conductor. Power is coupled into the cavity via either direct conduction/displacement, or radiation, with high power RF oscillators or transmitters tuned to the cavity's resonant frequency. A remove receiver, also tuned to this resonant frequency, then extracts this power wirelessly. The propagation loss in the earth-ionosphere cavity increases with frequency but, at the fundamental frequency, is about 11% less than the equivalent loss on a 200KV power line. The wireless concept described here differs from that used in microwave wireless power transmission in that the latter beams power along a line of sight path, normally from outer space to earth [32]

PROMISES OF WORLDWIDE WIRELESS POWER TRANSFER

The benefits of wireless power transfer have not changed since originally described by Tesla in 1900 [9] and 1904 [10]. A cheap, efficient means of distributing energy would revolutionize development and improve access to new energy sources. Energy could be coupled into the cavity at the source, eliminating the need for the costly and time-consuming process of constructing and maintaining power transmission lines. The system would enable better utilization of remote sources of energy and would facilitate power transfer to remote users worldwide. While Tesla primarily proposed supplying power for lighting in conjunction with his high frequency single electrode lighting systems, he also envisioned "...energy of a waterfall made available for supplying light, heat and motive power anywhere – on sea, or land or high in the air..."[10]. Of course, the economic viability of such a system depends upon either 1) a technical means for controlling/measuring the supply and use of wireless power around the world, or 2) a very low cost source energy.

Nikola Tesla

Nikola Tesla was a prolific inventor best known for the AC induction motor and AC polyphase distribution system which are the basis for our present AC power system. His other inventions include the Tesla coil, high frequency generators, the Tesla Magnifying Transmitter, key elements of radio, single electrode high frequency, the single electrode x-ray tube, a viscous turbine, and remote control. Following his developments in low frequency AC machines and power distribution systems , Tesla experimented with single electrode, high frequency, high voltage lamps utilizing rarefied gases, the forerunner of present fluorescent lights. Initially he utilized patented high frequency alternators with 384 poles to produce the necessary 20 KHz power, but subsequently invented the disruptive discharge high voltage transformer, a.k.a. Tesla Coil, in 1891 [33].

In a Tesla Coil, low frequency AC power is amplified in voltage with a conventional transformer. The output of this transformer feeds the Tesla Coils' resonant LC primary circuit through a spark gap. The spark gap creates a broad spectrum of energy, components which resonate the primary and secondary circuits of the Tesla coil. The secondary of the Tesla Coil is tuned to be electrically ¼ wavelength long, with one terminal grounded, and acts as a "slow wave" device to resonantly amplify the voltage further.

Tesla found that the high frequency output from the Tesla coil could readily power lights and motors utilizing a single wire with a ground return. Tesla presented these results in this lecture to the IEE in London in 1892[7]. Following the work of Kelvin and Crookes, Tesla also noted that slightly rarefied gases were excellent conductors, leading him to propose a system for " ...transmitting intelligence or perhaps power, to any distance through the earth or environing medium". [34] In February 1893, at his lecture on high frequency currents before the Franklin Institute of Philadelphia (repeated in March in St Louis.) Tesla proposed to determine the capacitance of the earth and the period of oscillations resulting from a disturbance of the earth's charge . After subsequent patented improvements to the Tesla Coil, Tesla patented the single wire power distribution system in March., 1897, [35] and patented the wireless power distribution 6 month later [36,37]. In the wireless system , the single wire conductor was replaced by a conductive path through a slightly rarefied gas coupled to bodies of large surface area, or open capacitors, connected to the high tension terminals of the transmitter and receiver, thus forming an open resonator circuit between the body and the earth. In his patent, Tesla claimed the use of the conductive layers in the upper atmosphere as the conductive path.

In the 1892 lecture in London, Tesla noted that " It is quite possible, however, that such 'no wire' motors, as they might be called, could be operated by conduction through the rarefied air at considerable distances. Alternate currents, especially of high frequencies, pass with astonishing freedom trough even slightly rarefied gases. The upper strata of the air are of difficulties of a merely mechanical nature. There is no doubt that with the enormous potential is obtainable by the use of high frequencies and oil insulation, luminous discharges might be passed through many miles of rarefied air, and that by thus directing the energy of many hundreds of thousands of horsepower, motors or lamps might be operated at considerable distances from stationary sources. But such schemes are mentioned merely as possibilities. We shall have no need to transmit powers in this way. We shall have no need to transmit powers at all. Ere many generations pass, our machinery will be driven by a power obtainable at any point of the universe..."[38] Tesla demonstrated plasma conduction in a glass tube with rarefied air surrounding a central axial platinum electrode, he observed that the wire was heated only at the ends, and not in the middle. He also observed that the pressure at which the gas becomes conducting is directly related to the applied voltage.

Colorado Springs Laboratory

Tesla moved to Colorado Springs in May 1899, after reaching the limits of his New York Laboratory with Tesla Coils operating at 4 million volts. The dry, electrostatic filled air at the 2000 m facility in Colorado Springs facilitated his developments. His primary and secondary coils were 51 ft. in diameter, and it was here that he developed the concept of an extra coil placed in series with the secondary but with loose inductive coupling so as to enable large resonant amplification of voltage. In addition to the development and improvement of the high power Tesla coil, Tesla concentrated on the development of sensitive receivers necessary for detecting communication signals. On July 3, 1899, using these devices, Tesla monitored the progression of a passing thunderstorm, observing electrical standing waves which he attributed to the storm's disturbance of the earth's electrical charge and a corresponding propagation of this disturbance around the conductive globe. Tesla also experimented with his single electrode x-ray tubes. The oscillator reportedly operated at frequencies between 45KHz and 150 KHz, at voltages between 12 MV and 18MV, and with secondary currents as high as 1100A [1,12,12,39].

Wardenclyffe Laboratory

Funded principally by J.P. Morgan, Tesla proceeded with the construction of a system of "World Telegraphy" at Wardenclyffe on Long Island upon his return from Colorado Springs in 1900 [12,13]. While he intended to use the facility publicly for communications, Tesla's secret aim was to implement wireless power transfer. The facility featured at 187 ft. wooden tower designed to support a 68 ft. diameter copper hemisphere, which was not completed because of Tesla's difficulty in obtaining funding following Marconi's success in demonstrating transoceanic wireless communication with much simpler equipment (albeit using Tesla's patents in the process) The transmitted was to have operated at 30 MV, which Tesla claimed was sufficient for worldwide power distribution; however, the transmitter was designed to handle up to 100MV. Aside from its toroidal elevated capacitor, patent 1,119,732 [40] filed in 1902 shows the Wardenclyffe configuration of the transmitter, which incorporated the 'extra coil" from the Colorado Springs experiments.

Tesla's Concept of Worldwide Wireless Power Transfer

Tesla outlined the requirements for wireless power distribution in patent 787,412, describing the earth as "….behaving like a perfectly smooth or polished conductor of inappreciable resistance with capacity and self-induction uniformly distributed along the axis of symmetry of wave propagation"[41]. He described reflections of signals from antipodes, the points on the globe diametrically opposite from the transmitter, as being similar to those from the end of a conducting wire, thus creating stationary waves on the conductive surface. He provided three requirements for resonance: 1) the earth's diameter should be equal to an odd number of quarter wavelengths, 2) the frequency should be less than 20 KHz to minimize Hertzian radiation; and 3) most critical, the wave train should continue for a minimal period, which he estimates to be 1/12 second, and which represents the period of time for a wave to propagate from and return to the source at a mean speed of 471,240 Km/sec. Tesla conceived the wave as propagating through the earth along a straight line path, the effect on the outside surface being that of concentric rings expanding to the equator and then contracting until reaching the opposite pole. Tesla also applied a fluid analogy to the earth and the water level representing the earth's state of charge at any given point. While his earlier work emphasized ground currents as the mechanism for transferring power, he later indicated that he had conclusively demonstrated that "…with two terminals maintained at an elevation of not more than thirty thousand to thirty five thousand feet above sea level, and with an electrical pressure of fifteen to twenty million volts, the energy of thousand of horse-power can be transmitted over distances which may be hundreds, and, if necessary, thousands of miles. In am hopeful , however that I may be able to reduce very considerably the elevation of the terminals now required…"[42].

Summary of Tesla's Proof of Concept

Tesla claimed to have observed the effects of the Colorado Springs transmitter at a distance of up to 600 miles. An advertising brochure for the World Telegraphy system claims the transmission of power around the globe in sufficient quantity to light incandescent lamps (50watts). Others report that a bank of 200 watt lamps, 50 watts each, were lit at a distance of 26 miles [12, 13]. The article in Century magazine shows photographs of an isolated extra coil powering and incandescent lamp as evidence of "...electrical vibrations transmitted to it through the ground from the oscillator..." [43]. However, this extra coil was most likely within the inductive field of primary transmitter, with the ground serving as a return path.

Rationale for a Renewed Interest in Wireless Power Transfer

Given Tesla's firm and unending belief in the feasibility of wireless power transfer, yet his inability, after considerable expenditure of time and money, to conclusively demonstrate its viability, the reader may question why there is a renewed interest in demonstrating the feasibility of wireless power transfer. Aside from the benefits outlined initially, the best reason probably lies in both 1) the legacy of Tesla himself, and 2) the fact that because of insufficient funding, Tesla was never able to teat a facility that had been developed strictly for power transfer, and thus hi wireless power transfer concept remains to be proven.

The legacy of Tesla speaks for itself in terms of his many and varied significant inventions, his insightful pioneering understanding of physics and electrical engineering, his tremendous drive and creative energy enabling him to constructively, work long hours on a protracted basis guided by a keen sense of vision, his ability to visualize and test concepts in his mind enabling him to achieve good results with little trail and error, and his genuine concern for improving the condition of humanity. The breath of his accomplishments at Colorado Springs with less than 8 months exemplifies these. The Colorado Springs experiments focused primarily in the development of wireless communications, i.e., radio rather than wireless power transfer. As indicated by Marincic [18], 56% of his time was spent on developing the Tesla Coil, 21% on receivers for small signals, 16% on measuring the capacity of the vertical antenna, and 6% on miscellaneous other research, including fireballs. Wireless power transmission experiments were limited to small distances.

While Tesla shared much with the world in the form of his patents, publications, lectures, he was also a very secretive person, and never fully documented his intended configuration for the wireless power system, even though he was confident there would be a workable solution. He believed that that his Magnifying Transmitter (Tesla Coil w/extra coil designed to excite the earth) would ultimately be recognized as his greatest invention [11], and felt that there would be no problem in wireless disturbing the earth's energy. He also believed the universe to be so full of energy that, ultimately, wireless distribution would not be necessary. Modern day researchers attempting to follow his path, must also be part detective. Tesla's belief and confidence in wireless power transfer is clear, however, so too was Edison's belief in magnetic ore separation, which, like Tesla's experience with Wardenclyffe, left him in deep financial debt. [44]

Recent Developments

In recent years, there has been a renewed interest in Tesla's work on high voltage, high frequency phenomena. Beginning in 1968, R. Golka formed Project Tesla to measure, under Air Force Contract, aircraft susceptibility to lighting discharge and to repeat Tesla's ball lighting experiments for application to laser fusion. In the process, he replicated Tesla's Colorado Springs transmitter and succeeded in operating it at twice Tesla's original power levels [25,26]. In 1986, Golka and Grotz proposed the application of this device to artificially resonating the earth-ionosphere waveguide [45].

Cheney reports on wireless power projects that had been planned and some partially implemented circa 1977-1980 in Canada, Central Minnesota and Southern California. [13]

Wait indicated how Tesla's early wireless experiments were the forerunner of modern developments in ELF. He observed that Tesla's fluid analogy for the process is faulty in its assumption that all of the signal energy would propagate through the fluid medium, i.e. the earth. Also faulty was Tesla's notion that energy propagates to the antipode via the center of the earth, although it is not known if Tesla had viewed this as a conceptual model as opposed to a physical model as presently interpreted.

Marincic, in his annotations of Tesla's Colorado Springs Notes [1,2] and his excellent review of Tesla's wireless work [18] applies results from recent ELF experimental data to show that the transfer of power via ELF radiation would be extremely inefficient. He indicates that for a typical gridded ELF antenna, 106 m. total length, that the antenna operating efficiency would be only 0.026% and for both receiving and transmitting antennas, the total efficiency would be (0.026%), not to mention the path loses, which are as low as 0.25 dB/Mm at 10Hz and 0.8dB/Mm at 50Hz. For a fixed size antenna, efficiency increases with operating frequency, but so do path losses, so that for long distance power transfer, the overall efficiency of a radiation-based system will be low.

Corum and Corum [27-31] also replicated some of Tesla's Colorado Springs fireball experiments but with much smaller scale equipment. This work extended to a critical engineering evaluation of Tesla's wireless power concept.[20-23], showing how the current moment in the tower of Tesla's transmitter could be used to excite the Schumann resonances in the earth-ionosphere cavity. They also hypothesized that Tesla intended to use his single electrode x-ray to both ionize a current path to the sphere of elevated capacitance and to rectify the RF energy enabling the sphere to be electrostatically charged at RF rates [20,21] The sphere would then be discharged to ground, either naturally or via a second x-ray device, at a Schumann resonance frequency. Corum and Corum have also verified that Tesla's electrical measurements such as the attenuation constant, phase velocity, cavity resonant frequency and Q are consistent with modern measurements [23] and that the loses due to glow discharge around the transmitter would be small [21].

J. F. Corum patented a toroidal helical antenna [46,47] one of whose applications could be a waveguide probe for either ELF communications or wireless power transfer. This antenna is physically small while reportedly possessing good radiating efficiencies with vertical polarization. Since the propagating Schumann modes are **primarily** vertically polarized, a vertically polarized antenna would have a distinct advantage over the horizontally polarized example presented by Marincic. However, in applying Corum's design formula to the 8Hz example presented in his patent, one finds that an antenna with a 6 km major radius (0.0002 free space wavelengths) would require a virtual continuum of 43, 200 semicircular loops each 600 m in diameter, with a total conductor length equivalent to half the circumference of the earth.

The Q of the earth-ionosphere cavity is generally reported to be about 6-8 but Corum and Spaniol [48] indicate that a low Q cavity does not necessarily limit the practicability of wireless power. However, Sutton and Spaniol [49] found that the previously measured Q values were limited by instrument noise and using modern equipment they measured levels as high as 1000, which they say were also confirmed by others. [50].

In 1986-1988, Nash, Smith, Craven and Corum of UWV utilized a ¼ wave coaxial resonator to develop a high frequency "Tesla Coil" and proposed coupling this device to a Tesla single electrode x-ray tube to generate ionizing radiation with possible application to wireless power transfer [53].

THE KEY ELEMENTS OF WORLDWIDE WIRELESS POWER TRANSFER

The key elements of worldwide wireless power transfer consist of:
1. source/transmitter
2. path
3. receiver
4. system considerations
5. environmental impact
6. economic viability

Each of these will now be explained in more detail, along with their subgroups.

Source/Transmitter

The source/transmitter, consisting of Tesla's Magnifying Transmitter is the most highly developed elements of the system, as evidenced by the standard terminology of "wireless power transmission". In this paper, the term "transfer" emphasizes the importance of other system elements as well. The Tesla Coil is remarkable efficient power processing element, and Corum and Corum have shown that Tesla's Colorado Springs Transmitter operated at power levels high by even modern standards, with peak average power levels some four orders or magnitude higher that those of the Stanford Linear Accelerator. [21]

Earth-Electrosphere/Ionosphere Cavity with Dielectric Atmosphere

The path comprises the earth (ground) and the atmosphere. The ground is a good conductor at lower frequencies, conductivity decreasing with frequency due to the skin effect. The lower atmosphere is normally a good insulator. At higher altitudes the air becomes conductive due to ionization caused by cosmic rays. The conductive layer, termed the electrosphere, [54] provides an electrostatic shield and an equipotential surface due to its high conductivity relative to the ambient currents. Lord Kelvin, in 1860 [55] originally postulated the existence of such a conductive layer based upon the fact that rarefied gases act as good conductors, and he thus postulated that this conductive layer together with the earth and intervening insulating atmosphere forms a capacitor. The potential of the electrosphere is about 300KV. The ionosphere, located above the electrosphere, is caused by ionizing solar radiation, different ionospheric layers (D,E,F) being attributed to different components of the radiation. The ionosphere is that part of the earth's atmosphere which reflects radio waves [54,56] . The properties of the path are normally measured under conditions (voltage, current, frequency) quite different from those expected for wireless power transfer, and this should be considered before drawing conclusions on the suitability of the path for such purposes. Also, the effects of weather on conductivity and the effects of magnetic storms must be considered.

Spherical Cavity Modes

The spherical cavity between the ground and the ionosphere resonates at specific modes as predicted by Schumann [57,58] and discussed by Wait[59] and Galejs[60]. The transverse electric field mode (TE) is cutoff below 1.5 KHz, so for the ELF frequencies normally considered for wireless power transfer, the cavity will only support transverse magnetic TM waves, [61]. The first seven Schumann resonances are naturally excited by lightning and this fact has been used to track lightning strikes around the globe. [61-67]. The polarization and ellipticity of the waves vary diurnally. Waves propagating in the cavity are attenuated with distance due to the finite conductivity of the conductive and dielectric layers, and the attenuation increase exponentially with frequency, increasing from 0.25 dB/Mm at 10 Hz to 20 dB/Mm at 1 KHz. (compared with 1.15 dB/Mm for a conventional 200KV power line [24]. Tesla has indicated that very little power is required to maintain a state of resonance in the cavity [21].

Waveguide Coupling

The key issue in wireless power transfer is how to couple power into and out of the cavity with minimal, or at least acceptable loss. Corum and Corum have indicated that Tesla more likely created the necessary current moments to excite the cavity by electrostatically charging an isolated capacitance at RF rates via a single electrode x-ray tube and then suddenly discharging this capacitance at a resonant frequency of the cavity [20-21]. They reported that the currents measured by Tesla would have been sufficient to generate relatively weak ELF global field strengths . Tesla noted that the discharge tended to pass upward away from ground, which he attributed to either electrostatic repulsion, or convection of the heated air. However, with such an electrically short tower, radiation into the cavity at cavity resonant frequencies would not be sufficiently efficient for technical or commercial viability. And while a resonating cavity would have purely reactive fields, and hence zero point radiation resistance together with non-stationary fields would be required for power transfer within the cavity. A radiative coupling approach appears to be infeasible for reasons stated above by Marincic.

Transmission Line Coupling

A second method for coupling power into the cavity would be via direct conduction/displacement with the conductive surfaces of the waveguide, which appears to be Tesla's original concept dating back to 1892. Several mechanisms could be considered as follows: 1) Recall that, in 1900, he proposed using balloons at 30-35 thousand feet of elevation. Conceivably the power could be conducted to these via an ionization path, created by a single electrode x-ray tube driven by the transmitter. 2) The conducting path formed by ionizing radiation might be used to couple directly into the electro sphere without the elevated conductive sphere. 3) An approach might also be borrowed from those used in present ionospheric modifications experiments [68]. 4) Perhaps with the extremely high operating voltages that Tesla had proposed, the displacement coupling with the atmospheric conduction path would be direct, as apparent from an artist's rendition of wireless power distribution from Tesla's Wardenclyffe facility [69]. Tesla originally indicated that the atmosphere could be made conductive at lower elevations with either high voltage or high frequency so this should be studied further. . With such a direct coupling approach, the power transfer mechanism would then be a spherical "transmission line", rather than a spherical wave guide.

Ground Currents

The ground currents in Tesla's Colorado Springs experiments were reported to have caused sparks within the ground, and to have shocked horses through their metal shoes within ½ mile from his transmitter. [70]. As an aside, ground currents were separately exploited for communications during WW I, when conversations over the then prevalent single wire telephone systems were susceptible to enemy interception by differentially amplifying the signals extracted from two separate and displaced ground plates. The phenomenon of magnetospheric plasma whistler waves was first noticed with these receivers, but was not identified until later [71].

Power Loss

Power loss can occur in all elements of the path, which have finite conductivity: the ground, the dielectric lower atmosphere, and the conductive upper atmosphere. Elaborate and extensive ground planes are often constructed with antenna systems in order to minimize resistive power loss to the ground. Since the ground is an intrinsic conductive element, losses are inevitable, but can be reduced by operating at lower frequencies and/or establishing distributed area contacts at the transmitter and receiver sites. The poor conductivity of the Colorado Springs soil appears to have caused Tesla some difficulty[1]. At Wardenclyffe, Tesla was planning to use saltwater filled with viaducts under the transmitter to establish a good ground connection. Similar to the ground, atmospheric losses can be reduced by operating at lower frequencies. This appears to conflict with Tesla's notion that gases conduct better at high frequencies, but could be explained by higher dielectric losses. One important feature to the wireless system is the possibility of storing power in the resonating fields within the earth-ionosphere cavity, however, the feasibility of doing this will be dependent upon the Q of the cavity and upon the relative amount of excess power being stored therein. As Tesla had indicated, the power losses are reduced with higher operating voltage since power would then be distributed at lower current levels. Precipitation can dramatically change the conductivity of the atmosphere, and the effects of this on power coupling need to be considered further.

Receiver

The receiver is the least understood element of the system, and one that is most crucial to the system's success. For system using a radiative coupling mechanism, an antenna's efficiency and size both benefit from higher operating frequencies which, as noted above, increased the system's path losses. A transmission line approach would require conductive/displacement coupling into the electrosphere, which requires invention and development.

Tesla expressed confidence in being able to extract power for both individual and home use as well as for powering ground and air transportation vehicles, as illustrated in an artist's rendition [69]. He indicated in patent 649,621: "Obviously the receiving coils, transformer, or other apparatus may be movable – at, for instance, when they are carried by a vessel floating in the air or by a ship at sea. In the former case the connection of one terminal of the receiving apparatus to the ground might not be permanent, but might be intermittently or inductively established without departing from the spirit of my invention. IT is to be noted here that the phenomenon here involved in the transmission of electrical energy is one of true conduction and is not to be confounded with the phenomenon of electrical radiation which have heretofore been observed and which from the very nature and mode of propagation would render practically impossible the transmission of any appreciable amount of energy to such distances as are of practical importance [36].

Tesla separately described the utilization of energy from ionized air, in connection with his description of the art of telautomatics; "Most generally I employed receiving circuits in the form of loops, including condensers, because the discharges of my high-tension transmitter, ionized the air in the hall so that even a very small aerial would draw electricity from the surrounding atmosphere for hours. Just to give an idea, I found for instance, that a bulb 12 inches in diameter, highly exhausted, and with one single terminal to which a short wire was attached, would deliver well on to one thousand successive flashes before all charge of the air in the laboratory was neutralized..." [72]

Systems Considerations

A wireless system would entail a multiplicity of transmitters and receivers each coupling into a common propagation and storage cavity, each requiring proper phasing and balance.

Safety

A wireless power system would expose the entire biosphere to ELF fields of varying intensity. The 78 Hz Seafarer/Sanguine/ELF submarine communication system provoked health concerns, as do high-tension power lines. The fields of wireless and wire-based power transmission systems need to be compared for equivalent power levels. There is much speculation about the adverse effects of magnetic fields on health. However, recent reports from PACE indicate that ELF energy at the lower Schumann resonance frequencies constitute a natural biological clock [71]. The first four Schumann resonances frequencies are within the range of brain wave activity. The fundamental mode is coincident with the theta wave spectrum, which ranges from 4 to 8 Hz, and is attributed to a normally unconscious state with enhanced mental imagery and a high level of creativity.[72] The next three Schumann modes are coincident with the beta wave spectrum which ranges from 13 to 26 Hz, and is associated with the normal conscious state.

Environmental Impact

Operating at high voltages and surrounded by a glow discharge, the transmitter could be a source of pollutants, including ozone, NO and nitric acid, as reported by Tesla during his experiments and steps would have to be taken to mitigate any such hazards if they exist.

Electromagnetic Interference (EMI) and Radio Frequency Interference (RFI)

The operating frequencies of a wireless system could be expected to be low enough so as to not interfere with present communications of electronic systems. The FCC does not make frequency allocations below 9Khz and Tesla had predicted the operating frequency to be below 20 kHz. Circuit interrupters in conventional Tesla coils could be expected to create a significant amount of wide-band EMI; however, modern transmitters could be expected to utilize more advanced switching devices which, together with shielding, could minimize radiated EMI/RFI. The glow discharge surrounding the high transmitter could also be a source of EMI/RFI.

Weather Modification

Since the potential of the electrosphere is about 300 KV relative to the earth, and the wireless system as proposed by Tesla was designed to operate at 30-100MV, there is a significant potential for electrically disturbing the atmosphere. It is not know whether this would be beneficial or harmful. Vonnegut [75] has suggested that the destructive effects of tornadoes may result from atmospheric electrical effects; however, Wilkins [76] concluded from laboratory model vortex experiments that the electrical effects were the effect, rather than the cause, of tornadoes.

Economic Viability

Given technical feasibility and safety, the wireless power transfer system must still be economically viable in order to succeed. Multiple transmitter could conceivably be phased to control the location of antinodes form which power could be extracted, however, this could be at best, a short term solution, unless wireless is constrained to a relatively few large scale facilities that will be expensive and technically difficult to construct. The worldwide regulation and control of wireless power distribution will be difficult if physically constrained to operate at selected resonant frequencies.

CONCLUSION

Times have changed since Tesla's initial investigations of wireless power. Tesla originally envisioned a distributed network of relatively low level suppliers and users of wireless power, and thought it would benefit remote users the most, although he also envisioned large scale power distribution. Our power needs have dramatically increased over the past 100 years, as have their complexity. Tesla expressed great confidence in the viability of wireless power distribution, yet was unable to see its fruition after nearly 50 years of effort. The fulfillment of his vision was undoubtedly impeded by limitation on funds and resources. Tesla demonstrated that the earth can be electrically resonated. The key challenge to feasible worldwide wireless power distribution is whether a means can be found for efficiently coupling power into and out of the cavity formed by the earth, the atmosphere, and the electrosphere/ionosphere. Radiative coupling does not appear to be viable . A conductive approach is proposed which is consistent with Tesla's original wireless concepts; this requires, however, further invention and development. The receiver is the element requiring the most development to make wireless power transfer feasible.

REFERENCES

1. Tesla, N Colorado Springs Notes 1899-1900 with commentaries by A Marincic , 1978 Nolti. Yugoslavia.

2. Ratzlaff, J.T., and Jost, F.A. Dr. Nikola Tesla (I English Serbo-Croatian Diary Comparisons, II Serbo-Croatian Diary Commentary by A. Marincic III Tesla/Sherff Colorado Springs Correspondence 1899-1900), 1979 Tesla Book Co., Millbrae, CA.

3. Ratzlaff, J.T. Dr Nikola Tesla, Complete Patents, 1983 Tesla Book Co. Millbare CA

4. Nikola Tesla Museum Nikola Tesla Lectures, Patents, Articles, 1956, Nolit, Beograd, 1973, Health Research, Mokelume Hill Ca.

5. Martin, T.C. The inventions Researches and Writings of Nikola Tesla , 1894, The Electrical Engineer , New York, 1986, Angriff Press, Hollywood Ca.

6. Tesla, N. Lecture delivered before the American Institute of Electrical Engineers at Columbia College, NY May 20, 1891 Reference [5] pp145-197 Reference 4 pp L-15-L47

7. Tesla, N. Lecture delivered before the Institution of Electrical Engineers in London February 1892, Reference 5 pp;.198-293. Reference 4 pp. L-48.

8. Tesla, N, Lecture delivered before the Franklin Institute Philadelphia Pa February 1893 and before the National Electric Light Association St. Louis March 1893, Reference 5 pp 294-373 Reference 4 pp L-107-L-155

9. Tesla, N "The Problem of Increasing Human Energy" The Century Illustrated Monthly Magazine, June 1900 also in Reference 4 pp A-109-A-152

10. Tesla, N. "The Transmission of Electric Energy without Wires", Electrical World and Engineering, March 5, 1904, also in Reference [4] pp. A-153-A-161.

11. Tesla, N., "My Inventions: The Autobiography of Nikola Tesla,", (introduction by B. Johnston), 1919, Hart Brothers, Williston, Vermont, 1982.

12. O'Neal, J.J, "Prodigal Genius" The Life of Nikola Tesla. David McKay Co., New York, 1955, also Angriff Press, Hollywood, Ca.

13. Cheney, M, "Tesla: Man out of Time, Dell New York, 1981.

14. Friedlander, G.D., "Tesla Eccentric Genius", IEEE Spectrum June 1972 pp.26-29

15. Trinkas, G. Tesla: The lost Inventions, High Voltage Press, Portland Oregon, 1988

16. Wait, J.R. "Historical Background and Introduction to the Special Issue on Extremely Low Frequency (ELF) Communication" IEEE Transactions on Communications. Vol. COM_22 No. 4, April 1974, pp.353-354.

17. Wait, J.R. "Propagation of ELF Electromagnetic Waves and Project Sanguine/Seafarer, IEEE J. Oceanic Engr. Vol OE-2 No.2 April 1977, pp. 161-172

18. Marincic, A.S. "Nikola Tesla and the Wireless Transmission of Energy" IEEE Trans.on Power Apparattus and Systems Vol. PAS-101, No.10 October 1982, 4054-4068

19. Corum, J.F., and Corum K.L., "A Technical Analysis of the Extra Coil as a slow wave Helical Resonator". Proceedings of the Second International Tesla Symposium, Colorado Springs, Colorado, 1986.

20. Corum, J.F. and Corum, K.L. "A Technical Analysis of the Extra Coil as a Slow Wave Helical Resonator", Proceedings of the Second International Tesla Symposium, Colorado Springs, Colorado, 1986.

21. Corum, JF. And Corum K.L., "Critical Speculations Concerning Tesla's Inventions and Applications of Single Electrode X-Ray Directed Discharges for Power Processing, Terrestrial Resonances and Particle Beam weapons." Proceedings of the Second International Tesla Symposium, Colorado Springs, Colorado, 1986.

22. Corum, J.F. and Aidinejad A, "The transient Propagation of ELF Pulses in the Earth-ionosphere Cavity". Proceedings of the Second International Tesla Symposium, Colorado Springs, Colorado, 1986.

23. Corum J.F. and Corum K.L., "A Physical Interpretation of the Colorado Springs Data" Proceedings of the Second International Tesla Symposium, Colorado Springs, Colorado, 1986.

24. Corum J. F. and Smith, J.E., "Distribution of Electrical Power by Means of Terrestrial Cavity Resonator Modes" Proposal submitted to Planetary Association for Clean Energy Inc, December 1986.

25. Golka R.K. " Long Arc Simulated Lightning Attachment Testing using a 150 kWh. Tesla Coil (unknown publication status),

26. Golka, R.K. "Project Tesla" Radio, Electronics, February 1981, 48-49, also see Reference [13] pp 282-284

27. Corum, J.F., and Corum K.L. "Laboratory Generation of Electric Fire Balls" (unknown publication status).

28. Corum J.F. & Corum K.L. "The laboratory Production of Electric Fire Balls" (unknown publication status).

29. Corum, J. F & Corum, K.L "Production of Electric Fire Balls" (unknown publication status).

30. Corum, J.F., Edwards J.D. and Corum K.L, "Further experiments with Electric Fire balls (unknown publication status).

31. Michrowski, A., "Laboratory Generation of Electric Fireballs" Planetary Association for Clean Energy Newsletter" Vol.6, No.1 July 1990, pp.21-22

32. Glasser, P.E. Solar Power from Satellites" Physics Today, February 1977, pp.30-37, summarized in Reference [13] pp. 284-285

33. Tesla, N. Patent 462-418, Method and Apparatus for Electrical Conversion and Distribution " Application filed on February 14, 1891, Reference [3] p. 211, Reference [4], p; P-221

34. Reference [8]; Reference [5] p. 349

35. Tesla, N. Patent 593,138, "Electrical Transformer" Application filed on March 20, 1897, Reference [3] p. 301 Reference [4] p. P-252

36. Tesla N. Patent 649, 621, " Apparatus for Transmission of Electrical Energy". Application filed on September 2, 1897, Reference [3] p. 311

37. Tesla N. Patent 645,576 "System of Transmission of Electrical Energy" Application filed on September 2, 1897 Reference [3] p. 311.
38. Reference [7] Reference [5] p. 235
39. Tesla N. "Possibilities of Electro-Static Generators" Scientific American March 1934, 115, 132-134, 163-165 April 1934, 205.
40. Tesla N. Patent 1,119,732 "Apparatus for Transmitting Electrical Energy", application filed on September 2, 1897, Reference [3] p. 397, Reference [4] p. P-331
41. Tesla N Patent 787,412, " Art of Transmitting Electrical Energy through the Natural Media" Application field on May 16, 1900 Reference [3] p. 435 Reference [4] p. P-357
42. Reference [9] Reference [4] p. A-150
43. Reference [9], Reference [4] p. A-123
44. Peterson, M "Thomas Edison Failure" Inventions and Technology, Vol, 6 NO. 3 Winter 1991, pp.8-14
45. Golka R.K and Grotz, Toby "Proposal: Project Tesla The demonstration of Artificial Resonating the Earth's Ionosphere Waveguide, a precursor for the wireless transmission of vast amounts of electrical power using the Earth's Schumann's Cavity" October 28, 1986
46. Corum J.F. Patent 4,622,558 "Toroidal Antenna" Application filed on November 7, 1985
47. Corum J.F. Patent 4,751,515, "Electromagnetic Structure and Method" Application field on July 23, 1986
48. Corum J.F. Corum K.L and Spaniol C. "Concerning Cavity Q" Proceedings of the International Tesla Symposium 1988, summarized in Reference [49]
49. Sutton, J.F. and Spaniol C. "A Measurement of the Magnetic Earth-Ionosphere Cavity Resonances in the 3-30 Hz range", Presented at the International Tesla Symposium 1988.
50. Personal correspondence
51. Michrowski, A. The Planetary Association for Clean Energy Newsletter Vol. 5 Nos 3 &4 December 1987, p. 6
52. Reference [31] p.3
53. Nash, M Smith J.E, and Craven R P,M. "A Quarter-Wave Coaxial Cavity as a Power Processing Plant" p. 285 in Michrowski A New Energy Technology, Planetary Association for Clean Energy. 1990.
54. Chalmers, J.A., Atmospheric Electricity, Pergamon Press, N.Y. pp. 13, 33-35
55. Kelvin, Lord "Atmospheric Electricity" Royal Institute Lectures, Pap or Elec and Mag. Pp.208-226, summarized in Reference [54]
56. Davies, K, Ionospheric Radio Propagation, US Dept of Commerce, NBS, Monograph, 80, 1965
57. Schumann W.O. "On the radiation free selfoscillations of a conducting sphere which is surrounded by an air layer and an ionospheric shell" (in German) Z Naturfosch, 72, 1952, 145-154, summarized in Reference [59,60,61]
58. Schumann W.O. "On the damping of electromagnetic selfoscillations of the system earth-air-ionosphere (in German), Z. Naturforsch , 72, 1952, 250-252 summarized in references [59,60,61]

59. Wait, J.R. Electromagnetic Waves in Stratified Media, Pergamon Press, New York 1970

60. Galejs, J. Terrestrial Propagations of Long Electromagnetic Waves, Pergamon Press, New York, 1972

61. Sentman D.D. "Magnetic elliptical polarization of Schumann resonances" Radio Science, Vol. 22, No. 4, July-August 1987, pp.595-606

62. Sentman D.D. "PC Monitors Lightning Worldwide" Computers in Science. Premier, 1987, page 25-34

63. Coroniti S. and Huges, J. Planetary Electrodynamics Vol. 2 Gordon and Breach, New York, 1969, summarized in Reference [61]

64. Jones, D.L. and Kemp, D.T. "Experimental and Theoretical observations on the transient excitation of Schumann resonances" Journal of Atmospheric and Terrestrial Physics, Vol 32, 1970 pp 1095-1108

65. Kemp, D.T. "The global location of large lightning discharges from single station observations of ELF disturbances in the Earth Ionosphere cavity" Journal of the Atmospheric and Terrestrial Physics Vol.33, 1981, pp. 919-927.

66. Jones, D.L.,"Extremely low frequency (ELF) Ionospheric Radio Propagation Studies Using Natural Sources" IEEE trans on Communications, Vol Com-22 No 4, April 1974, pp. 477-484.

67. Mitchell, V. B. "Schumann resonance – some properties of discrete events" Journal of Atmospheric and Terrestrial Physics, Vol 38, 1976, pp. 77-78

68. Eastlund, B.J and Ramo, S. Patent 4,712,155 "Method and Apparatus of Creating an Artificial Electron Cyclotron Heating Region of Plasma" Application filed on January 28, 1985

69. Reference [11] p. 89

70. Reference:[13] p. 138.

71. Stix, T.H. "Waves in Plasmas: Highlights from the past and present" Phys. Fluids B 2 (8) August 1990, 1729-1743

72. Reference:[11] p.107.

73. Reference: [31] p. 4.

74. Allen W. G. Overlords and Olympians, Health Research, 1974, Mokelume Hill, California, p. 12

75. Vonnegut, B. "Electrical theory of Tornadoes" J Geophys, Res. 65, 1950-203-212 summarized in Reference [54]

76. Wilkins, E.M. "The role of electrical phenomena associated with tornadoes" J. Geophys Res. 69, 1964, 2435-47 summarized in reference 54.

Kurt Van Voorhies holds patents #5,442,369, and #6,239,760 and can be reached at Vortekx, Inc., DeTour Village, MI, vortekx@sault.com

14 Distribution of Electric Power by Means of Terrestrial Cavity Resonator Modes

James Corum, PhD and James E. Smith, PACE Conference, Dec. 5, 1986

The Earth-Ionosphere cavity can be used as a means to distribute electrical energy for industrial purposes at extremely low frequencies. The technology which will permit the wireless distribution of electrical power to or from remote geographical regions is now available for research and development.

It is advanced that the Earth-Ionosphere cavity possesses electrical properties which are appropriate for the wireless distribution of electrical energy to any point on Earth on an industrial scale. Such a remarkable proposition, though seemingly a fanciful concept, is actually no more profound than the notion advanced in the early 1950's to use a cavity resonator to wirelessly distribute microwave power to process food, i.e., a microwave oven.

Electrical energy at the appropriate frequency may be introduced at one point in a cavity resonator and efficiently collected at another by devices tuned to the same frequency. The resonator itself serves as a two port reactive distribution system. The ELF (extremely low frequency) resonator formed by the cavity between the Earth and the lower E-region of the ionosphere is a natural resource that will actually permit the terrestrial distribution of electrical power across a continent, without the necessity of an interconnecting land-line grid of high tension transmission lines.

History

From an historical standpoint, it is significant that Nikola Tesla long ago envisaged such a global power distribution system. A flag of caution should be raised here. It has been common in the past to discard Tesla's far-sighted vision as baseless. We believe that such depreciation has stemmed from critics who were, in fact, uninformed as to Tesla's techniques, measurements and physical observations. After reviewing Tesla's technical disclosures, it is our considered judgement that not only is industrial scale power transmission practical, but that Tesla's actual data is consistent with the very best experimental data available today. It could have only been gotten as a result of authentic terrestrial resonance and power transmission measurements.

Tesla proposed that the Earth itself could be set into a resonant mode at frequencies on the order of 7.5 Hz. From his notes, his private correspondence, his diary and his patent disclosures, it is clear that Tesla's physical explanation and interpretations were erroneous. However, as is often the case, significant explorations or inventions have been made on the basis of faulty physical concepts. Experiments were done, demonstrations performed and data taken. Let us review some of the history behind this early research.

Tesla's ELF Enterprise

In May of 1899, Tesla arrived at Colorado Springs, Colorado with $100,000. This is the same Tesla whose patented AC power system, purchased by the Westinghouse Corporation which was selected and installed in the original Niagara Falls electric power project of the 1890's. Perhaps it is not unremarkable that almost a century later most of the civilized world still employs a power generation and distribution system in virtually the same form as his early disclosures.

Within three months of his arrival at Colorado Springs, he and his associates constructed a laboratory which housed a prodigious RF signal generator. The primary and secondary were wound on a circular fence 51 feet in diameter and had an input power in excess of 250 KW provided by the Colorado Springs Electric Power Company. The secondary was used to drive a helical resonator, or extra coil, 10 feet high, wound with 100 turns of 6 gauge wire on a coil form about 8 feet in diameter. Emanating from the midst of the extra coil was a tower about 150 feet high, capped with a copper sphere 3 feet in diameter. The resonant frequencies of the driving transmitter have been variously reported as between 50 KHz to 150 KHz. This transmitter, we believe, was used as one component in a recently uncovered process to produce significant currents in the vertical tower and its attachments at pulse frequencies of 7.5 Hz to 15 KHz.

A very colorful account of the first time Tesla fired up his equipment is given in O'Neill's now classic, though somewhat unreliable, biography. Tesla, on various occasions actually said that, he had created sparks 150 feet in length. His experiments in Colorado Springs lasted nine months and cost in excess of $200,000.

Tesla returned to New York on January 21, 1900 and soon received the financial backing of J. P. Morgan, Thomas Fortune Ryan, John Jacob Astor and others. His patent application of January 18, 1902 reveals his intention to construct a massive Tesla coil driven generator for global power distribution.

The installation was subsequently constructed at Wardenclyffe, Long Island in 1902. The tower was 154 feet tall and the cap sphere was 50 feet in diameter. It was never completed, however, and was destroyed during WW I. Similar towers were to have been built at Niagara Falls, in Australia and in Europe. Tesla, however, had to abandon the Wardenclyffe project when his financial backers withdrew their support.

Physical Operation

Tesla had proposed that the Earth itself could be set into resonant electrical oscillations which he experimentally determined to be no lower than 6 Hz and no greater than 20 KHz. He claimed to have resonated the Earth in this frequency range by using a huge spark gap transmitter energized by the standard secondary of his monstrous Tesla Coil. His patent application of February 19, 1900, entitled "Apparatus for Transmission of

Electrical Energy" is probably the closest description available of the equipment used at Colorado Springs the previous summer. Assuming Tesla's claimed demonstration of distant power transmission without wires as a working hypothesis, then a plausible physical explanation is that the discharges from the electrode at the top of his giant tower would have significant spectral components at the Schumann resonance frequencies and excite a standing wave mode in the Earth-Ionosphere cavity. These physical issues have been addressed in recently presented technical publications. The overwhelming documented technical evidence clearly substantiates the above position.

Schumann Resonances

In 1952, the German physicist, W. O. Schumann recognized the possibility that a somewhat unusual example of a resonant cavity might be provided by the Earth itself as one boundary surface, and the ionosphere as the other. These two concentric spheres could then form the boundaries of a resonant electromagnetic cavity. (Sea water has a conductivity of 4 Siemens/meter[1] while the ionosphere has an effective conductivity on the order of 1 milliSiemens/m. Evidently, the structure can easily support damped oscillations.)

Determination of the cavity resonant frequencies follows from a solution of Maxwell's Equations subject to the given boundary conditions. At extremely low frequencies (ELF), where the wavelength is large compared to the effective height of the ionosphere, the electric field is essentially radial, and its amplitude distribution varies as the cosine of the polar angle measured from the position of the source antenna. Amplitude distributions for the first and second modes of oscillation of the Schumann cavity are as shown in Figure 1, when the Earth-Ionosphere cavity is excited by a source which launches vertically polarized electromagnetic waves from the North Pole.

Measured Electrical Properties

There are a variety of electrical properties of the Earth-Ionosphere cavity which have been experimentally determined over the past twenty years and are now well documented in scientific literature.

(a) Spectral Response

The resonant frequencies of the cavity have been predicted and observed. One would expect natural phenomena to excite cavity oscillations. This, indeed, does happen. The cavity is set into oscillation by solar flares, for example. But, by far the dominant natural phenomenon exciting cavity resonances is thunderstorm activity occurring world-wide. The power density spectrum of a lightning stroke is very broad, containing a wide band of frequencies. Electrically, the Earth-Ionosphere cavity behaves like a multiply tuned LC network driven by an impulse generator, and oscillations are excited at the natural resonant frequencies of the network. Thunderstorm activity is more or less continuously present on Earth, with the main centers of activity being Southeast Asia, the Congo and

---

[1] Ed. Note: 1 Siemens = 1/ohm, also called "mho" or "mhos" which is a measure of conductivity

the Amazon Basin. Consequently, experimental measurements of the atmospheric noise power density spectrum would be expected to reveal peaks at the cavity resonant frequencies, should Schumann's hypothesis be correct. Figure 2 is a typical measurement of the atmospheric noise spectral density vs. frequency. The first few cavity resonances reported above are quite evident. This is how the measured values were determined.

These sorts of measurements have been reported by many observers over the past 20 years. The spectra are frequently skewed about the center frequency and may undergo variations up to about 1 Hz in periods on the order of a minute or so.

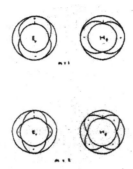

Figure 1. Radial electric field E_r and azimuthal magnetic field intensity H_ϕ for the first two cavity resonator modes of the Earth-ionosphere shell.

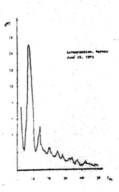

Figure 2. Typical spectrum of cavity noise. Prominent Schumann resonances at 8, 14, 20 and 26 Hz are visible. Peaks at 32, 37 and 43 Hz are apparent.

(b) Cavity Q

An important practical question associated with the Earth-Ionosphere cavity is its ability to store or contain energy without dissipating it by heating up the Earth or the ionosphere boundaries. In electrical and microwave circuit theory, a quantity called the Q of the resonant cavity is determined as a ratio between the stored energy and the energy loss per cycle in the cavity,

$$Q = \omega_o \; \frac{\text{stored energy}}{\text{power lost}}$$

where ω_o is the resonant angular frequency assuming no losses. The Earth-Ionosphere cavity Q has been measured and documented experimental data places it in the range between 3.8 to 7.8.

(c) Propagation Attenuation Constant

While the above Q is relatively low for a tuned circuit, it does indicate that the waveguide propagation losses are surprisingly small. For electromagnetic propagation on a transmission line or in a waveguide, a forward traveling wave attenuates as

$$E(\ell) = E_o \, e^{-\alpha \ell}$$

where α is the attenuation constant in nepers per meter[2]. The measured value of the attenuation constant for ELF waves propagating in the Earth-Ionosphere cavity has been experimentally determined to be on the order of one quarter of a dB per thousand kilometers. By way of comparison, single circuit 200 kV 60 Hz overhead power transmission lines have attenuation constants on the order of 1.15 dB per thousand kilometers. Experimentally established transmission and distribution losses are 23% less for the Schumann Cavity than for conventional power transmission lines.

The issue of the practicality of the proposed distribution system does not rest upon the efficiency of the transmission medium. Rather, the technical issue to be faced concerns the electromagnetic coupling mechanism to be used. This issue, we believe, was addressed by Tesla and the experimental results which he disclosed testify to his conspicuous success.

The Earth-Ionosphere cavity is, indeed, capable of being artificially excited into oscillation and the cavity can be employed as a medium for the global distribution of electrical power.

---

[2] Ed. Note: 1 neper = 8.6 dB, which is a unit of attenuation, close to 10 dB where 1 dB $\cong$ 20% loss

What is required is the creation of a practical engineering capability to efficiently launch electrical power into the cavity and to couple energy from the cavity.

It is absolutely astonishing that Tesla's public disclosures and technical publications match the electrical properties of the Schumann Cavity fifty years before there was even a theoretical model to predict rough values.

The conclusion should be obvious: Tesla could have only obtained these numbers by successfully stimulating the cavity. Tesla had to have solved the problem of launching energy into the Earth-Ionosphere waveguide and coupling energy from the cavity. We believe that the technical aspects of his apparatus have been sufficiently disclosed, in his patents, to be able to replicate his cavity stimulation and power transmission experiments. This experimental investigation should be carried out immediately. Clearly, whoever executes a sound and careful program of research along these lines will develop a technology capable of distributing electrical energy on a vast scale without the necessity of a land-line network.

It is evident that we are advocating one of the most visionary energy distribution systems ever conceived. And yet we maintain that it is technically sound and can be swiftly inaugurated at a fraction of the capital investment required by the only other alternative electrical power distribution system - high voltage overhead power transmission lines.

Tesla was aware of this and could clearly see through to the logical conclusion. When he returned to New York City in 1900, he wrote:

> "Men could settle down everywhere, fertilize and irrigate the soil with little
> effort, and convert barren deserts into gardens, and thus the entire globe
> could be transformed and made fitter abode for mankind."

This program will inevitably have an even broader impact upon the entire civilized world. The electrical power industry will experience a major innovation. The global economics of today, which is so petroleum dominated, would be transformed overnight to reflect the importance of those nations which are happily endowed with natural resources appropriate for the generation of electrical energy.

Such research will not only revolutionize the areas of energy, transportation, agriculture and commerce, but, in all probability, could even inspire significant alterations in the present structure of world governing bodies. We are referring to the consequences initiated by a global diffusion of energy. International society could perhaps be on the verge of a metamorphosis comparable in magnitude to the great agitation, evolution and achievement which so characterized the European Renaissance and the forward progress of civilization to which it gave birth.

During the last century, natural science seemed, for all intents and purposes, to have reached its maturity. From our vantage point today, that period is called "The Golden Age of Classical Physics". Yet, almost a hundred years ago, remarkable discoveries began to be made which would engender profound modifications of classical physics. It was the experimental science of the 1890's which would soon give birth to what, today, we call modern physics. It was a renaissance no less than the transition which had occurred several centuries earlier in art, literature and natural philosophy.

It has been observed that, standing on the threshold of the 1890's, only a writer of science fiction could have dreamed of the revolution on physical thought which was to occur over the next few years. And even the poets and writers of that day were unable to grasp the impact which the new science would soon have on industry, the military and the political life of the entire planet, which we have observed during the twentieth century.

Today, we stand on a similar threshold. But now it is technology which is experiencing such radical growth. We submit that the "high tech" society which we enjoy today may be but a destitute and primitive shadow of the flourishing civilization which could soon emerge across the threshold of the 21st century.

The power distribution system which we are proposing will surely require careful and considered investigation. There are no simple engineering answers.

Engineering has been called "that profession which utilizes the resources of the Earth for the benefit of mankind." We are proposing the initial step in what eventually will be an engineering project the scale of which civilization has never endeavored to attain before. But never since the days of Columbus, could so much be gotten for so small a financial investment. Never before in recorded history has it been within the grasp of the technical community to so dynamically influence the advancement of civilization.

The Engineering Challenge

There is a need for a practical waveguide probe capable of exciting the Earth-Ionosphere cavity at 8 Hz where the wavelength is about 37.5 million meters. Poor radiation efficiency and physical size limitations for such probes in previously known technology have been overcome with our inventions, patented in the U.S. and other nations (patents 4,751,515 and 4,622,558). With these, a contrawound structure waveguide probe of reasonable size can be built which can excite the Earth-Ionosphere cavity. It employs the earth as an image current source and has a maximum dimension of 0.001 free space wavelengths (with much smaller sizes possible), designed to launch vertically polarized, omnidirectional energy efficiently into the cavity at its primary resonant frequency, or sufficiently close to a resonance frequency so as to be within the resonance frequency bandwidth. Because propagation losses are so low at the primary Schumann resonance frequency, signals at that frequency may be transmitted to any point of the earth without significant attenuation.

An important element of the inventions is that the path inhibit propagation, thereby creating slow waves, and provide an electromagnetically closed path so that a standing inhibited-velocity wave, or resonant operation, can be established in response to the flow of electrical current through the path. One half of the electrically conducting path may be eliminated in embodiments of the structure by employing the image theory technique. Thus, a conducting image surface electrically supplies the missing portion of the path.

209

The image surface may be a conducting sheet, a screen or wires arranged to act electrically as a conducting sheet, or may be the earth, in accordance with the improvement disclosed in the patents of the known electromagnetic theory.

COMPARISON OF PHYSICAL PARAMETERS

Physical Parameter	Accepted Experimental Values	Predicted from Tesla's Disclosures
Attenuation Constant (dB/Mm)	$.20 \leq \alpha \leq .30$	.26
Resonant Frequency (Hz)	$6.8 \leq f_o < 7.8$	6
Cavity Q	$3.8 \leq Q \leq 7.8$	$3.2 \leq Q \leq 6.4$
Coherence Time (sec.)	no data available	0.08484
Phase Velocity	$.71 \leq V_f \leq .83$	0.8
Cavity Mode Structure	$P_n (\cos \theta)$	"Projections of all the stationary nodes onto the earth's diameter are equal."
Cavity Thickness (Km)	$35 \leq h \leq 80$	{ "greater than 8 Km" / "about 20 Km" }

Table 1. Documented numerical evidence that Tesla excited terrestrial resonances in 1899. Additionally there is a host of descriptive evidence.

Nikola Tesla, *The American Magazine*, April, 1921

15 The Real Tesla Electric Car

Mike Gamble
mike.gamble.retired@gmail.com

ABSTRACT

There are many stories both good (real) and bad (hoax) about Tesla's fabled electric powered Pierce-Arrow car. If these stories are indeed fact then sound electrical engineering should be able to reproduce it. This paper documents my efforts in trying to reverse engineer Tesla's electric car motor. So as not to loose anyone in the details, it starts with a tutorial review of radio communications and resonant circuits. Then it proceeds to reverse engineer the system design by extracting possible motor and controller specifications from those stories. The following presentation is my effort in this endeavor.

INTRODUCTION

This paper on the "First Tesla Electric Car Motor" was reverse engineered using sound (standard) electronic and electrical engineering design practices.

For those of you that haven't met me, I am Mike Gamble. Presently, I am retired from (BR&T) Boeing Research and Technology. As a Boeing electrical engineer I put in 30 years doing electronic designs and running an R&D laboratory specializing in the design of nitinol and piezo actuators.

BASIC RADIO COMMUNICATIONS

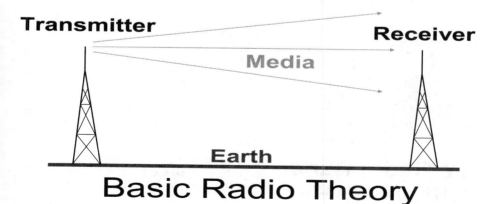

This paper starts with a tutorial on basic radio communications theory as background material for understanding Tesla's resonant electrical work. Radio communications is divided into three major parts or segments. First, is the "Transmitter" or signal generator which is the wave source. Second, is the "Transmission Media" or wave conductor which provides the signal path from the transmitter to the receiver. Third, is the "Receiver" or wave sink which reproduces the transmitter's signal. For clarity this figure only shows a single receiver, however one transmitter usually powers multiple receivers. The radio engineering definition is if the receiver is working correctly it gives a good reproduction of the transmitter's signal. However, most people just turn "ON" the receiver; the rest is all "Black Magic".

TRANSMISSION MEDIA

The second part of radio communications is the transmission media. Over time, for many years and centuries this media has been known by various titles, to name a few of the more common ones: Aether, Essence, Space, Cosmic carrier field and Electromagnetics. Tesla used the term Aether; I prefer using electromagnetics. Also, on the religious side of house it is known by the names: Heavens and The Water. Basically it is a lossless elastic media of infinite strength through which both matter and energy can pass. As mentioned it has both a scientific and religious aspect. However, only some of its properties are measurable the others are more esoteric. Its electrical properties are as follows:

a) ε_o (permittivity) $= 8.854 \times 10^{-12}$ Farad/meter

b) μ_o (permeability) $= 4\pi \times 10^{-7}$ Henry/meter

c) R_o (resistance) $= 377$ Ohms

CRYSTAL RADIO RECEIVER

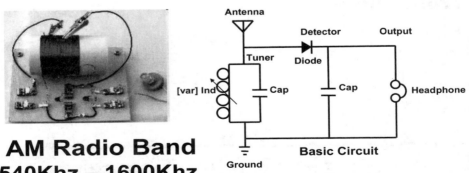

AM Radio Band
540Khz – 1600Khz

The third part of radio communications is the receiver shown here in both picture and schematics form. The most basic radio circuit is a "Crystal Radio" which gets it name from the "germanium" crystals that were used as the detector. Now days it's built using silicon diodes; I have made many of these sets when I was a kid. A basic radio receiver is composed of four (4) parts or stages: Antenna, Tuner (resonant circuit), Detector and Output (headphone).

MATLAB (TUNER) RESONANT SIMULATION AT 1MHz

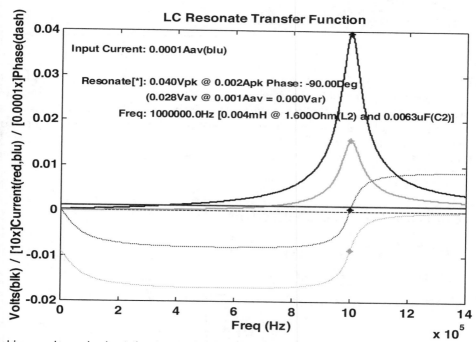

Taking a closer look at the tuner or resonant circuit. This MatLab simulation of a 1 MHz resonant signal is approximately the center of the AM radio band. The solid lines are the voltage and current signals while the dotted lines are their phasing. The antenna inputs a traveling wave. The tuner converts this traveling wave to a standing wave by phase shifting it 90 degrees and amplifying it through resonance. The LC resonant circuit values for this example are:

L2 = 4uH @ 1.6 Ohm and C2 = 6.3nF

With an input (antenna) current of 100uA the circuit generates a resonant output voltage (Vres) of 40mV producing a voltage gain of 400. Similarly it outputs a current (Ires) of 2mA producing a current gain of 20.

REGULAR RADIO

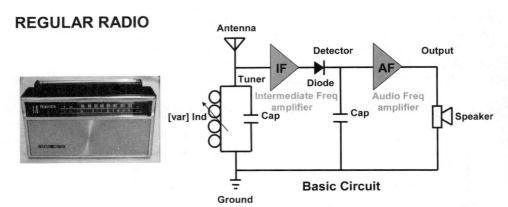

A normal radio design then adds two more stages to this basic system. First, is an "Intermediate Frequency" amplifier located between the tuner and detector circuits. Second, is an "Audio Frequency" amplifier located between the detector and output circuits. These modifications make a real transistor radio (pictured on the left) from the basic crystal radio circuit. These added amplifier circuits produce enough signal power to drive a speaker rather than just a small headphone.

TESLA'S RADIO MODIFICATION

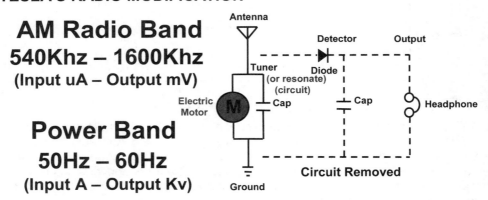

AM Radio Band
540Khz – 1600Khz
(Input uA – Output mV)

Power Band
50Hz – 60Hz
(Input A – Output Kv)

However, Tesla did not go in that direction. He made four (4) different modifications to the basic radio circuit. First, he removed the detector and output stages; he only used the antenna and tuner or resonant circuit stages. Second, he replaced the inductor in the resonant LC circuit with an electric motor. Engineering Note: Motors are rotating inductance machines. Third, he reduced the circuits operating frequency from the radio band (1 MHz) down to the power band (60 Hz). Fourth, he increased the circuit's power level from "mWatts" up to "kWatts". Engineering Note: Low frequency resonant circuits tend not to be self starting, therefore a driver or starter circuit (pulser) maybe required to get it running.

TESLA PATENT #685,957 (11/05/1901)

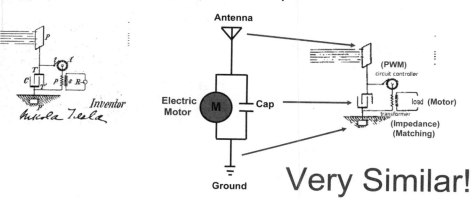

On Nov 5, 1901 Tesla applied for and received patent (#685,957) for his "Radiant Energy Device". A closer examination between the initial (first) cut of the modified radio tuner and Tesla's radiant energy patent shows they are "Very Similar"! However, there are some differences: First, his patent uses the (old) term "condenser" rather than (modern) "capacitor". Second, he used mica as the capacitor dielectric. I would use a "polypropylene" capacitor; both have good high voltage and temperature stability. Third, is the type of controller circuit used to start it. Tesla used what he called an interrupter. This was a mechanical device with a motor turning a "make and break" commutator switch; I would use a PWM (chopper) circuit to do the same thing. Fourth, you may have to add an impedance matching transformer between the motor and LC resonant circuit (as pictured).

DETAILED ANALYSIS - LC CIRCUIT LOOP EQUATIONS

$$I_{res} = \frac{V_{res}}{1 / CS}$$

$$(I_{in} - I_{res}) = \frac{V_{res}}{LS + R}$$

$$[S = -2\pi(F_{res})j]$$

This analysis is started by writing loop equations of the different current paths through the circuit. As shown this circuit has two (2) paths. The first equation is the resonant current (Ires) which only flows between the Inductor (L) and the

217

capacitor (C). The second equation is the input current (Iin) which only flows straight through the inductor. Engineering Note: The inductor carries the difference between the input and resonant currents (Iin – Ires). Also, shown is that a "real" inductor contains both inductance (L) and resistance (R). Combining and rearranging these two loop equations generates the following two transfer functions both based only on the input current (Iin). The first is for the resonant current (Ires) and the second is for the resonant voltage (Vres).

S-PLANE TRANSFER FUNCTIONS

$$I_{res} = \frac{LCS^2 + RCS}{LCS^2 + RCS + 1} I_{in} = \frac{CS(LS + R)}{LCS^2 + RCS + 1} I_{in}$$

$$V_{res} = \frac{LS + R}{LCS^2 + RCS + 1} I_{in} = \frac{1}{CS} I_{res}$$

By inspection of these two equations it is shown that the difference between the resonant voltage (Vres) and resonant current (Ires) equations is the term 1 / C*S which is capacitor impedance.

MATLAB RESONANT CIRCUIT SIMULATION AT 60 Hz

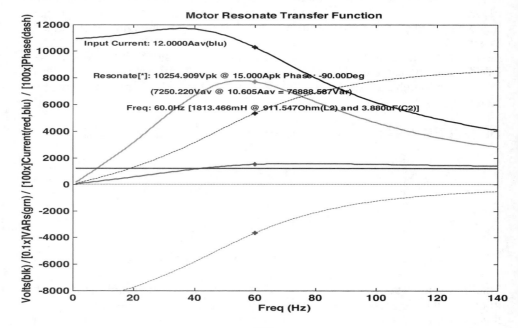

This is the MatLab Bode plot of a low frequency LC resonant circuit's Gain and Phase frequency sweep. As shown all the variable maximums: Voltage (35 Hz), Current (80 Hz) and VARs (55 Hz) occur at different frequencies. Therefore, the 60 Hz operating point is a compromise position as compared to the 1 MHz resonant plot where they all lined up. Engineering Note: That's why low frequency resonant circuits may not be self starting. Also, shown is that the resonant current's phase tracks the voltage's phase by 90 degrees generating a standing wave.

LC RESONANT CIRCUIT @ 76.9 kVAR

	Iin	Hz	Vres	Ires	L	R	C
1)	4A	60	21.75 kV	3.54A	16321mH @	8.20K	0.43uF
2)	7A	60	*12.43 kV*	6.19A	5329mH @	2.68K	1.32uF
3)	10A	60	8.70 kV	8.84A	2611mH @	1.31K	2.69uF
4)	12A	60	*7.25 kV*	10.61A	1813mH @	912	3.88uF
5)	14A	60	6.21 kV	12.37A	1332mH @	670	5.28uF

This table of resonant circuit values was generated by running MatLab simulation for different values of input current (Iin) keeping the frequency at 60Hz and the power at 76.9 kVAR. It gives the corresponding resonant voltage (Vres), current (Ires), inductance (L), resistance (R) and capacitance (C) values. Engineering Note: The *7.3 kV* and *12.4 kV* numbers (italics) are standard utility AC power distribution voltages.

SIMULINK LC RESONANT CIRCUIT MODEL

L2= 1813mH

Lr2= 912 Ohm

C2= 3.9uF

Iin = 12A

fres= 60Hz

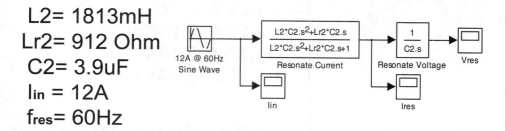

SIMULINK OUTPUTS

I_{in}= 12A$_{pk}$ @ 60Hz V$_{res}$= 10.25KV$_{pk}$ @ I$_{res}$= 15.0A$_{pk}$

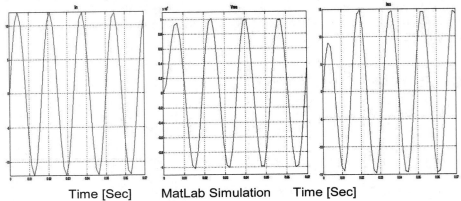

Time [Sec]   MatLab Simulation   Time [Sec]

Running MatLab Simulink for this same LC resonant circuit with an input current (Iin) of 12A and given component values generates a large resonant voltage (Vres) of 10.25 kV and a smaller current (Ires) of 15A. As shown a resonant circuit has a large voltage gain, but only a small current gain.

HIGH VOLTAGE RESONANT CAPACITORS

Depending on the input current these are the resonate cap specs:

Iin	Cap	Volts
14A =	5.28uF @	20Kv
12A =	3.88uF @	20Kv
10A =	2.69uF @	25Kv
7A =	1.32uF @	50Kv
4A =	0.43uF @	60Kv

From the proceeding LC resonant circuit table of component values these are the required AC capacitor (C) specifications for the different values of input current (Iin). The picture shows approximate (HV) High Voltage capacitor sizing.

LC RESONANT (TANK) CIRCUIT CONCLUSIONS

The resonant voltage (Vres) and current (Ires) are solely determined by the input current (Iin). A resonant circuit generates a large voltage gain, but only a small current gain. This voltage gain is greater than the current gain by a factor of at least a hundred to one (>100:1). The circuit's resonant frequency is only determined by the inductance (L) and capacitance (C) values. Depending on what the motor impedance values are, you may need to use an impedance matching (step-down) transformer to properly match it to the resonant tank circuit.

TESLA'S ELECTRIC CAR MOTOR STORIES

It is rumored in books and on the web that Tesla built and drove an electric powered Pierce Arrow car back in the early 1930s. I picked a few of those stories off the web to see if there was any truth to them and to correlate that data to see if it was even close to a "real" motor design.

"Electrical power is everywhere present in unlimited quantities and can drive the world's machinery without the need for coal, oil or gas" – Nikola Tesla, 1892

I came across this picture and could not resist starting the reverse engineering section with it. It's a picture of a 1931 Pierce Arrow; I know it's been photo shopped just look at the tires. However, the quote by Tesla himself lends some credibility to the story's truth; he certainly thought it was possible!

The following four (4) Tesla car stories are all similar, but each details different points. In the stories I highlighted (italics) any information about either the motor or the controller. You may get tired of reading the "same" story over and over again (four times), but repetition does have a way of making a point!

(1 of 4) Nikola Tesla's 'Black Magic' Touring Car

In the summer of 1931, Nikola Tesla, the inventor of alternating current and the holder of some 1200 other U.S. patents, along with his nephew Peter Savo, installed a box on the front seat of a brand new *Pierce-Arrow* touring car at the company factory in Buffalo, New York. The box is said to have been *24 inches long, 12 inches wide and 6 inches high*. Out of it protruded a *1.8 meter long antenna and two ¼ inch metal rods*. Inside the box was reputed to be some *dozen vacuum tubes -- 70-L-7 type* -- and other electrical parts. *Two wire leads ran from the box to a newly-installed 40 inch long, 30 inch diameter AC motor* that replaced the gasoline engine.

As the story goes, Tesla inserted the *two metal rods* and announced confidently, "We now have power" and then proceeded to drive the car for a week, "often at speeds of up to 90 mph." One account says the motor developed *1,800 rpm* and got fairly hot when operating, requiring a *cooling fan*. The "converter" box is said to have generated enough electrical energy to also power the lights in a home.

(2 of 4) Nikola Tesla's Wireless Electric Automobile Explained

Nikola Tesla proved in 1931 that it is possible to power our vehicles without a drop of fossil fuel. He removed the gasoline engine of a *Pierce Arrow* and replaced it with an *electric motor* and drove for hours, at speeds as high as 90 mph. Today, 81 years later, it is still possible to convert any gasoline engine vehicle into an all-electric vehicle and it will operate for hours – without having to stop and recharge. Not a drop of oil, gasoline, hydrogen fuel, natural gas or water. No combustion engine. No exhaust system. No pollution.

Supported by the Pierce-Arrow Co. and General Electric in 1931, Tesla took the gasoline engine from a new Pierce-Arrow and replaced it with an *80-horsepower alternating-current (AC) electric motor* with no external power source. At a local radio supply shop he bought *12 vacuum tubes*, some wires and assorted resistors, and assembled them in a circuit *box 24 inches long, 12 inches wide and 6 inches high, with a pair of 3-inch rods* sticking out. Getting into the car with the circuit box in the front seat beside him, he pushed the rods in, announced, "We now have power," and proceeded to test drive the car for a full week, often at speeds of up to 90 mph. His car was never plugged into any electrical receptacle for a recharge. As it was an *alternating-current motor* and there were no batteries involved, where did the power come from?

Tesla used the collection of **vacuum tubes** (also called a valve amplifier), wires and assorted resistors to build a radio wave receiver/amplifier **24 inches long, 12 inches wide and 6 inches high, with a pair of 3-inch rods 1/4" in diameter** sticking out. The pair of rods that Tesla pushed in were used to close (complete) the circuit – like an on/off switch. The rod ends were most likely the positive and negative leads (connections) between the car antenna and the radio wave receiver/amplifier. By pushing them into the box containing the radio wave receiver/amplifier the connection was completed allowing the **radio waves that were received from the air by the antenna to flow through the receiver/amplifier to the electric motor.**

My Lineout: Tesla certainly did not work for Edison's "General Electric" Co; he worked for George "Westinghouse".

(3 of 4) Tesla's Black Box Car

Mr. Savo reported that in 1931, he participated in an experiment involving **aetheric power**. Unexpectedly, almost inappropriately, he was asked to accompany his uncle on a long train ride to Buffalo. A few times in this journey, Mr. Savo asked the nature of their journey. Dr. Tesla remained unwilling to disclose any information, speaking rather directly to this issue.

Taken into a small garage, Dr. Tesla walked directly to a **Pierce Arrow**, opened the hood and began making a few adjustments. In place of the engine, there was an **AC motor. This measured a little more than 3 feet long, and a little more than 2 feet in diameter**. From it trailed **two very thick cables**, which connected with the dashboard. In addition, there was an ordinary **12-volt storage battery**.

The motor was rated at **80 horsepower**. Maximum rotor speed was stated to be **30 turns per second**. A **6-foot antenna rod** was fitted into the rear section of the car.

Dr. Tesla stepped into the passenger side and began making adjustments on a "power receiver" which had been built directly into the dashboard. The receiver, no larger than a short-wave radio of the day, used **12 special tubes, which Dr. Tesla brought with him** in a box-like case.

Mr. Savo told Mr. Ahler that Dr. Tesla built the receiver in his hotel room, a **device 2 feet in length, nearly 1 foot wide, a 1/2 foot high**.

These **curiously constructed tubes** having been properly installed in their sockets, Dr. Tesla pushed in **2 contact rods** and informed Mr. Savo that power was now available to drive. Several additional meters read values, which Dr.

Tesla would not explain. No sound was heard. Dr. Tesla handed Mr. Savo the ignition key and told him to **start the engine**, which he promptly did.

Yet hearing nothing, the accelerator was applied, and the car instantly moved. Tesla's nephew drove this vehicle without other fuel for an undetermined long interval. Mr. Savo drove a distance of 50 miles through the city and out to the surrounding countryside. The car was tested to speeds of 90 mph, with the speedometer rated to 120.

(4 of 4) Nikola Tesla had an *Electric Pierce Arrow* back in 1931

The ICE engine was replaced with an **Electric Motor**. The power source was a **black box of radio tubes**, in the glove compartment. The **box had an antenna** sticking out. Tesla would fool with some tuners and tune in the right frequency and got **240 volts delivered through the air** to his car. The car ran almost silent.

Here is the story: It was a **Pierce Arrow**, one of the luxury cars of the period. The engine had been removed, leaving the clutch, gearbox and transmission to the rear wheels undisturbed. The gasoline engine had been replaced with a round, completely **enclosed electric motor of approximately 1m in length and 65cm in diameter, with a cooling fan in front**. Reputedly, it has no distributor. Tesla was not willing to say who had manufactured the engine. It was possibly one of the divisions of Westinghouse.

The **"energy receiver"** (gravitational energy converter) had been **built by Tesla himself**. The dimensions of the **converter housing were approximately 60 x 25 x 15cm**. It was installed in front of the dashboard. Among other things, the converter contained **12 vacuum tubes, of which three were of the 70-L-7 type**. A **heavy antenna approximately 1.8 meters long, came out of the converter**. This antenna apparently had the same function as that on the Moray converter (see chapter on Radiant Energy). Furthermore, **two thick rods protruded approximately 10cm from the converter housing**.

Tesla pushed them in saying "Now we have power." The motor achieved a maximum of **1800rpm**. Tesla said it was fairly hot when operating, and therefore a **cooling fan** was required. For the rest, he said there was enough power in the converter to illuminate an entire house, besides running the car engine. The car was tested for a week, reaching a top speed of 90 miles per hour effortlessly. Its performance data were at least comparable to those of an automobile using gasoline. At a stop sign, a passerby remarked that there were no exhaust gasses coming from the exhaust pipe. Petar answered "We have no motor." The car was kept on a farm, perhaps 20 miles outside of Buffalo, not far from Niagara Falls.[1]

PROPOSED TESLA MOTOR SPECIFICATIONS

1) *80 Hp AC induction*   **59.7 kW [74.6 kVAR]**

2) *1800 RPM*   **4-pole [60Hz]**

3) *Fan cooled*   **Yes - external**

4) *Enclosed housing*   **Yes - option**

5) *Two (heavy) leads*   **Single phase HV / Hi current**

6) *Not self starting*   **Run coil only**

7) **Power factor (PF)**   **0.8 [.75 - .85]**

8) *240 Vac*   **310.8 Aac**
      (480 Vac   **155.4 Aac)**

9) *Dimensions: 1m(40")L x 65cm(25.5")Dia.*

Using the highlighted information from the four stories along with "real" electric motor engineering data, I pieced together this proposed motor design. The story data (italics) is on the left and the engineering specs (bold) are on the right. There are many inconsistencies in the stories as to who was driving, where things were located and how they worked. Even, the story's motor and controller details are spotty but consistent; no one story has all the numbers. However, none of the stories mention the motor's "Power Factor". I assumed a PF value of 0.8 based on the spread (0.75 – 0.85) for single phase motors. Engineering Note: Usually electric motors are not run with the current (310.8A) higher then the voltage (240V) because of line losses. That is why I included the 480V data.

ELECTRIC MOTOR DESIGN EQUATIONS

Resistance R = Volt(PF) / Amp
= 240(0.8) / 310.8 = 0.6178

---

[1] Some attribute an old car with men on the roof with a tall antenna as the Tesla car but it is from the book by Archibale Williams, called the *Romance of Modern Invention*, (Lippincott, 1910), which appears to be a wirelessly powered vehicle but actually was a "travelling station for wireless telegraphy" attributed to Marconi. – Ed. Note

Inductance L = (Volt)sin(Acos(PF)) / 2π(Hz)Amp
= 240(0.6) / 2π(60)310.8 = 1.229mH

Hp = Volts(Amps)PF / 746
= 240(310.8)0.8 / 746 = 79.99Hp

RPM = 60(Hz) / N (N= number of pole pairs)
= 60(60) / 2 = 1800RPM

As a sanity check I ran all these proposed motor numbers through standard electric motor design equations. No ringers; they come up as a valid (REAL) motor design.

MOTOR RUNNING IMPEDANCE

Volts R [ohm] L [60Hz]
240V   618mΩ 1.229mH
Z240V = 0.618 + 0.463j = 0.771Ω

480V   2.477Ω 4.929mH
Z480V = 2.477 + 1.858j = 3.096Ω

Plugging in the numbers for resistance and inductance you get a "REAL" motor run impedance (Z) for both 240V and 480V power.

1800RPM MOTOR OPERATION

Poles	N	Freq
4	*2*	*60Hz*
6	3	90Hz
8	4	120Hz
12	6	180Hz

This is a table of different motor frequencies and pole combinations that will run at 1800RPM. I highlight 60Hz as the most likely combination that Tesla used as it was his design frequency.

PRESENT DAY AC INDUCTION MOTOR - BALDOR

Note: No one builds
a 1-PH motor
over 15Hp

75Hp 3-PH External Fan Enclosed Baldor

BALDOR·RELIANCE Product Information Packet: CEM4316T - 75HP,1780RPM,3PH,60HZ,365TC,A36068M,TEFC

Nameplate NP2383L										
CAT.NO.	CEM4316T	SPEC NO.	A36-1117-1816							
HP	75	AMPS	169/84.9	VOLTS		230/460	DESIGN	B		
FRAME	365TC	RPM	1780	HZ		60	AMB	40	SF	1.15
DRIVE END BEARING	65BC03J30X	PHASE	3	DUTY		CONT	INSUL.CLASS	F		
OPP D.E. BEARING	65BC03J30X	TYPE	P	ENCL		TEFC	CODE	G		
SER.NO.		POWER FACTOR	87	NEMA-NOM-EFFICIENCY		95.4				
		MAX CORR KVAR	14	GUARANTEED EFFICIENCY		94.5				
NEMA NOM/CSA QUOTED EFF										
		SUIT FOR 208V AT 186 AMPS		MOTOR WEIGHT		907				

This is the closest present day electric motor I could find for doing a comparison with Tesla's motor design. It is a Baldor 75Hp, 3-Phase, 1800RPM, Fan cooled, Fully enclosed motor. However, the comparison is not 100% as no one builds a 1-Phase motor bigger than 15Hp and today's standard motor sizing is 50, 75, 100Hp. Looking at the "Nameplate" numbers it has a Power Factor of 0.87; 3-Phase motor operation is much more efficient. It runs on both 230V @ 169A and 460V @ 84.9A; much lower current levels. And it weighs in at 907lbs; about double that of a gas engine.

BALDOR MOTOR MECHANICAL DIMENSIONS

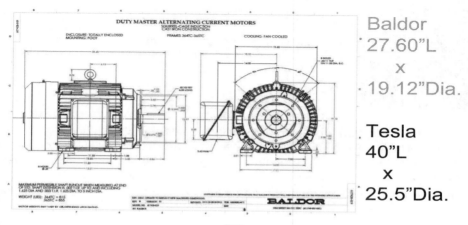

Baldor
27.60"L
x
19.12"Dia.

Tesla
40"L
x
25.5"Dia.

Looking at the Baldor motor's mechanical dimensions it is 27.6" long and 19.12" in diameter. Compared to Tesla's motor of 40" long and 25.5" in diameter. The diameters are similar. The 6 inch difference in diameter could be frame sizing, coil winding style, insulation material, bigger cooling fins or just older technology. However it is about 12 inch shorter than Tesla's motor. This would certainly be enough room for an impedance matching transformer and the heavy current wiring would be a very short run.

TESLA MOTOR CONCLUSIONS

In comparing Tesla's motor design to a present day Baldor his specs are close enough to be a "REAL" motor. This certainly adds more credibility that the story is true. However, his motor was not self starting; it only had a 4-Pole run coil winding. A start coil winding would have changed the motor impedance and thrown the circuit out of resonance. Its longer length might have included a transformer or just be older technology. Finally, we are left with three (3) assumptions or loose ends: First, is the exact input operating voltage (*240V /* 480V). Second is the operating frequency (*60Hz*). And third is the Power Factor (*0.8*). The highlighted numbers are my assumptions.

UTILITY STEP-DOWN TRANSFORMERS (TESLA DESIGN)

Standard Specs:
(30:1) *7.25Kv* to 240V @ 75Kva
(51:1) *12.47Kv* to 240V @ 75Kva

If Tesla used (needed) an impedance matching step-down transformer they were readily available. Remember he also designed the AC power system grid and the transformers that run it. Standard utility power distribution voltages are *7.25 kV* and *12.47 kV*. At the time Westinghouse was building these 240V (center tapped) step-down transformers in various sizes (5 kVA to 100 kVA) so Tesla could have used a 75 kVA unit right off the assembly line!

IMPEDANCE MATCHING WITH STEP-DOWN TRANSFORMERS

Primary   Secondary $V_{out} = V_{in} / N$

(N:1)

$I_{out} = I_{in} * N$

$Z_{out} = Z_{in} / N\text{^}2$

N = Turns Ratio

This section is a short tutorial on step-down transformer design. Starting with the input side there are three (3) variables: the input voltage (Vin), the input current (Iin) and the input impedance (Zin). Similarly the output side also has three variables: the output voltage (Vout), the output current (Iout) and the output impedance (Zout). The transformer's input and output sides are connected together by the "turns ratio" (N). The picture shows a step-down transformer schematics and three design equations:

Output Voltage = Input Voltage / turns ratio
Output Current = Input Current * turns ratio
Output Impedance = Input Impedance / turns ratio squared

STEP-DOWN TRANSFORMER TURNS RATIO

$$N = \left[\frac{(7.25Kv)(310.8A)}{(10.61A)\ (240V)} \right]^{0.5} = 29.75\ (30{:}1)$$

Turns Ratios

$$N = \left[\frac{(7.25Kv)(155.4A)}{(10.61A)\ (480V)} \right]^{0.5} = 14.87\ (15{:}1)$$

As shown the "turns ratio" (N) calculation for a step down transformer is equal to the square root of (input voltage * output current) / (input current*output voltage). Examples are for a **7.25 kV** step down to both 240 V (30:1) and 480 V (15:1) power output levels.

229

MATLAB (30:1) TRANSFORMER SIMULATION AT 60Hz

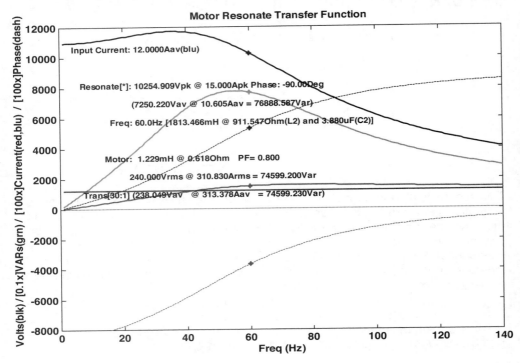

This is the MatLab simulation bode plots of Gain and Phase for a 7.25 KV resonant design (30:1) step-down transformer coupled motor operating at 240V.

STEP-DOWN TRANSFORMER

Input   N   Output [real]
7.25kV / 30   = 238V **[240V]**
10.61A * 30   = 313A [311A]
911.55Ω / 30^2 = 0.98Ω [0.77Ω]

Input   N   Output [real]
7.25kV / 15   = 476V **[480V]**
10.61A * 15   = 159A [155A]
911.55Ω / 15^2 = 3.93Ω [3.09Ω]

Plugging the numbers into the step-down transformer equations for both 240V and 480V generates this table. It shows that the calculated numbers compared to the "real" motor values are close enough to run, providing a good impedance match!

SIMULINK LC RESONANT CIRCUIT MODEL
(WITH (30:1) STEP-DOWN TRANSFORMER)

$L2 = 1813mH$
$Lr2 = 912\ Ohm$
$C2 = 3.9uF$
$I_{in} = 12A$
$f_{res} = 60Hz$

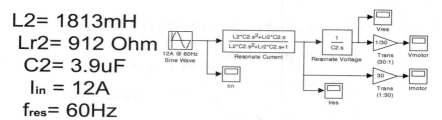

SIMULINK VOLTAGE OUTPUTS

Vres = 10.25KVpk   **Vm = 337Vpk**

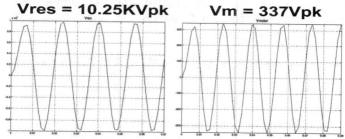

Transformer Voltage (30:1)

SIMULINK CURRENT OUTPUTS

Ires = 15Apk   **Im = 440Apk**

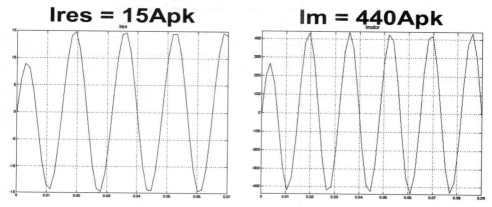

Transformer Current (1:30)

Running MatLab Simulink for this same LC resonant circuit with a (30:1) step-down transformer an input current (Iin) of 12A and the given component values generates a motor voltage (Vm) of 337V and a motor current (Im) of 440A. As shown the Simulink run provides good verification of the step-down transformer's impedance matching.

STEP-DOWN TRANSFORMER TURNS RATIO

$$N = \left[\frac{(12.5Kv)(310.8A)}{(6.16A)\,(240V)}\right]^{0.5} = 51.22\;(51{:}1)$$

Turns Ratios

$$N = \left[\frac{(12.5Kv)(155.4A)}{(6.16A)\,(480V)}\right]^{0.5} = 25.61\;(26{:}1)$$

Repeating the transformer matching process for the higher standard value resonant voltage. Examples are for a **12.5 kV** step down to both 240V (51:1) and 480V (26:1) power output levels.

MATLAB (51:1) TRANSFORMER SIMULATION AT 60Hz

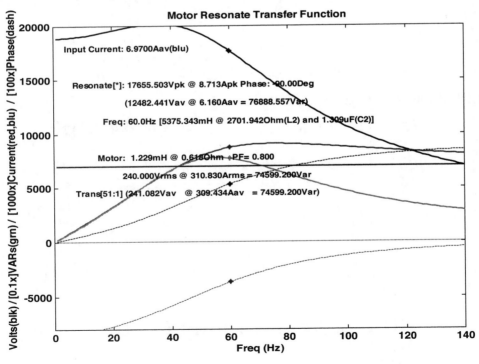

This is a MatLab simulation bode plots of Gain and Phase for a 12.5 kV resonant design (51:1) step-down transformer coupled motor operating at 240 V.

TRANSFORMER MATCHING CONCLUSIONS

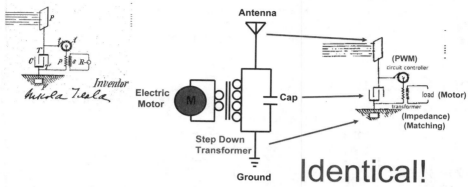

Impedance matching the motor to the resonant circuit with a step-down transformer produces an "Identical" system to that of Tesla's patent (#685,957). Simulations show that the system could run an 80Hp motor provided the antenna could supply a 12A input current (Iin). That leaves only the motor controller design to reverse engineer.

PROPOSED TESLA CONTROLLER SPECIFICATIONS

Tesla built box
1) Contained 12 tubes
   a) Tesla brought with him
   b) Curiously constructed tubes
   c) Three were 70L7

2) Two (3"x1/4"dia) movable rods

3) One 6ft (heavy) antenna

4) Connected to two (heavy) motor leads

5) 240Vac
  (480Vac)

6) Dimensions: 24"L x 12"W x 6"H

Custom Design

Non Standard?
Standard MFG

Iron / Ferrite

<<1/4λ @ 60Hz

HV / Hi Current

7.25 kV @ 10.61 A

Includes Cap (Cres)?

Using the highlighted information from the same four stories along with "real" electronic circuit design, I pieced together this proposed controller design. The story data (italics) is on the left and the engineering specs (bold) are on the right. Engineering Note: The control box dimensions are way too small to include a 75kVA step down transformer; therefore it must have been located in the motor housing. Therefore, the two leads connected to the motor would carry high voltage making the "heavy" term apply to thick HV insulation. However, the control box is large enough to contain the resonant capacitor.

CONTROLLER BLOCK DIAGRAM

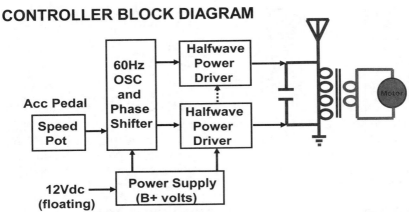

Based on the previous specifications chart this is what I get for the controller's block diagram design. Now comes the hard part of trying to build the circuits with vacuum tubes. I had to dig out my old college tube books for design references. Engineering Note: For each tube type used I picked the oldest numbered tubes found in a 1963 RCA tube book; they are all discontinued 8-pin octal tubes. As they were already discontinued by the early 60s I assumed these tubes were the standard (new) in the 1930s.

POWER SUPPLY SPECIFICATIONS

Starting with the system power supply; it's basically a DCDC converter that takes 12Vdc up to 70Vdc for the tube's B+ supply voltage. To design this circuit using tubes would require one (1) dual triode tube with a 12V heater for a multivibrator (osc) circuit, one (1) center tapped power transformer and two (2) half wave rectifier tubes with either a 6V or 12V heater.

Power Supply

12SN7 Specs:
High-mu Dual Triode
12V Heater
450Vmax Voltage
20mAmax Current

6X5 Specs:
Vacuum Rectifier
6V Heater
1250Vmax Voltage
245mAmax Current

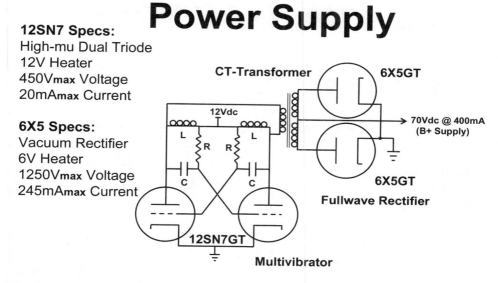

The power supply schematics would look something like this. One dual triode (12SN7) tube running as an oscillator driving the input of a center tapped transformer. The transformer output is connected to two half wave rectifier (6X5) tubes to provide full wave power rectification.

REFERENCE OSCILLATOR SPECIFICATIONS

Next is the reference oscillator which is also a multivibrator circuit. However, this oscillator runs at a frequency of 60Hz and outputs about 100V to the power driver section. To design it would require two (2) beam power tubes. One or both of the multivibrator tube grids would have to be phase shifted for speed control. The phase shifter circuit would also require one (1) beam power tube. The phase shifting or speed control signal would come from a potentiometer connected to the vehicle's accelerator (gas) pedal.

POWER DRIVER SPECIFICATIONS

Last is the power driver stage consisting of two half wave power driver circuits (current sources) that generated 12 amps of current from the 12V power supply. This is done by alternately charging and discharging two big inductors. The two 3" by ¼"dia. adjustable rods were the iron/ferrite cores of these power inductors. The system would be started (brought up to resonance) at a low power level and then the cores were "pushed-in" (adjusted) to generate the high power levels. The "curiously constructed" tubes Tesla used for the power drivers turn out to be

Westinghouse "Ignitrons". He may have had to parallel these tubes to get the high currents. I have not been able to track down any specs of these very early Westinghouse power tubes; the best to date is a 1938 WL-654 Ignitron rated at 480V @ 12.5A.

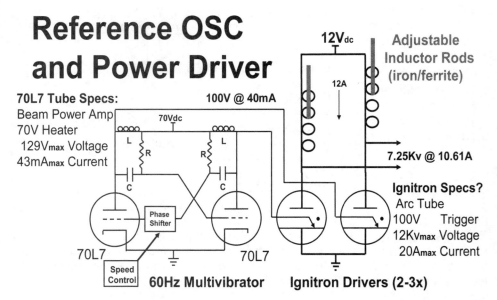

The schematic of the 60Hz reference oscillator, phase shifter and power driver circuits would look something like this. The multivibrator circuit was composed of two (2) 70L7 beam power tubes outputting about 100V each which would alternately fire (trigger) the two Ignitron's igniter thereby charging the inductors from the 12V supply. Tesla may have had to parallel these ignitron tubes to get the proper current rating. These half wave power inductors alternately drove the motor's step-down transformer. The phase shifter circuit was composed of another (1) 70L7 beam power tube which moved the oscillator's frequency off the 60Hz resonant point thereby reducing the system's power output providing speed control of the motor. This would be a high voltage / high power system which could be very dangerous to work with!

IGNITRON TUBES

Ignitron

From Wikipedia, the free encyclopedia, An ***ignitron*** is a type of gas-filled arc tube used as a controlled rectifier dating from ***1930***. Invented by Joseph Slepian while employed by **Westinghouse**. Westinghouse was the original manufacturer and owned trademark rights to the name "Ignitron".

Tesla would have had access to these curiously constructed "Ignitron" tubes. Ignitrons were the predecessors of modern day (SCRs) Silicon Controlled Rectifiers.

Ignitron tubes had a big pool of mercury in the bottom which vaporized when struck by an arc from the igniter electrode. They ran hot with a blue glow; some of the bigger ones were water cooled. If the EPA thinks a little bit of mercury vapor in fluorescent lights is bad they would really love these tubes.

TESLA MOTOR CONCLUSIONS

From an electrical engineering standpoint Tesla's AC motor drive system (antenna, LC tuner, step-down transformer, motor) would certainly "RUN"; he did not break any rules. However, being a high voltage / high power system it would be dangerous to work with. The controller circuit most likely gave him some problems as tube circuits are not all that stable; they tend to drift so some tweaking would be needed. Also, a 3-Phase power system would be far more efficient than a single Phase one.

References

http://energybytesla.org/electric-car/nikola-teslas-free-energy-car

http://waterpoweredcar.com/teslascar.html

http://www.tfcbooks.com/teslafaq/q&a_016.htm

https://teslamotorsclub.com/tmc/threads/nikola-tesla-pierce-arrow-car.2035/

16 HIGH Q RESONANT WIRELESS POWER TRANSFER

ROY DAVIS
ROYD@QTI.QUALCOMM.COM

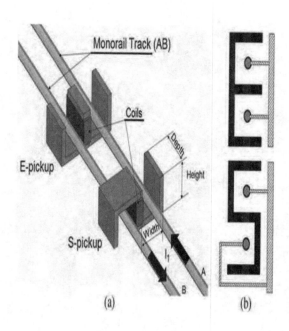

Adapted from a presentation at the USPTO, 6/9/15

PRESENTATION DESCRIPTION

- Wireless power transfer using inductively coupled resonators

    - Dates back at least to Tesla and his work in late 19th century

- This long history means

    - A variety of disparate explanations, articulations and characterizations of the related principles - complicated and often difficult to reconcile

- This presentation

    - Discusses the fundamentals of inductive power transfer

    - Introduces many ideas that are "in the works"

    - Reviews the history of wireless power transfer

    - Provides detailed technical introduction to resonator theory applicable to

        - ## Wireless power transfer applications

        - ## For charging or powering devices and vehicles

TRANSFORMER COUPLING

- When two inductors are brought close together magnetic flux from one cuts through the other

- Current flowing in first inductor (called a primary) induces current in other inductor (called secondary)

- Two inductors purposefully sharing magnetic flux is a common and useful electrical component called a transformer

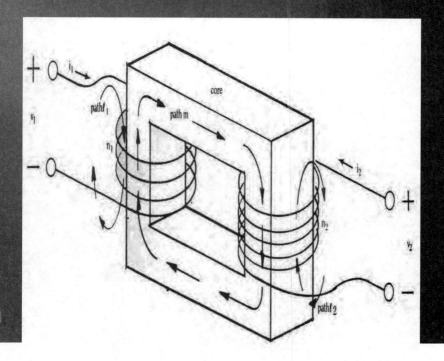

BASIC TRANSFORMER EQUATIONS

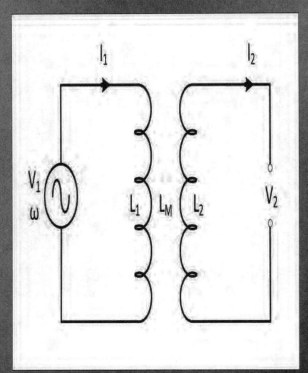

L — inductance of coil

L_M — mutual inductance

ω — frequency

I_{SC} — short circuit current

V_{OC} — open circuit Voltage

$$V_{OC} = j\omega L_M I_1 \quad I_{SC} =$$

$$V_{OC}/j\omega L_2$$

$$= L_M/L_2 L_1 \quad VA_{MAX} =$$

$$V_{OC} I_{SC}$$

COUPLING COEFFICIENT K

- Inductance that is not mutual is flux not shared between two inductors

    - Called leakage inductance

- Coupling coefficient **k** would be close to 1 for traditional transformer

- Wireless power transfer leakage inductance may exceed the mutual inductance

    - Coupling coefficient **k < 0.5**

    - Very loosely coupled (larger distance) **k < 0.1**

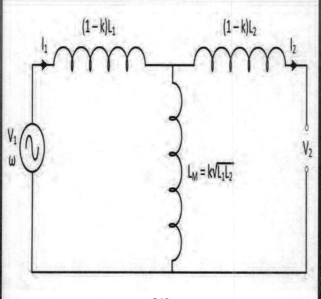

243

EARLY **UNIVERSITY** OF **AUCKLAND** DEMONSTRA-TION

WHAT IS A RESONATOR ?

A bell is a resonator

- Stores energy in oscillation between metal flexure (potential) and mechanical motion (kinetic)

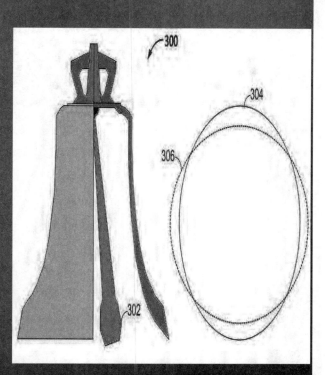

- Striking bell imparts energy that is stored and released over time as sound waves

  - The time a bell rings after being struck is a measure of quality – often called **Q**

PENDULUM RESONATOR

- ## Energy in pendulum resonator

  - ### Alternates between kinetic energy in motion of the weight and potential energy of the raised weight

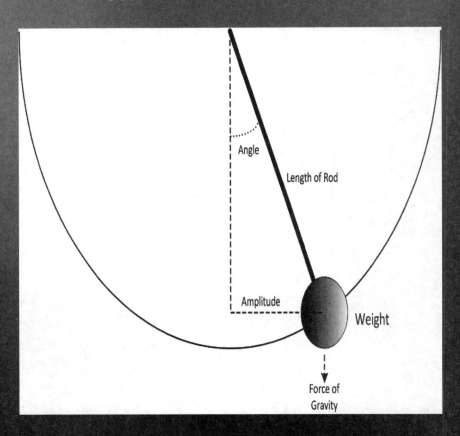

V & I PHASE RELATIONSHIP

- In a circuit where electricity is doing work the Voltage and current are in phase

- In an inductor the current is stored in the magnetic field, then released causing the current to lag the Voltage

- In a capacitor the current is stored in an electric field, then released causing the current to lead the Voltage

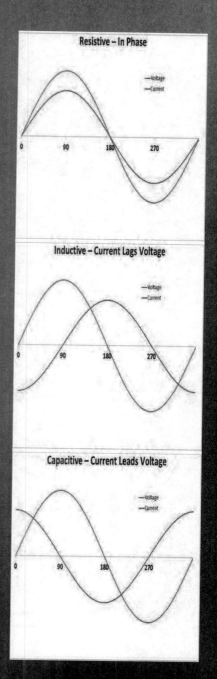

TANK CIRCUIT IN ACTION

- Stores energy in L-C resonator
  - Alternates between the magnetic field of inductor and electrical field in capacitor

TANK CIRCUIT
(ENGINEERING SLANG FOR ELECTRICAL RESONATOR)

- Stores energy by circulating current (M2) between inductor (L) and capacitor* (C)

- Energy is alternately stored in magnetic field of inductor and electric field of capacitor

- Resistor (R) represents loss that is unloaded Q

- If impedance of generator (E) is much higher than impedance at resonance of tank circuit, then loaded Q is high

*In electrical power engineering capacitor is often referred to as a compensating capacitor to correct the power factor to 1, which brings circuit to resonance

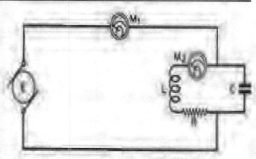

Figure 22
PARALLEL-RESONANT CIRCUIT

The inductance L and capacitance C comprise the reactive elements of the parallel-resonant (antiresonant) tank circuit, and the resistance R indicates the sum of the r-f resistance of the coil and capacitor, plus the resistance coupled into the circuit from the external load. In most cases the tuning capacitor has much lower r-f resistance than the coil and can therefore be ignored in comparison with the coil resistance and the coupled-in resistance. The instrument M, indicates the "line current" which keeps the circuit in a state of oscillation—this current is the same as the fundamental component of the plate current of a class-C amplifier which might be feeding the tank circuit. The instrument M, indicates the "tank current" which is equal to the line current multiplied by the operating Q of the tank circuit.

CIRCULATING TANK CURRENT EXAMPLE

*The tank current is very nearly the value of the line current multiplied by the effective circuit Q. For example: an r-f line current of 0.050 ampere, with a circuit **Q** of 100, will give a circulating tank current of approximately 5 amperes.*

$$I_{circulating} = I_{line}*Q = 0.05*100 = 5 \text{ amperes}$$

*From this it can be seen that both the inductor and the connecting wires in a circuit with a high **Q** must be of very low resistance, particularly in the case of high power transmitters, if heat losses are to be held to a minimum.*

PROTOTYPE FOR THUNDER MOUNTAIN RAILROAD

251

AUTO
ASSEMBLY
PLANT
ROBOT

QUALITY FACTOR (Q FOR SHORT)

- Quality factor **Q** represents effect of electrical resistance r_{AC} in wire that makes up circuit

- Inductor usually dominates the **Q** so capacitor loss is absorbed into r_{AC}

- r_{AC} is resistance of wire in the coil inductor at operating frequency ω

- At higher frequencies Skin Effect increases loss by reducing useful cross section of wire

- **Q** is calculated as $Q = \omega L / r_{AC}$

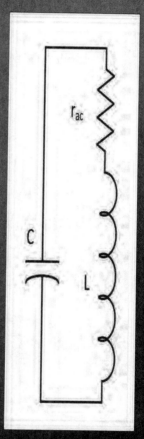

Patented May 17, 1927.

1,628,983

UNITED STATES PATENT OFFICE.

KENNETH S. JOHNSON, OF JERSEY CITY, NEW JERSEY, ASSIGNOR TO WESTERN ELEC.
TRIC COMPANY, INCORPORATED, OF NEW YORI, N. Y., A CORPORATION OF NEW
YORK.

ELECTRICAL NETWORK.

Application filed July 9, 1923. Serial No. 850,218.

The transmi ssion losses in such a structu re
will depend (1) upon the ratio Q of the
reactance to the resi stance of the coils
(..)   and upon the frequency

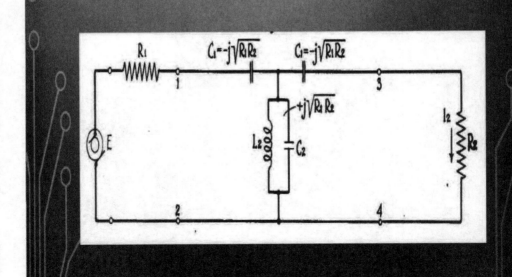

Q = SHARPNESS OF RESONANCE

Early engineers referred to Q as "sharpness of resonance"

- Good quality **L** and **C** with a 4.4Ω resistor resonates and magnifies Voltage in curve A

- Same **L** and **C** with a 9.4Ω resistor gives curve B

- Same **L** and **C** with 14.4Ω gives curve C

More resistance in resonant circuit reduces magnification factor of resonance

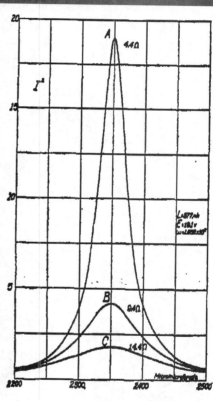

FIG. 19.—*Resonance curves for series circuit with different resistances*

SKIN EFFECT

- Electric current (I, flowing up in this example) creates magnetic field [106]

- Magnetic field creates eddy currents [108]

- Eddy currents oppose current I in center of wire

- Eddy currents aid current I close to surface of wire

- Effective current flow is confined to annular ring [104] near surface of wire

- Skin effect reduces effective cross section of wire

  - Increases effective resistance we call r_{AC}

- $r_{AC} = r_{DC}$ at low frequency where skin depth [104] > wire diameter [102]

- At higher frequency skin depth [104] < wire diameter [102] so $r_{AC} > r_{DC}$

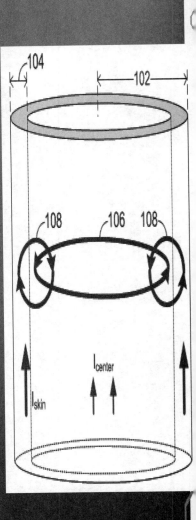

TABLE I.

Effect of Frequency up on Resistance of Straight Wires.

A Frequency of current	B Thickness of Annulus Copper μ=1	C Thickness of Annulus Iron μ=300
80	0.719	0.0976
120	.587	.0798
160	.509	.0691
200	.455	.0617
480	.293	.0399
800	.228	.0309
1,000	.203	.0276
1,800	.152	.0206
3,200	.114	.0154
4,000	.102	.0138
9,000	.068	.0092
16,000	.051	,0069
20,000	.045	.0062
26,000	.034	.0046
40,000	.032	.0044
		.0035

LITZ WIRE

- **Reginald Fessenden invented braided wire to reduce skin effect**

- **Now referred to as Litz wire after the German word Litzendraht for braided wire**

UNITED STATES PATENT OFFICE.

REGINALD A. FESSENDEN, OF BRANT ROCK, MASSACHUSETTS, ASSIGNOR TO NA-
TIONAL ELECTRIC SIGNALING COMPANY, OF PITTSBURGH, PENNSYLVANIA, A
CORPORATION OF NEW JERSEY.

HIGH-FREQUENCY ELECTRICAL CONDUCTOR.

1,039,717.

Specification of Letters Patent.

Patented Oct. 1, 1912.

Application filed January 7, 1911. Serial No. 601,380.

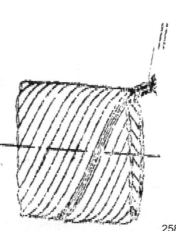
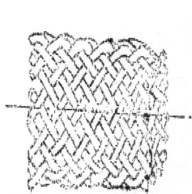

Fig. 3

EARLY IMPROVEMENT IN Q

- A pendulum resonator has two major sources of resistance

  - Air resistance

  - Friction at the pivot

- This illustration from Christiann Huygens' *Horologium oscillatorium* of 1673 shows improved aerodynamics of the weight

  - Cross section small in direction of motion

  - Streamlining shape by making ends pointed

- Improved aerodynamics

  - Reduces air resistance

  - Increased **Q** of resonator

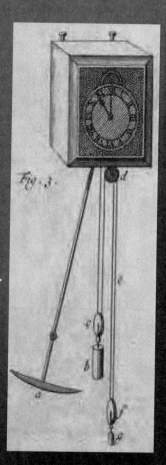

SILICON FAB WAFER MOVER ROBOT

WHAKAREWAREWA PEOPLE MOVER

Installed at the Whakarewarewa (Fuk ar ray wa ray wa)

Geothermal Park, New Zealand

BATTERY ELECTRIC VEHICLE

- Transports park visitors over 1.8 Kilometer road

- Developed 1995 by University of Auckland

- Commissioned 1998

PEOPLE MOVER CHARGING COILS

- Pickup coils on underside of vehicle
- Bipolar type coils pick up horizontal magnetic flux

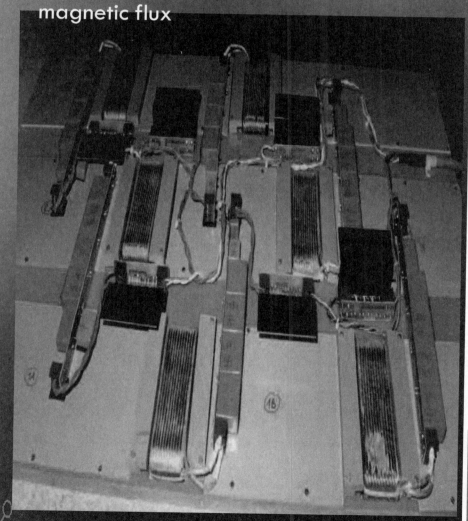

PEOPLE MOVER CHARGING STATION

- 20 Kilowatts wireless charger

- Primary charging resonator is mounted in concrete island

TURIN WIRELESSLY CHARGED BUS

Turin, ITALY

CITROEN ELECTRIC VEHICLE WITH WIRELESS CHARGING

ROLLS ROYCE WIRELESS POWER DEMONSTRATOR

Built 2010 by Halo IPT at Auckland University 7 KW Wireless Charger

20 KW WIRELESS CHARGER

VEHICLE CHARGING PAD WINDINGS

Double D coils for bipolar operation

Quadrature coil for single coil operation

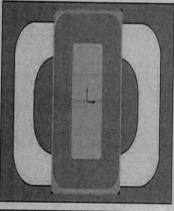

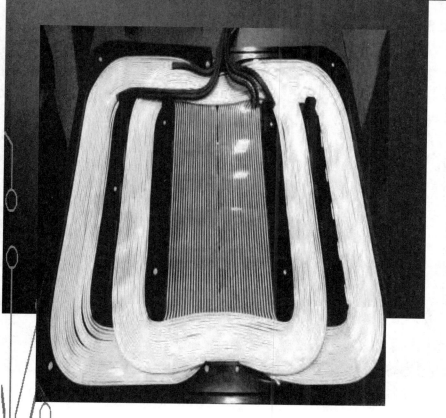

DOUBLE D FLUX COUPLING

- Flux exits primary at one pole, couples to opposing secondary pole, flows through ferrite backing and out the opposite pole back to the primary and through primary ferrite backing

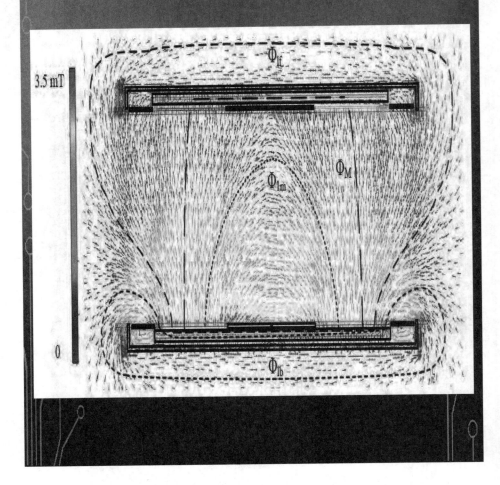

H-BRIDGE PRIMARY SIDE POWER SUPPLY

- AC Line input rectified to DC and stored in large capacitor
- Rectifier is usually a complex power factor correction circuit
- MOSFET switches form full-wave H bridge to chop DC into high frequency square wave
- Inductor forms a current source to feed resonant transmitter pad

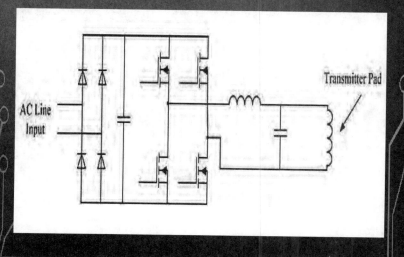

SECONDARY SIDE
POWER PICKUP

- Pickup coil feeds a full-wave bridge rectifier

- Inductor (and often bulk capacitor) smooths to DC

- Switch Mode Power Supply (SMPS) regulates output Voltage

- Recent innovation is to control impedance seen by resonator by adjusting SMPS for maximum power transfer

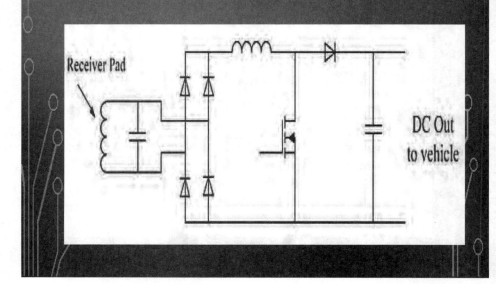

DYNAMIC ELECTRIC ROADWAY TEST BED

Preparing for FeAsiBility analysis and development of on-Road charging solutions future electric vehiCles

Test track for charging on the go ⟹

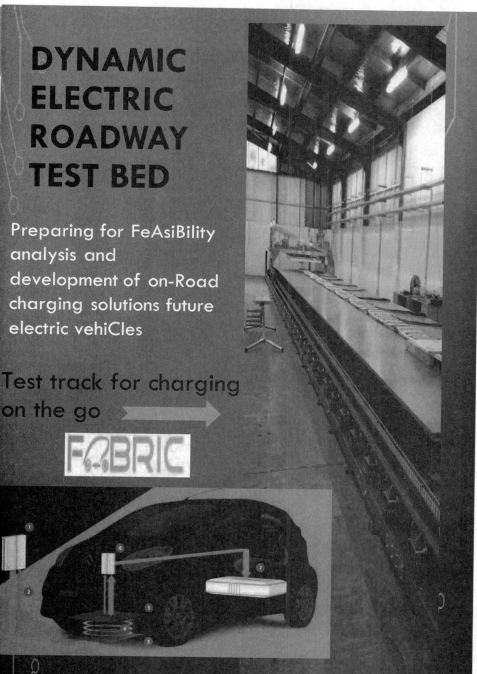

SOME NEW WIRELESS POWER

CHALLENGES AND SOLUTIONS

SWITCHABLE CURRENT DOUBLER

- Reducing range for impedance adjustment improves efficiency

- Current doubler acts as "gear change" to expand adjustment

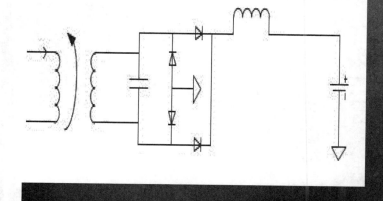

MULTIFILAR COIL WINDINGS

- Current balancing to improve efficiency

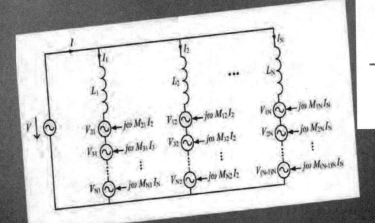

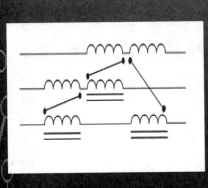

276

WIRELESS CHARGING FOR METAL CASE PHONES

- Wireless charging via magnetic fields for a smartphone with a metal case is a challenge

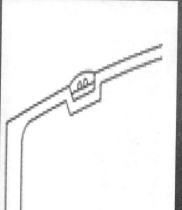

Coil wires embedded in groves in metal cover is one solution

Other solutions involve segmenting the metal cover and forming loops

FIELD CONTROL BY PARASITIC COILS

- Induced current in parasitic coils oppose field of primary coil inside the parasitic coil but enhance field outside parasitic coil

- Improved coupling efficiency for large primary/small secondary

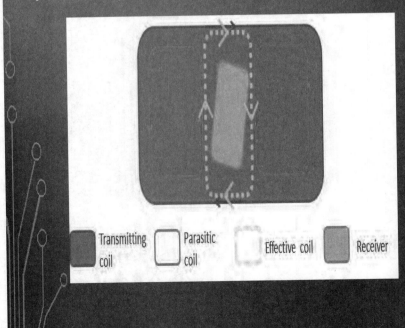

FLUX STEERING

- Steering magnetic flux only to verified wireless power receivers

- One large primary coil produces flux over entire surface

- Parasitic coils oppose flux in inactive regions

- Switching of parasitic coils requires lower current/Voltage switches

- Does not reduce efficiency of main primary coil

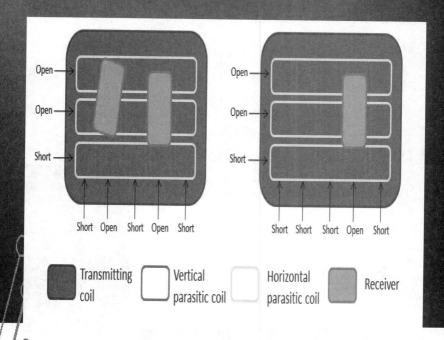

279

CLASS E DRIVER PROTECTION

- Switched mode drivers have catastrophic failure modes if stressed due to load variations

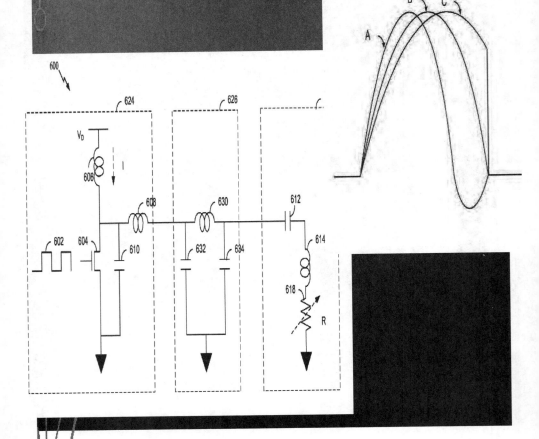

DUAL INDUCTOR

- Inductors in class-E drivers are the largest and most expensive components

- Dual winding inductor reduces number and size

- One winding feeds energy into magnetic field while other winding takes it out – reduces size of core to avoid saturation

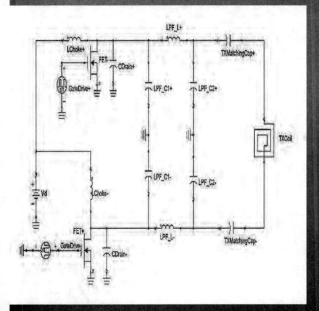

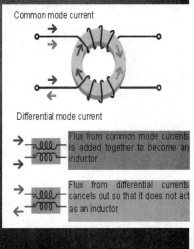

Common mode current

Differential mode current

Flux from common mode currents is added together to become an inductor

Flux from differential currents cancels out so that it does not act as an inductor

CROSS CONNECT PREVENTION

- Multiple charging stations in close proximity can have cross connection of control and power signals

- Typically use radio (such as Bluetooth) for control signaling

- Multi-level signaling, probes and protocols avoid problem

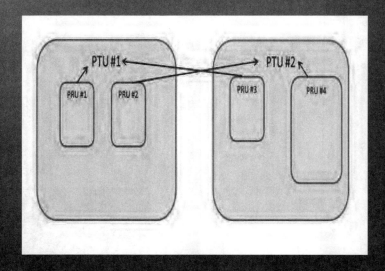

BRIEF HISTORY LESSON

MUSSCHENBROEK/CUNAEUS AND CAPACITORS

- Pieter van Musschenbroek and Andreas Cunaeus discovered the capacitor in form of a Leyden Jar in 1744

- Water inside jar forms one plate of capacitor

- Hand on outside of jar forms other plate of capacitor

- Glass of jar is insulator or dielectric.

MICHAEL FARADAY AND INDUCTORS

1791-1867

1839 publication *Experimental Researches in Electricity* explored magnetic induction

Faraday showed that

- An electrical current in a wire produces a magnetic field

- A changing (not static) magnetic flux cutting a wire produces an electrical current

- Making a coil of wire multiplies effect of changing magnetic flux

- Coils have "self induction" where interrupting an electrical current flow causes a "back electromotive force"

- A bundle of iron wires inserted into the center of coil intensifies the self induction

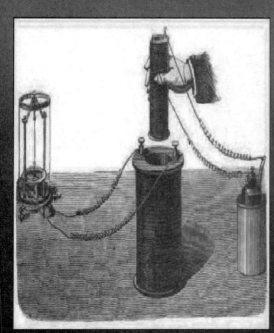

OLIVER LODGE AND SYNTONIC JARS

Olivier Lodge connected a square loop of brass wire to a Leyden Jar

- Square loop forms an inductor

- Leyden Jar is a capacitor

- Together they form an L-C tank circuit

When two of these devices are tuned to the same frequency

- By sliding the vertical wire

- Electrical energy introduced to one on right causes energy to be coupled to one on left

As indicated by sparks between two balls

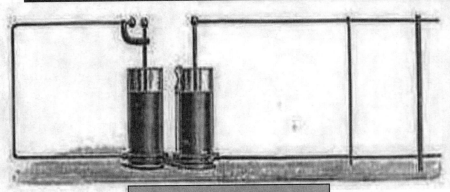

Oliver Lodge's L-C Circuit

HERTZ AND L-C RESONATOR

- Hertz combined

  - Inductor (**L**), formed by ring of wire, with

  - Capacitor (**C**) formed by two brass balls

- Combination of **L** and **C** forms resonator

- Hertz found in 1885 that two identical resonators couple energy from one to the other over a distance

FIG. 2. HERTZ RESONATOR.

TESLA'S COLORADO EXPERIMENTS (1899)

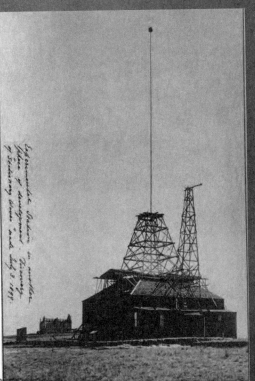

- Tesla coupled energy
  - from one high-**Q** resonator to
  - a second high-**Q** resonator

- Tesla's resonators were huge,
  - had very high **Q**
  - generated very high Voltage

- To early developers these were the same technology
  - Wireless power transfer
  - Wireless communications
  - Early receivers were powered by the transmitter

- Tesla's investors wanted him to develop wireless telegraph
  - Tesla focused on general wireless power transfer

TESLA'S WIRELESS POWER RECEIVERS

Tesla's "magnifying transformer" generated strong high frequency magnetic fields

The power receivers tuned to resonate at the same frequency as the magnifying transformer

Electric light bulbs glowing at a distance from his laboratory building

TESLA'S WIRELESS EXPERIMENTS AT COLORADO SPRINGS

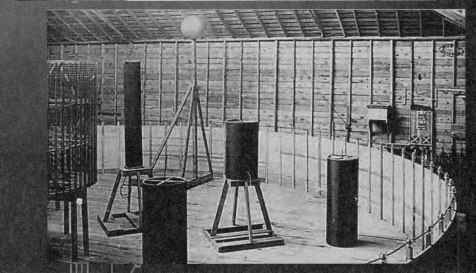

DESIGN
EXAMPLE

EXAMPLE: ELECTRIC VEHICLE CHARGING PAD DESIGN

Desired configuration:

- Outside diameter = 1000 mm

- Inside diameter = 400 mm

  - Dimensions accommodate physical constraint of fitting between wheels of a car

- Operating frequency = 40 Kilohertz

  - Operating frequency is trade- off between

    - More efficient coupling at higher frequency, and

    - More loss in drive electronics at higher frequency

- Inductance ≈ 40 uH

  - Inductance is selected to provide a reasonable load for power supply

- **Q** > 100

  - For loss in inductor < 1% relative to circulating current, component Q of inductor is selected to be Q>100

- Inductor design in this example will be further developed in other examples below

- Parameters and results closely match examples used in commercial products

NATIONAL BUREAU OF STATISTICS, CIRCULAR C74 (1924)

Provides all information needed to design and optimize inductors including the flat spiral example here

U. S. DEPARTMENT OF COMMERCE
NATIONAL BUREAU OF STANDARDS

RADIO INSTRUMENTS AND MEASUREMENTS

CIRCULAR C74

- Inductors are described on pages 14 to 18

- Resonance on pages 31 to 41 — page 37 defines "sharpness of resonance" which is the same as **Q**

- Inductive coupling on pages 57 to 59

- General inductor configurations on page 131

- Calculations for coil self-capacitance are on pages 132 to 136

- Definitions of dimensions of coil on page 257

- Flat spiral inductance specific calculations on pages 296 to 299

- Calculations for increase of resistance with frequency (skin effect) are on pages 299 to 300

CRITICAL DIMENSIONS OF FLAT SPIRAL FROM NBS C74

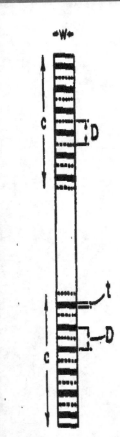

Fig. 184.—*Sectional view of flat spiral wound with metal ribbon*

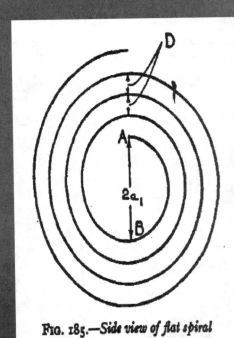

Fig. 185.—*Side view of flat spiral*

EXAMPLE: ELECTRIC VEHICLE CHARGING PAD DESIGN

Using calculations available based on NBS C74, following aspects of the charging coil are derived:

- Number of turns = 7
- Wire diameter = 5.189 mm (AWG #4)[3]
- Spacing between turns = 38 mm
- Wire length = 15.4 m
- Inductance = 38.7 μH
- Resistance (of wire at DC) r_{DC} = 0.0128 Ω

ELECTRIC VEHICLE CHARGING PAD R$_{AC}$ EXAMPLE

Reactance of inductor at operating frequency

- $X_L = 2 \pi f L$
  $= 2 * \pi * 40000 * 0.0000387 = 9.73 \ \Omega$

Degradation of resistance due to skin

AC resistance (r_{AC}) effect can be calculated from coil parameters derived in example above

Important parameters are repeated here:

- Inductance = 38.7 μH

- Resistance (of wire at DC)

$r_{DC} = 0.0128 \ \Omega$

- Wire diameter = 5.189 mm

- Frequency of operation = 40

- Skin depth defines annular ring of copper wire that is useful for electrical current flow

- Center area of wire is subtracted and increase of resistance at the operating frequency is calculated

  - Skin depth @ 40 Kilohertz = 0.813 mm

  - Wire cross-section area = 21.15 mm^2

  - Annulus area = 11.18 mm^2

  - Increase in resistance = r_{AC}/r_{DC}

  = 21.15/11.18

  = 1.89

  - AC resistance = r_{DC} * 1.89

    = 0.0128 * 1.89

    = 0.0242 Ω

  - $Q = X_L/r_{AC} = 9.73/0.0242 =$ **402**

RESONATOR BANDWIDTH

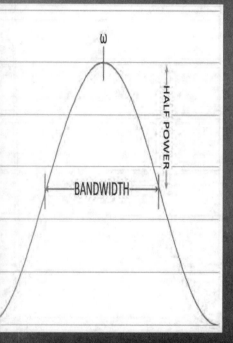

- Electrical resonators are usually excited with a continuous sine wave at a single frequency

- If sine wave frequency matches ω, then response is maximum

- If sine wave is lower or higher in frequency response is less

- If response is one-half of maximum it is considered cutoff or edge of bandwidth

$$\text{Bandwidth} = \text{Cutoff}_{upper} - \text{Cutoff}_{lower}$$

$$\text{Bandwidth} = \omega_{upper} - \omega_{lower}$$

- Q is related to bandwidth by

$$\Delta\omega = \omega/Q$$

$$Q = \omega/\Delta\omega$$

- Frequency can be normalized to $\omega = 1$

ELECTRIC VEHICLE CHARGING PAD BANDWIDTH EXAMPLE

- High **Q** resonators have narrow bandwidth

- Low **Q** resonators have wider bandwidth

- Adjusting frequency of resonator and drive sine wave is more critical with high-**Q** resonators

- If a resonator has

  - Peak response at 40 KHz

  - **Q** of 402

- $\Delta\omega = 40000$ **Hz/402 = 99 Hz**

- Lower cutoff is 39951 Hz

- Upper cutoff is 40049 Hz

- Value of **Q** is unitless

  - Units of Hertz carry to result

GAMMA AND BANDWIDTH

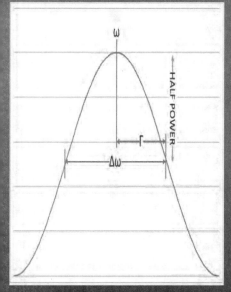

- Physicists like to use Greek letter gamma (**Γ**) for resonance width, or deviation from center frequency (half bandwidth)

$$Γ = Δω /2$$

- Relation with **Q** now becomes

$$Γ = ω/2Q$$

- Solving for **Q**

$$Q = ω/2Γ$$

- As above, **Γ** may be expressed as fractional bandwidth

  - Frequency disappears from equation

$$Q = 1/2Γ$$

- Physicists also call this **intrinsic loss rate**

PHYSICAL RESONATOR

- Time a bell rings after being struck is a measure of Q_{loaded}

    - Most of energy is released in form of sound waves

- Measuring bell in a vacuum

    - Reveals $Q_{unloaded}$

    - Only energy loss is heating from metal flexure

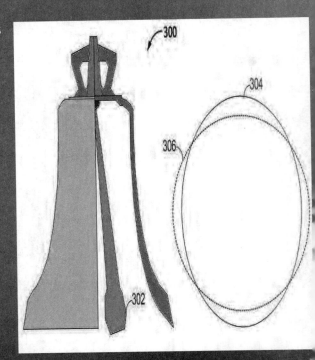

$$Q = 2\pi \times \frac{\text{Energy Stored}}{\text{Energy dissipated per cycle}}$$

UNLOADED VERSES LOADED Q

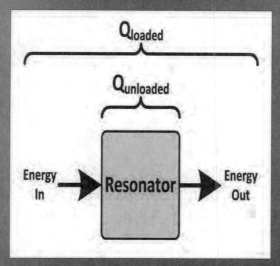

- As discussed above, resonators have Native **Q**, Intrinsic **Q**, Natural **Q**, Unloaded **Q**
  - $Q_{unloaded}$ of circuit standalone
  - $Q_{unloaded}$ defines losses in components
- Operating **Q**, Loaded **Q** mean the **Q** with energy coupled into and out of circuit
  - Q_{loaded} defines conditions when delivering power to load

$$Q_{loaded} = X_L / (r_{AC} + r_{LOAD})$$

FREQUENCY ALIGNMENT OF COUPLED RESONATOR

- For coupling to be efficient

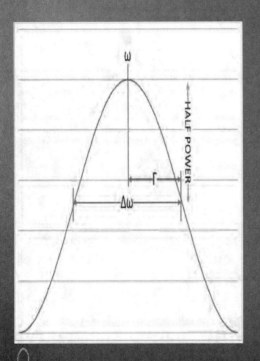

  it is important that frequency of resonance of each of the resonators be closely aligned

- Using half-bandwidth of resonance expression

$$\omega_1 - \omega_2 << \Gamma_1$$

- Difference of resonance frequencies much less than resonance bandwidth

$$\omega_1 - \omega_2 << \Gamma_2$$

- For both resonators

$$\omega_1 - \omega_2 << \sqrt{(\Gamma_1 - \Gamma_2)}$$

- Less than geometric mean of resonance bandwidths

EXAMPLE OF LOOSELY COUPLED RESONATORS

To operative efficiently coupled
resonators should satisfy

$$k/\sqrt{(\Gamma_1 \Gamma_2)} > 1$$

Substituting for Γ

$$k/\sqrt{((1/2Q_1)(1/2Q_2))} > 1$$

If coupling factor $k = 0.05$ and resonator
unloaded Qs $= 100$

$$0.05/\sqrt{((1/2*100)(1/2*100))} = 10$$

Which satisfies the requirement to be greater than 1

- One might see in literature this expression
as a system figure of merit:

$$Q = \sqrt{(Q_1 Q_2)} \text{ Geometric Mean of Unloaded } Q\text{s}$$

- This condition of critical coupling is discussed in a
later section where a practical means of operation
in this condition is explained

$$Q = \sqrt{(Q_1 Q_2)} \text{ Geometric Mean of Loaded } Q\text{s}$$

POWER FACTOR

- Electrical power engineers relate coupling of power into and out of a circuit as the power factor

  - Power factor is simply the ratio of true power to apparent power

Power factor = true power/apparent power

- Power being delivered to do useful work is true power

- Power circulating in the circuit because of reactance is apparent power

- Phase angle between Voltage and current increases with increased reactance

- Increased difference in phase angle causes losses to increase

POWER FACTOR

- Power factor (PF) can be expressed as the cosine of the angle between Voltage and current

PF = cos θ, power factor of alternating current

- Electrical power engineers think in terms of power factor of a load

  - They also use a different term for the capacitor that resonates an inductor — compensating capacitor

- Coupling of resonators is critical

    - Resistance of the load is reflected to the source with good power factor

RADIATION RESISTANCE

- Radiation loss is unintended leakage of electromagnetic (EM) wave

  - Also called antenna effect

- Radiation loss is represented as another resistor ($r_{radiation}$) in series with r_{AC}

- Do not confuse $r_{radiation}$ with energy transferred via magnetic field coupling

- Called non-radiative wireless energy transfer

- Energy is constrained to the alternating magnetic field by recapture of field energy by primary and secondary coils

  - Very little energy escapes as a radiating electromagnetic wave

- The $Q_{unloaded}$ of an inductor with all losses taken into account would be

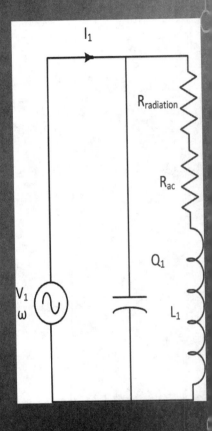

$$Q = XL/(r_{AC} + r_{radiation}) = 2\pi fL/(r_{AC} + r_{radiation}) = \omega L/(r_{AC} + r_{radiation})$$

ELECTRIC VEHICLE CHARGING PAD RADIATION RESISTANCE EXAMPLE

The equation for calculation of the radiation resistance from the reference is

$$r_{radiation} = (177*N*S / \lambda^2)^2$$

S = surface area of coil

N = number of turns

λ = wavelength

- Small loops are of interest to the radio communications field so calculations of $r_{radiation}$ are available

- In a WEVC system $r_{radiation}$ is very small and of interest only for emissions compliance

- Assume a 1 meter diameter coil.

- **Surface area calculation**

$$S = \pi (d/2)^2 = \pi (1/2)^2 = 0.785 \text{ m}^2$$

- **Wavelength calculation**

$$\lambda = 3 \times 10^8 / f = 3 \times 10^8 / 80 \times 10^3 = 3750 \text{m}$$

- **Therefore**

$$R_{radiation} = ((177 * 7 * 0.785) / 3750^2)^2$$

$$= 4.79 \text{ n}\Omega \ (10^{-9} \ \Omega)$$

- **For our example charging pad inductor**

$$X_L \gg R_{AC} \gg R_{radiation}$$

$$9.73 \ \Omega \gg 0.0242 \ \Omega \gg 0.00000000479 \ \Omega$$

TEST SETUP

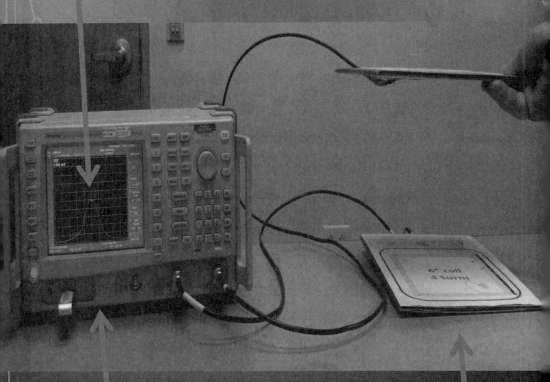

Operating
Frequency

Secondary
Coil

Spectrum analyzer
with tracking
generator

Primary Coil

- Tracking generator
  sweeps frequency

- Spectrum analyzer displays
  amplitude vs. frequency

- Operating frequency
  at center of display

- Distance between coils is
  varied

HIGH Q VERSES LOW Q DEMONSTRATION

- Illustrates difference between high Q_{loaded} verses low Q_{loaded}

    - $Q_{unloaded}$ of resonators same for both cases

    - High Q_{loaded} illustrates loose coupling operation

    - Low Q_{loaded} illustrates tight coupling operation

- Spectrum analyzer with a built-in tracking generator

    - Source is tracking generator generates sine wave that is swept across a frequency band

    - Load is spectrum analyzer that measures output by sweeping a receiver across band

    - Spectrum analyzer displays amplitude verses frequency

    - Operating frequency set at center of display

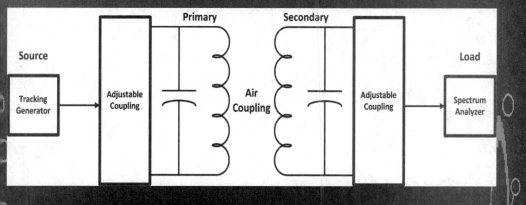

COMPARISON OF
RESONATOR BANDWIDTHS

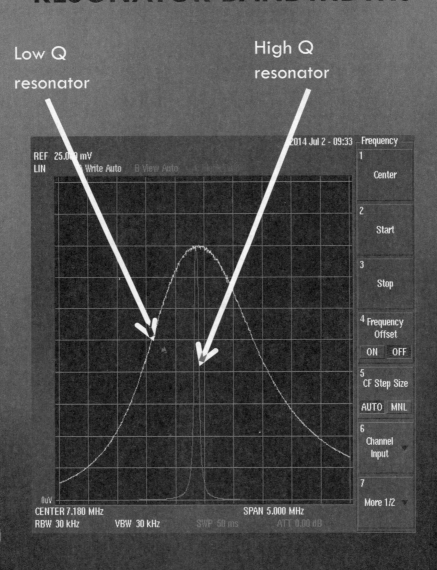

RESONATOR Q MEASUREMENT

$$Q = f_{center}/\Delta f \quad \text{(One definition of Q)}$$

- This is the operating Q of the coupled resonators
- Notice span is 200 KHz on left and 10 MHz on right

$$Q = 6.78/(6.81 - 6.74) = 97 \qquad Q = 6.93/(8.16 - 6.12) = 3.4$$

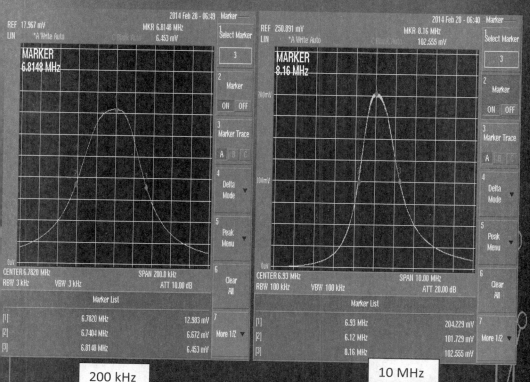

200 kHz

10 MHz

HIGH Q TEST – RANGE OF SEPARATION

Operating Q = 97

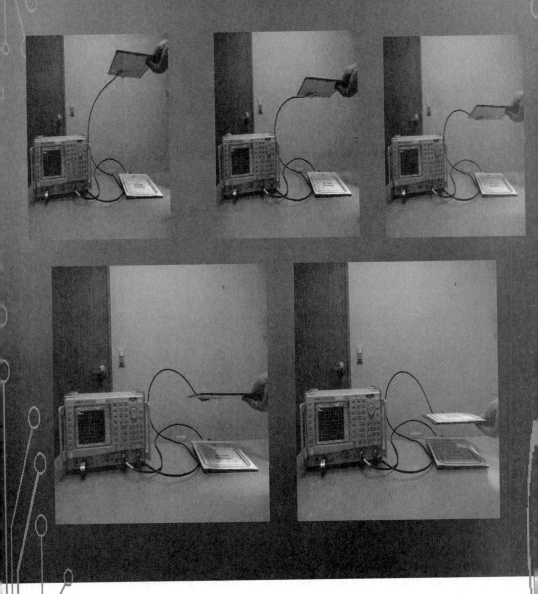

HIGH Q TEST – SINGLE FREQUENCY

Measuring only single operating frequency
At highest position (1) good coupling, better at 2, then
degrades to lowest position (6)

HIGH Q TEST – FREQUENCY SWEEP

1   2   3

4   5   6

- With frequency sweep we can see what is really going o
- Highest position (1), critical coupling (2) then over couplin
  (3 to 6) as coils get too close

LOW Q RESONATORS

Q = 3.4

Separation of coils much less

Highest position (1)

Critical Coupling (2)

Over coupling (3 & 4)

Coils are touching (4)

1

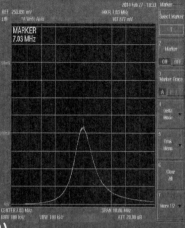

2

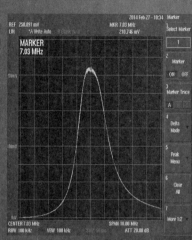

3

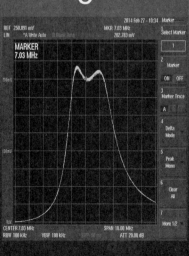

4

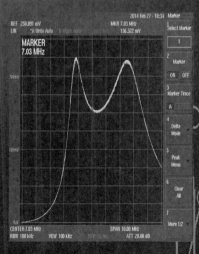

COORDINATED ADJUSTMENT OF COUPLING

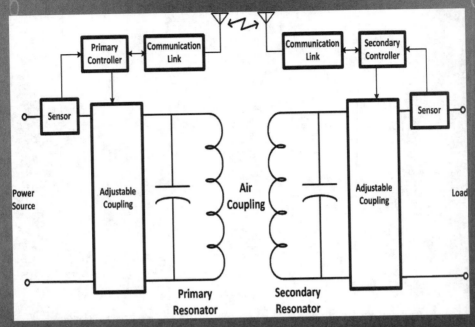

For high-Q wireless power transfer to be efficient

- Adjustment of coupling from power source to primary resonator must be coordinated with adjustment of coupling from secondary resonator to load

- When the air coupling or the load changes, the coupling at each stage must be adjusted to compensate

- A communications link between primary and secondary is necessary for coordination

ADJUSTABLE COUPLING METHODS

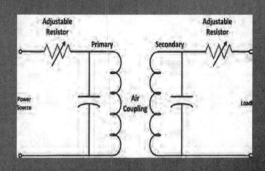

- Simplest adjustable coupling is adjustable resistors

- Adjustable inductors or capacitors can be used

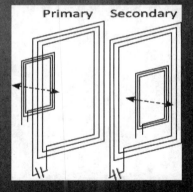

- Physical adjustment of inductive coupling is popular for demonstrations but not practical for products

TAPPED COUPLING LOOP

- Tapped coupling loop overlays but is electrically isolated from resonator

    - Resonator is made up of loop inductor connected across a capacitor with no other connections

    - Coupling is only by shared flux

- Minimal effect on resonant frequency of resonator

    - Changing taps does not alter resonator

    - Avoids retuning resonator when coupling is changed

- Strong circulating current in resonator isolated from tap switch

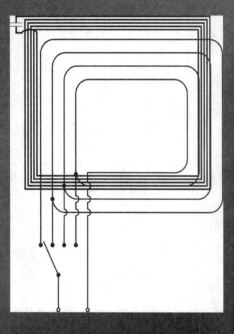

GEOMETRIC TAPPED COUPLING LOOP

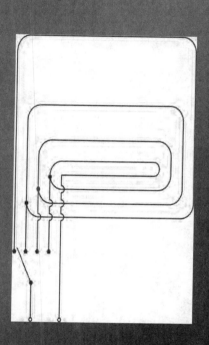

- <u>Tapped coupling loop</u> with larger and smaller loops provides larger range of coupling adjustment

    - Smallest loop gives very loose coupling for very high-**Q** distant operation

    - Using all loops gives tight coupling for low-**Q** close operation

319

SUMMARY

- Techniques and methods described have been developing and used for more than a century

- Basic concepts of the coupling of high-frequency, high-**Q** resonators known years before the term **Q** was used.

- Early developers in this field produced very high-**Q** L-C resonators
  - Coupled energy over considerable distance
  - Resonators tuned to be precisely aligned in frequency

- Modern semiconductors efficiently produce Kilowatts to drive resonators
  - Wireless power transfer by High-**Q** resonators are now becoming available as commercial products
  - This technology continues to improve, but the core concepts are the same

- Consumer acceptance of electric vehicles and portable electronic devices is improved by eliminating inconvenient and dangerous electrical connections

- Low power wireless charging enables portable devices to be always on and serving our needs – Charging can take place anywhere we set our portable device down

- Efficient and safe wireless charging of BEVs encourages take-up
  - No longer will drivers have to wrestle with a heavy power cord

OTHER TOPICS

FIELD

- Radio antenna engineers use the terms near field and far field to describe the regions where fields are converted to waves

- **Reactive Near Field $< \lambda/2\pi$**
  - Magnetic field dominates

- $\lambda/2\pi <$ **Transition Zone** $< 2D^2/\lambda$
  - Some magnetic field energy couples to an electric field
  - Creates an EM wave
  - Also called the Fresnel region

- $2D^2/\lambda <$ **Radiating Far Field**
  - Pure magnetic field has decayed
  - EM wave well established and contains most of the energy that was not returned to the coil
  - Also called the Fraunhofer Zone

- **D** is largest dimension of the resonator

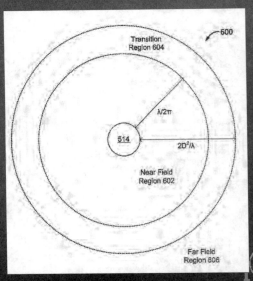

ACOUSTIC ANALOGY – NEAR FIELD

- Illustrates what happens when a source of energy is in a confined volume

  - Like the **Reactive Near Field**

- Imagine a room that is smaller than the wave length of sound signal

  - When the cone moves out, the air pressure of room increases

  - When the cone moves in, air pressure of room decreases

- Wave length of the sound signal is longer than the room

  - The air pressure in whole room changes at the same time

- No wave is produced – Non-radiative

  - All of the air pressure in the room is increased or decreased as the speaker **cone moves**

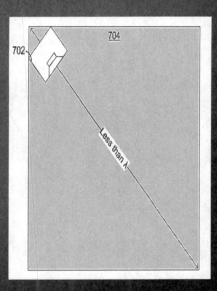

702

704

Less than λ

ACOUSTIC ANALOGY – FAR FIELD

- Illustrates what happens when a source of energy is in an unconfined volume
  - Such as **Far Field**

- Now imagine a room is much larger than wave length of sound signal
  - When speaker cone moves out air pressure immediately in front of cone is increased
  - Before increased pressure can propagate to the rest of room, cone moves in, reducing pressure immediately in front of cone

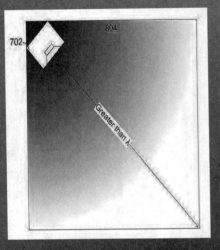

- Air pressure farther away from speaker cone cannot keep up with changes cone makes
  - Pressure waves propagate away from speaker

SOUND WAVES

- As speaker cone moves out pressure builds to a peak, decays to neutral, builds to a negative peak, then decays back to neutral

- If one were to follow the peak of pressure, it would appear to move away from the speaker at a speed of about 1 meter in 3 milliseconds

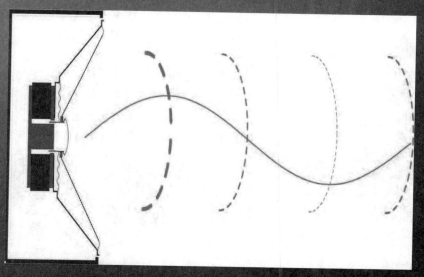

EVANESCENT WAVES

- Occur when electromagnetic waves interact with a boundary

- Practical wireless power systems operate at a small fraction of a wavelength so no waves are formed

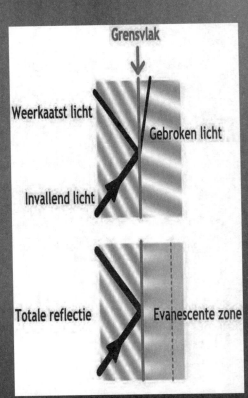

Wireless Power System Components

1	Power Supply	Power Supply
2	DC-to-higher frequency conversion	Amplification and voltage conversion
3	Impedance Matching	Tuning, switching in and out elements
4	Signaling/system control	In-band and out-of-band
5	Antennas/Coils	Repeaters
6	WP Field	Evenness; interference mitigation; emissions; safety; foreign object detection
7	Load Management	Conveying or disconnecting WP to the load/battery
8	User Interface	User Interface
9	Interoperability	Connectivity to other solutions e.g. wired
10	Application/Use case	E.g. charging in public places

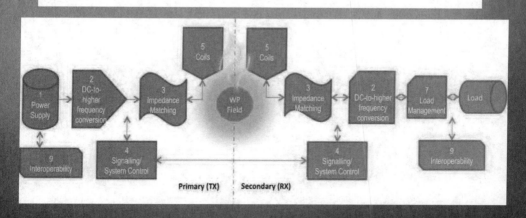

Q CHEAT SHEET

- Relates **Q** to ratio of energy stored in magnetic field of inductor to resistive loss

$$Q = \omega L / r_{AC}$$

- Relates Q to ratio of energy stored in inductor to resistive and radiation

  losses

$$Q = \omega L / (r_{AC} + r_{radiation})$$

- Relates **Q** to ratio of frequency of oscillation to half-power bandwidth

$$Q = \omega / \Delta\omega$$

- Relates **Q** to ratio of frequency of oscillation to resonance width (1/2 of bandwidth)

$$Q = \omega / 2\Gamma$$

- $Q_{unloaded}$ is resonator standalone defined by losses in components
- Q_{loaded} is resonator delivering power to load

$$Q_{loaded} = \omega L / (r_{AC} + r_{LOAD})$$

- Resonant frequency of coupled resonators should be matched within a resonance width

$$\omega_1 - \omega_2 << \sqrt{(\Gamma_1 - \Gamma_2)}$$

- Efficiency of coupled resonators is approximately the ratio of unloaded to loaded **Q**

$$\eta \approx Q_{unloaded} / Q_{loaded}$$

SUMMARY OF TERMS

Inductor An electrical component that stores energy in a magnetic field. Usually consists of a coil of wire. May include   material with high magnetic permeability to increase and contain magnetic flux.

Capacitor An electrical component that stores energy in an electric field. Usually consists of a pair of electrically conductive   plates separated by an insulator.

Resonator A circuit or mechanism that stores energy in an oscillation.

R_{DC} Resistance at DC.

R_{AC} Effective resistance at a frequency. Skin effect reduces useful cross section of a conductor and increases effective   resistance.

ω Angular frequency in units of radians per second. $\omega = 2\pi f$

Q Unitless quality factor for a component or a resonator. Ratio of stored to dissipated energy. Also ratio of bandwidth to resonant frequency. Also called resonance ratio or sharpness of resonance.

Γ Resonance width. Also called intrinsic loss rate. Half bandwidth to resonant frequency ratio. Another way of expressing quality factor. $Q = 1/(2\ \Gamma)$

$Q_{unloaded}$ Unloaded Q of a resonator, also known as **intrinsic Q, natural Q**. Ratio of energy stored in resonator to energy dissipated by components of resonator. Makes no allowance for coupling energy into or out of resonator.

Q_{loaded} Loaded Q of a resonator, also known as **operating Q, system Q**. Ratio of energy stored in resonator to energy dissipated and coupled into and out of resonator. Usually dominated by coupled energy.

M Mutual coupling factor. Fraction of flux from one inductor that cuts windings of a second inductor.

Non-Radiative Coupling by magnetic field with little conversion of that field to an electromagnetic wave.

Near Field Volume close to a resonator where most of the energy in oscillation is returned to resonator at each cycle.

Far Field Volume distant from a resonator where the energy in oscillation has escaped in form of a wave.

Transition Zone Volume between near field and far field where some of energy is returned to resonator and some of energy is converted to a wave that propagates energy away from resonator.

REFERENCES

[1] The Theory, Design and Construction of Induction Coils, H. Armagnat,
Translated by Otis Allen Kenyon, McFraw Publishing Co, 1908
Available as a reproduction:
http://www.amazon.com/Theory-Design-Construction-Induction-
Coils/dp/116519158X/ref=tmm_hrd_swatch_0?_encoding=UTF8&sr=8
_1&qid=1403138556

[2] The Theory and Design of Inductance Coils, V.G.Welsby, Macdonald & Co.
Ltd., 2nd edition, 1960

[3] Principles of Electricity applied to Telephone and Telegraph Work, American
Telephone and Telegraph Company, 1953 edition

[4] Radio Handbook, William Orr, 20th edition, 1975, ISBN 0-672-24032-

7 [5] Reference Data for Radio Engineers, IT&T, sixth edition, 1975

[6] US Patent 8,410,636, Low AC resistance conductor designs

[7] Tesla's Oscillator and Other Inventions, Thomas Commerford Martin,
The Century Magazine, April 1895
http://www.tfcbooks.com/tesla/1895-04-00.htm

[8] The New Telegraphy, Adolphus Slay, The Century Magazine, April
1898 http://earlyradiohistory.us/1898sla.htm

[9] Experimental Researches in Electricity, volumes I-III, Michael Faraday,1839
http://www.gutenberg.org/ebooks/14986

[10] TN -261 — Safety Code 6 (SC6) Radio Frequency Exposure
Compliance Evaluation Template (Uncontrolled Environment Exposure
Limits), Industry Canada, Spectrum Management and
Telecommunications, Appendix —
General Antenna Theory
https://www.ic.gc.ca/eic/site/smt-gst.nsf/eng/sf10112.html

[11] Evanescent Waves, Andrew Cmu, Carnegie Mellon University,
http://www.andrew.cmu.edu/user/dcprieve/Evanescent%20waves.htm

REFERENCES CONTINUED

[12] www.antenna-theory.com/antennas/smallLoop.php

[13] Design of a contact-less battery charging system for people movers at Whakarewarewa geothermal park, IPENZ Engineering TreNZ transactions, January 2003 https://www.ipenz.org.nz/ipenz/forms/pdfs/TreNz1.pdf

[14] IPT Fact Sheet Series: No. 2, Magnetic Circuits for Powering Electric Vehicles, Department of Electrical and Computer Engineering, The University of Auckland

[15] Field Service Memo: *Electromagnetic Radiation and How It Affects Your Instruments*, United States Department of Labor, Occupational Safety & Health Administration, May 20, 1990 https://www.osha.gov/SLTC/radiofrequencyradiation/electromagnetic_fieldmemo/electromagnetic.html

[16] *Radio Instruments and Measurements*, U.S. Department of Commerce, National Bureau of Standards, Circular C74, 1937 reprint http://terac.org/wp-content/uploads/2014/04/circ74-rescan.pdf

[17] *The Influence of Frequency Upon the Self-Inductance of Coils (August 1906)*, Bulletin of the Bureau of Standards, Vol. 2, 275-296, 1906, Scientific Paper 37 https://archive.org/details/influe227529619063737coff

[18] *Electrical Network*, US Patent 1,628,983, Kenneths S. Johnson, 1923

[19] *The Story of Q*, Estill I. Green, Bell Telephone System Monograph 2491, also published in American Scientist, Vol. 42, pages 584 to 594, October 1955

[20] High Frequency Electrical Conductor, US Patent 1,039,717, Reginald A. Fessenden, 1911

[21] UCSB ECE218b class notes, Resonant Circuits http://www.ece.ucsb.edu/Faculty/rodwell/Classes/ece218b/notes/Resonators.pdf

17 Wireless Energy Transfer with Efficient Scalar Waves

Prof. Dr.-Eng. Konstantin Meyl

Introduction

It will be shown that scalar waves, normally remaining unnoticed, are very interesting in practical use for information and energy technology for reason of their special attributes. The mathematical and physical derivations are supported by practical experiments. The demonstration will show:

1. the wireless transmission of electrical energy,
2. the reaction of the receiver to the transmitter,
3. free energy with an over-unity-effect of about 3,
4. transmission of scalar waves with 1.5 times the speed of light,
5. the inefficiency of a Faraday cage to shield scalar waves.

Tesla radiation

Here is shown extraordinary science, five experiments, which are incompatible with textbook physics. Following my short lecture I will present you the transmission of longitudinal electric waves.

It is a historical experiment, because already 100 years ago the famous experimental physicist Nikola Tesla measured the same wave properties, as I have. He holds a patent concerning the wireless transmission of energy (1900)[1]. Since he also found out that at the receiver arrives very much more energy, than the transmitter sends out, he spoke of a "Magnifying Transmitter". By the effect back on the transmitter Tesla found the resonance of the earth and that lies according to his measurement at 12 Hz. Since the Schumann resonance is a wave, which goes with the speed of light, but at 7.8 Hz, Tesla comes to the conclusion, that his wave has 1.5 times the speed of light[2].

As founder of the diathermy Tesla already has pointed to the biological effectiveness and to the possible use in medicine. The diathermy of today has nothing to do with the Tesla radiation; it uses the wrong wave and as a consequence hardly has a medical importance.

The discovery of the Tesla radiation is denied and isn't mentioned in the textbooks anymore. For that there are two reasons:

1. No high school ever has rebuilt a "Magnifying Transmitter". The technology simply was too costly and too expensive. In that way the results have not been reproduced, as it is imperative for an acknowledgement. I have solved this problem by the use of modern electronics, by replacing the spark gap generator with a function generator and the operation with high-tension with 2-4 Volts

low-tension. I sell the experiment as a demonstration-set so that it is reproduced as often as possible. It fits in a case and has been sold more than 100 times. Some universities already could confirm the effects. The measured degrees of effectiveness lie between 140 and 1000 percent.

2. The other reason, why this important discovery could fall into oblivion, is to be seen in the missing of a suitable field description. The Maxwell equations in any case only describe transverse waves, for which the field pointers oscillate perpendicular to the direction of propagation.

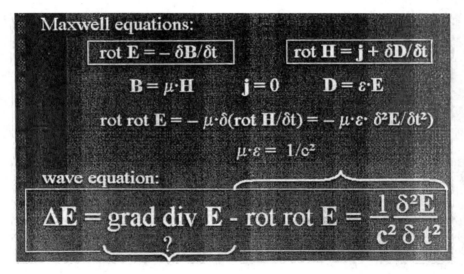

Figure 1. The vectorial part of the wave equation (derived from Maxwell equations)[*]

Wave Equation

By using the Laplace operator the well-known wave equation, according to the rules of vector analysis, can be taken apart in two parts: in the vectorial part (rot rot E) which results from the Maxwell equations and in a scalar part (grad div E), according to which the divergence of a field pointer is a scalar. We have to ourselves which properties has the wave part, which has the scalar wave?

If we derive the field vector from a scalar potential φ, then this approach immediately leads to an inhomogeneous wave equation, which is called plasma wave. Solutions are known, for example, the electron plasma waves, which are longitudinal oscillations of the electron density (Langmuir waves).

---

[*] Ed. Note: The German calculus convention of "rot" corresponds to the Western "curl" which is symbolized usually with an upside down delta and the cross product: "▼×"

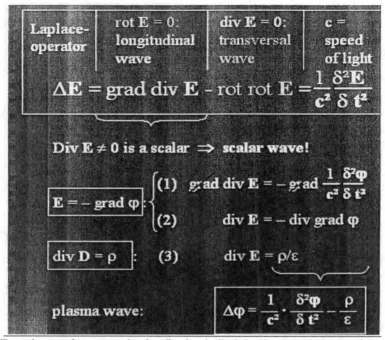

Laplace-operator	rot E = 0: longitudinal wave	div E = 0: transversal wave	c = speed of light

$$\Delta \mathbf{E} = \text{grad div } \mathbf{E} - \text{rot rot } \mathbf{E} = \frac{1}{c^2}\frac{\delta^2 \mathbf{E}}{\delta t^2}$$

Div E ≠ 0 is a scalar ⟹ scalar wave!

$\mathbf{E} = -\text{grad } \varphi$:
- (1) $\text{grad div } \mathbf{E} = -\text{grad } \dfrac{1}{c^2}\dfrac{\delta^2 \varphi}{\delta t^2}$
- (2) $\text{div } \mathbf{E} = -\text{div grad } \varphi$

$\text{div } \mathbf{D} = \rho$: (3) $\quad \text{div } \mathbf{E} = \rho/\varepsilon$

plasma wave: $\quad \Delta \varphi = \dfrac{1}{c^2}\cdot\dfrac{\delta^2 \varphi}{\delta t^2} - \dfrac{\rho}{\varepsilon}$

Figure 2. The scalar part of wave equation describes longitudinal electric waves (derivation of plasma waves)

Vortex model

The Tesla experiment and my historical rebuild however show more. Such longitudinal waves obviously exist even without plasma in the air and even in vacuum. The question thus is asked, what the divergence **E** describes in this case? How is the impulse passed on, so that a longitudinal standing wave can form? How should a shock wave come about, if there are no particles which can push each other?

I have solved this question, by extending Maxwell's field theory for vortices of the electric field. These so-called potential vortices are able to form structures and they propagate in space for reason of their particle nature as a longitudinal shock wave. The model concept bases on the ring vortex model of Hermann von Helmholtz, which Lord Kelvin made popular. In my books[3] the mathematical and physical derivation is described.

In spite of the field theoretical set of difficulties every physicist at first will seek for a conventional explanation. He will try two approaches:

Resonant circuit interpretation

Tesla had presented his experiment among others to Lord Kelvin and 100 years ago he talked about a vortex transmission. In Kelvin's opinion however, it was not a wave but a radiation. He had recognized clearly, that every radio technical interpretation had to fail, because the course of the field lines is a completely different one. It assumes a resonant circuit, consisting of a capacitor and an inductance.

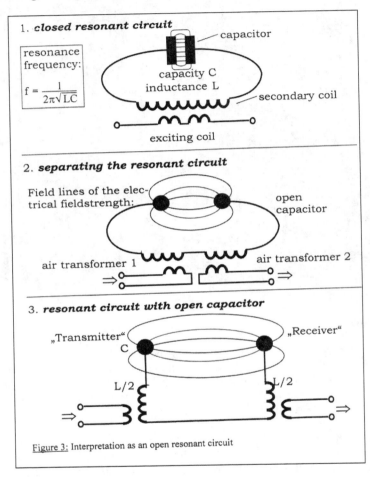

1. **closed resonant circuit**

capacitor

resonance frequency:

$$f = \frac{1}{2\pi\sqrt{LC}}$$

capacity C
inductance L

secondary coil

exciting coil

2. **separating the resonant circuit**

Field lines of the electrical fieldstrength:

open capacitor

air transformer 1

air transformer 2

3. **resonant circuit with open capacitor**

„Transmitter"

C

L/2

„Receiver"

L/2

Figure 3: Interpretation as an open resonant circuit

If both electrodes of the capacitor are pulled apart, then between both is stretching an electric field. The field lines start at one sphere, the transmitter, and they bundle up again at the receiver. In that way, a higher degree of effectiveness and very tight coupling can be expected. In this way some of the effects can be explained unequivocally, but not all.

The inductance is split up in two air transformers, which are wound completely identical. If fed in sinusoidal tension, voltage is transformed up in the transmitter, and then it is again transformed down at the receiver. The output voltage should be smaller or at maximum equal the input voltage - but it is substantially bigger! There can be drawn and calculated an alternative wiring diagram, but in no case the measurable result comes out, that light-emitting diodes at the receiver glow brightly (U>2Volt), whereas at the same time the corresponding light-emitting diodes at the transmitter go out (U<2Volt)! To check this both coils are exchanged.

The measured degree of effectiveness lies despite the exchange at 1000 percent. If the law of conservation of energy should not be violated, then only one interpretation is left: The open capacitor withdraws field energy from its environment. Without consideration of this circumstance the error deviation of every conventional model calculation lie at more than 90 percent. There one should do without the calculation.

It will concern oscillating fields, because the spherical electrodes are changing in polarity with a frequency of approx. 7 MHz. They are operated in resonance. The condition for resonance reads: identical frequency and opposite phase. The transmitter obviously modulates the field in its environment, while the receiver collects everything what fulfils the condition for resonance.

Also in the open question for the transmission velocity of the signal the resonant circuit interpretation fails. But the HF-technician still has another explanation at the tip of his tongue.

Near field interpretation

In the near field of an antenna effects are measured, which on one hand go as inexplicable, because they evade normal field theory; on the other hand they come as very close scalar wave effects. Everyone knows a practical application: e.g.: the entrance at department stores, where the customer has to go through scalar wave detectors.

In my experiment the transmitter is situated in the mysterious near zone. Also Tesla always worked in the near zone. But for those who ask why, they will discover that the field effect is nothing else but the scalar wave part of the wave equation. My explanation goes as follows:

> The charge carriers which oscillate with high-frequency in the antenna rod form longitudinal standing waves. As a result also the fields in the near zone of a Hertzian dipole are longitudinal scalar wave fields. The picture shows clearly how vortices are forming and how they come off the dipole.

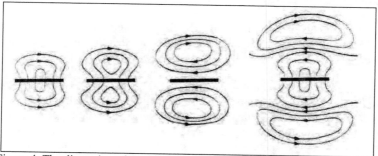

Figure 4. The dispersion of the dipole's electric field.

Like charge carriers in the antenna rod, the phase angle between current and tension voltage amounts to 90 degrees, occurring in the near field and also in the electric and magnetic field phase shifting 90 degrees. In the far field however the phase angle is zero. In my interpretation the vortices are breaking up, they are decaying, and inverse radio waves are formed.

Vortex interpretation

The vortex decay however depends on the velocity of propagation. Calculated at the speed of light the vortices already have decayed within half the wavelength. The faster the velocity, the more stable they get, to remain stable above 1.6 times the velocity. These very fast vortices contract in the dimensions. They now can tunnel. Therefore speed faster than light occurs at the tunnel effect. Therefore no Faraday cage is able to shield fast vortices.

Since these field vortices with particle nature following the high-frequency oscillation permanently change their polarity from positive to negative and back, they don't have a charge on the average over time. As a result they almost unhindered penetrate solids. Particles with this property are called neutrinos in physics. The field energy which is collected in my experiment, according to that stems from the neutrino radiation which surrounds us. Because the source of this radiation, all the same if the origin is artificial or natural, is far away of my receiver, every attempt of a near field interpretation goes wrong. After all does the transmitter installed in the near field zone supply less than 10% of the received power. The 90% however, which it concerns here, cannot stem from the near field zone!

Experiment

At the function generator I adjust frequency and amplitude of the sinusoidal signal, with which the transmitter is operated. At the frequency regulator I turn so long, till the light-emitting diodes at the receiver glow brightly, whereas those at the transmitter go out. Now an energy transmission takes place.

If the amplitude is reduced so far, till it is guaranteed that no surplus energy is radiated, then in addition a gain of energy takes place by energy amplification.

If I take down the receiver by pulling out the earthing, then the lighting up of the LED's signals the mentioned effect back on the transmitter. The transmitter thus feels, if its signal is received.

The self-resonance of the Tesla coils, according to the frequency counter, lies at 7 MHz. Now the frequency is ran down and see there, at approx. 4.7 MHz the receiver again glows, but less bright, easily shieldable and without discernible effect back on the transmitter. Now we unambiguously are dealing with the transmission of the Hertzian part and that goes with the speed of light. Since the wavelength was not changed, does the proportion of the frequencies determine the portion of the velocities of propagation. The scalar wave according to that goes (7/4.7=) 1.5 times the speed of light!

If I put the transmitter into the aluminum case and close the door, nothing should arrive at the receiver. Expert laboratories for electromagnetic compatibility indeed cannot detect anything despite the receiver lamps glowing! By turning off the receiver coil it can be verified that an electric and not a magnetic coupling, is present, although the Faraday cage should shield the electric fields. The scalar wave obviously overcomes the cage with a speed faster than light, by tunneling!

Literature

1. Nikola Tesla: Apparatus for transmission of electrical energy. US-Patent No. 645,576, N.Y. 20.3.1900.
2. Nikola Tesla: Art of transmitting electrical energy through the natural mediums, US-Patent No. 787,412, N.Y. 18.4.1905.

3. Konstantin Meyl: Elektromagnetische Umweltvertraglichkeit,
   Teil 1: Umdruck zur Vorlesung, Villingen-Schwenningen 1996, 3.Aufl. 1998
   Teil 2: Energietechnisches Seminar 1998, 3. Auflage 1999,
   Teil 3: Informationstechnisches Seminar 2002, auszugsweise enthalten in:
   K. Meyl: Skalarwellentechnik, Dokumentation fur das Demonstrations-Set,
   INDEL-Verlag, Villingen-Schwenningen,
4. Konstantin Meyl: Scalar Waves, INDEL-Verlag.
   (see: http://www.k-meyl.de).

Address
   Prof. Dr.-Ing. Konstantin Meyl,
   TZA (Transferzentrum der Steinbeis-Stiftung)
   LeopoldstraBe 1,
   D-78112 St. Georgen/Schwarzwald (Germany)
   Tel.: 0049-/0- 7724-1770, Fax.: 0049-/0- 7721-51870 (Mobil: 0172-7413378),
   Email: mevKgk-meyl.de

• More information in the internet: http://www.k-meyl.de

Electrodynamics with the scalar field

K.J. van Vlaenderen

koen@truth.myweb.nl

A. Waser

aw@aw-verlag.ch

October 27, 2001

Abstract

The theory of electrodynamics can be cast into biquaterion form. Usually Maxwells equations are invariant with respect to a gauge transformation of the potentials and one can choose freely a gauge condition. For instance, the Lorentz gauge condition yields the potential Lorenz inhomogeneous wave equations. It is possible to introduce a scalar field in the Maxwell equations such that the generalised Maxwell theory, expressed in terms of the potentials, automatically satisfy the Lorenz inhomogeneous wave equations, without any gauge condition. This theory of electrodynamics is no longer gauge invariant with respect to a transformation of the potentials: it is electrodynamics with broken gauge symmetry. The appearence of the extra scalar field terms can be described as a conditional *current* regauge that does not violate the conservation of charge, and it has several consequences:

- the prediction of a longitudinal electroscalar wave (LES wave) in vacuum.
- superluminal wave solutions, and possibly classical theory about photon tunneling.
- a generalised Lorentz force expression that contains an extra scalar term.
- generalised energy and momentum theorems, with an extra power flow term associated with LES waves.

A charge density wave that only induces a scalar field is possible in this theory.

1 Introduction: quaternions and biquaternions

The theory of electrodynamics can be formulated very efficiently in *biquaternion* form . Hamilton's quaternions [1] were used also by J.C. Maxwell in his Treatise on Electricity and Magnetism [2] [3]. A. Waser has used the biquaternion form also in [4] . The quaternion is very suitable to express the typical 4-qualities in physics, such as 4-position, 4-speed, 4-momentum, 4-force, 4-potential and 4-current [5]. A quaternion is defined as next:

$$X = x_0 + ix_1 + jx_2 + kx_3 \tag{1}$$

where i, j, k are hypercomplex roots of -1, and x_0, x_1, x_2, x_3 are real numbers.

$$ii = jj = kk = -1 \qquad ij = -k \qquad jk = -i \qquad ij = -k \tag{2}$$

The scalar part of the quaternion is represented by x_0, while the vector part is represented by x_1, x_2 and x_3. If we define $\vec{x} = (x_1, x_2, x_3)$ and $\vec{i} = (i, j, k)$ then we can notate a quaternion in a short scalar vector form by the use of the internal vector product:

$$X = x_0 + \vec{i} \cdot \vec{x} \tag{3}$$

This notation makes the use of prefixes S or V, for indicating the scalar or vector part of a quaternion, redundant. Now the internal and external vector products and the vector itself can be used in quaternion equations. The quaternion sum and product are as follows:

$$X + Y = (x_0 + y_0) + \vec{i} \cdot (\vec{x} + \vec{y}) \tag{4}$$

$$XY = (x_0 y_0 - \vec{x} \cdot \vec{y}) + \vec{i} \cdot (x_0 \vec{y} + y_0 \vec{x} + \vec{x} \times \vec{y}) \tag{5}$$

This product is the consequence of the product rules of the Hamiltonian numbers i, j, k as defined in (2). Note that $XY \neq YX$, because $\vec{x} \times \vec{y} = -\vec{y} \times \vec{x}$. The quaternion product is associative and distributive:

$$X(YZ) = (XY)Z \qquad X(Y + Z) = XY + XZ \tag{6}$$

The conjugate and the length of a quaternion are defined as follows:

$$X^* = x_0 - \vec{i} \cdot \vec{x} \tag{7}$$

$$|X| = \sqrt{XX^*} = \sqrt{x_0^2 + \vec{x} \cdot \vec{x}} = \sqrt{x_0^2 + x_1^2 + x_2^2 + x_3^2} \tag{8}$$

By replacing the real numbers x_0, x_1, x_2, x_3 for complex numbers:

$$X = x_0 + iy_0 + \vec{i} \cdot (\vec{x} + i\vec{y}) \tag{9}$$

a *complex* quaternion is obtained [6]. A complex quaternion is also called a *biquaternion*. The imaginary number i, should not be confused with $\vec{i} = (i, j, k)$. In general, a single biquaternion equation

$$(a_0 + ib_0) + \vec{i} \cdot [\vec{a} + i\vec{b}] = (c_0 + id_0) + \vec{i} \cdot [\vec{c} + i\vec{d}] \tag{10}$$

is a compact notation of two scalar equations and two vector equations:

$$\begin{aligned} a_0 &= c_0 \\ b_0 &= d_0 \\ \vec{a} &= \vec{c} \\ \vec{b} &= \vec{d} \end{aligned} \tag{11}$$

For instance, the four Maxwell equations can be expressed by just one biquaternion equation.

2 Minkowski space and biquaternion operators for physics

The four dimensional Minkowski space can be expressed in biquaternion form:

$$\mathbf{X} = (ict + \vec{i} \cdot \vec{x}) \qquad (12)$$

where X is the position quaternion. Notice that $x = 0$, $\vec{y} = \vec{0}$ and $y = ct$. In general a biquaternion represents 8-dimensional space. In case we want to represent 4-dimensional Minkowskian space in quaternions we set $x = 0$ and $\vec{y} = \vec{0}$. The lenght of X is invariant under transformation between inertial systems:

$$|\mathbf{X}| = \sqrt{(ict)^2 + \vec{x} \cdot \vec{x}} = \sqrt{x_1^2 + x_2^2 + x_3^2 - c^2t^2} \qquad (13)$$

The biquaternion velocity is defined as:

$$\mathbf{V} = \frac{d\mathbf{X}}{dt} = \frac{d(ict)}{dt} + \vec{i} \cdot \frac{d\vec{x}}{dt} = ic + \vec{i} \cdot \vec{v} \qquad (14)$$

The length of V is:

$$|\mathbf{V}| = \sqrt{\mathbf{v}^2 - c^2} = ic\sqrt{1 - \frac{\mathbf{v}^2}{c^2}} \qquad (15)$$

In stead of using V and dt we might define a relativistic time differential $d\tau$ (time dilatation):

$$d\tau = \frac{|\mathbf{V}|}{ic} dt = \sqrt{1 - \frac{\mathbf{v}^2}{c^2}} dt \qquad (16)$$

which is known as the "proper time", and define a "relativistic biquaternion speed":

$$\mathbf{U} = \frac{d\mathbf{X}}{d\tau} = \frac{ic}{|\mathbf{V}|} \frac{d\mathbf{X}}{dt} = \frac{ic\mathbf{V}}{|\mathbf{V}|} \qquad (17)$$

but then it is obvious that $|\mathbf{U}| = ic$ and this makes no sense [7]. So either we use the four-dimensional velocity V and the time differential dt, or we use the concept of time-dilatation $d\tau$ and a three-dimensional velocity $\frac{d\vec{x}}{d\tau}$. We choose for the first option, because it does not make sense to mix 4-dimensional qualities with 3-dimensional qualities in one theory. Now that a physical space has been described, three biquaternion operators for the use of physics are defined as follows:

$$\text{Nabla} \quad : \quad \nabla = \frac{i}{c}\frac{\partial}{\partial t} + \vec{i} \cdot \vec{\nabla}, \qquad \vec{\nabla} = \left(\frac{\partial}{\partial x_1}, \frac{\partial}{\partial x_2}, \frac{\partial}{\partial x_3}\right) \qquad (18)$$

$$\text{d'Alembert} \quad : \quad \square = -|\nabla|^2 = -\nabla\nabla^* = \frac{1}{c^2}\frac{\partial^2}{\partial t^2} - \vec{\nabla} \cdot \vec{\nabla} \qquad (19)$$

3 Electrodynamics in biquaternion form

The quaternion electromagnetic potential and the quaternion current are defined as follows:

$$\mathbf{A} = \frac{i}{c}\Phi + \vec{i} \cdot \vec{A} \qquad (20)$$

$$\mathbf{J} = ic\rho + \vec{i} \cdot \vec{J} = ic\rho + \vec{i} \cdot \rho\vec{V} = \rho\mathbf{V} \qquad (21)$$

By applying the quaternion product we can determine the differential of the electromagnetic potential, ∇A:

$$\nabla\mathbf{A} = \left(\frac{i}{c}\frac{\partial}{\partial t} + \vec{i} \cdot \vec{\nabla}\right)\left(\frac{i}{c}\Phi + \vec{i} \cdot \vec{A}\right) =$$
$$-\left(\frac{1}{c^2}\frac{\partial\Phi}{\partial t} + \vec{\nabla} \cdot \vec{A}\right) + \vec{i} \cdot \left[\vec{\nabla} \times \vec{A} + \frac{i}{c}\left(\frac{\partial\vec{A}}{\partial t} + \vec{\nabla}\Phi\right)\right] \qquad (22)$$

If one defines the electric vector field, magnetic vector field, and an extra scalar field as follows:

$$\vec{E} = -\frac{\partial \vec{A}}{\partial t} - \vec{\nabla}\Phi \tag{23}$$

$$\vec{B} = \vec{\nabla} \times \vec{A} \tag{24}$$

$$S = \frac{1}{c^2}\frac{\partial \Phi}{\partial t} + \vec{\nabla} \cdot \vec{A} \tag{25}$$

then

$$\nabla A = -S + \vec{i} \cdot \left[\vec{B} - \frac{i}{c}\vec{E}\right] \tag{26}$$

Usually the Maxwell equations are defined by (in biquaternion form):

$$-\nabla^*(\nabla A + S) = \mu J \quad \text{or by} \tag{27}$$

$$\Box A - \nabla^* S = \mu J \quad \text{or by} \tag{28}$$

$$-\nabla^* \left(\vec{i} \cdot \left[\vec{B} - \frac{i}{c}\vec{E}\right]\right) = \mu J \tag{29}$$

It is easy to verify that these three equations are identical. Equation (31) can be expanded by applying the quaternion product:

$$\left(-\frac{i}{c}\frac{\partial}{\partial t} + \vec{i} \cdot \vec{\nabla}\right)\left(\vec{i} \cdot \left[\vec{B} - \frac{i}{c}\vec{E}\right]\right) =$$

$$-\vec{\nabla} \cdot \vec{B} + \frac{i}{c}\vec{\nabla} \cdot \vec{E} + \vec{i} \cdot \left[\vec{\nabla} \times \vec{B} - \frac{1}{c^2}\frac{\partial \vec{E}}{\partial t} - \frac{i}{c}\left(\frac{\partial \vec{B}}{\partial t} + \vec{\nabla} \times \vec{E}\right)\right] \tag{30}$$

The biquaternion equation is a short hand notation of the famous Maxwell equations in seperated scalar and vector form:

$$\frac{1}{c^2}\frac{\partial^2 \Phi}{\partial t^2} - \vec{\nabla}^2\Phi - \frac{\partial}{\partial t}\left(\frac{1}{c^2}\frac{\partial \Phi}{\partial t} + \vec{\nabla} \cdot \vec{A}\right) = c^2\mu\rho = \frac{\rho}{\epsilon} \tag{31}$$

$$\frac{1}{c^2}\frac{\partial^2 \vec{A}}{\partial t^2} - \vec{\nabla}^2\vec{A} + \vec{\nabla}\left(\frac{1}{c^2}\frac{\partial \Phi}{\partial t} + \vec{\nabla} \cdot \vec{A}\right) = \mu\vec{J} \tag{32}$$

$$\vec{\nabla} \cdot \vec{B} = 0 \tag{33}$$

$$\vec{\nabla} \cdot \vec{E} = c^2\mu\rho = \frac{\rho}{\epsilon} \tag{34}$$

$$\vec{\nabla} \times \vec{B} - \frac{1}{c^2}\frac{\partial \vec{E}}{\partial t} = \mu\vec{J} \tag{35}$$

$$\frac{\partial \vec{B}}{\partial t} + \vec{\nabla} \times \vec{E} = 0 \tag{36}$$

The Lorentz 4-force density biquaternion, $F = \frac{i}{c}P + \vec{i} \cdot \vec{F}$, is defined by the following biquaternion equation:

$$F = J(\nabla A + S) \tag{37}$$

Applying the quaternion product:

$$J(\nabla A + S) = \rho V(\nabla A + S) = \rho(ic + \vec{i} \cdot \vec{V})\left(\vec{i} \cdot \left[-\frac{i}{c}\vec{E} + \vec{B}\right]\right)$$

$$= -\vec{J} \cdot \vec{B} + \frac{i}{c}\vec{J} \cdot \vec{E} + \vec{i} \cdot \left[(\rho\vec{E} + \vec{J} \times \vec{B}) + ic(\rho\vec{B} - \frac{1}{c^2}\vec{J} \times \vec{E})\right] \tag{38}$$

In seperated scalar and vector form, the Lorentz force equation is:

$$0 = \vec{J} \cdot \vec{B} \tag{39}$$

$$P = \vec{J} \cdot \vec{E} \tag{40}$$

$$\vec{F} = \rho \vec{E} + \vec{J} \times \vec{B} \tag{41}$$

$$\vec{0} = \rho \vec{B} - \frac{1}{c^2} \vec{J} \times \vec{E} \tag{42}$$

In the biquaternion Lorentz force equation, J can be eliminated by substituting the biquaternion Maxwell equation:

$$F = \frac{1}{\mu} \nabla^* (\nabla A + S)(\nabla A + S) \tag{43}$$

By expanding the imaginary scalar part and the real vector part of this equation one finds the well known energy and momentum theorems:

$$\mu(\vec{J} \cdot \vec{E}) = -\frac{\partial}{2\partial t}\left[\frac{1}{c^2}E^2 + B^2\right] - \vec{\nabla} \cdot (\vec{E} \times \vec{B}) \tag{44}$$

$$\mu\left(\rho\vec{E} + \vec{J} \times \vec{B}\right) = \left[\frac{1}{c^2}\left((\vec{\nabla} \cdot \vec{E})\vec{E} + (\vec{\nabla} \times \vec{E}) \times \vec{E}\right) + (\vec{\nabla} \times \vec{B}) \times \vec{B}\right]$$
$$- \frac{1}{c^2}\frac{\partial(\vec{E} \times \vec{B})}{\partial t} \tag{45}$$

This demonstrates that electrodynamics can be cast in biquaternion form. By considering a non-zero real scalar and imaginary vector part in the current biquaternion, one can introduce the magnetic monopole and magnetic current. A magnetic 4-current is imaginary with respect to the electric 4-current. The total current can be called an 8-current. One can also consider an 8-potential or an 8-Lorentz force with a non-zero real scalar and imaginary vector part in A or in F. The biquaterion mathematics enables a logical treatment of such extentions of the theory of electrodynamics.

4 A gauge asymmetrical theory of electrodynamics that includes scalar field S

No matter if S=0, which is called the *Lorenz gauge* condition, the biquaternion Maxwell equation is invariant with respect to a gauge transformation of the biquaternion potential:

$$A \rightarrow A' = A - \nabla^* \Gamma \tag{46}$$

where Γ is an arbitrary scalar function. The easiest way to prove this is by showing that only the S field is changed by such a gauge transformation:

$$\nabla A \rightarrow \nabla A' = \nabla A - \nabla \nabla^* \Gamma = \nabla A + \Box \Gamma \tag{47}$$

and since $\Box \Gamma$ is strictly a scalar function, only the S field is transformed:

$$S \rightarrow S' = S + \nabla \nabla^* \Gamma = S - \Box \Gamma \tag{48}$$

Electrodynamics is gauge invariant, because the expression $(\nabla A + S)$ in each equation is invariant. It is asssumed that $\vec{\nabla} \cdot \vec{A}$ can be chosen freely, and that any potential can be gauge transformed into another potential such that the two potentials are not different with respect to measurable physics. Potentials can be transformed into potentials that satisfy a condition, such as the Lorenz condition [8] [9]: $S' = 0$. The latter requires that $\Box\Gamma = S$. If we only consider potentials that satisfy the Lorenz condition (S=0) then the gauge transformation Γ has to satisfy $\Box\Gamma = 0$. It is not clear if the wave-nature of Γ has a physical meaning. In case S=0, then the Maxwell equation is defined also by $\Box A = \mu J$, which is called the Lorenz inhomogeneous wave equation.

Another type of gauge transformation exists that gives Lorenz's inhomogeneous wave equation:

$$\mu J \quad \rightarrow \quad \mu J' = \mu J - \nabla^* \Gamma \tag{49}$$

where Γ is an arbitrary scalar field. This is a current gauge transformation. The Maxwell equation is not invariant with respect to this transformation. A regauge of the current J is the special transformation $\Gamma = S$. After a current regauge, the resulting Maxwell equation automatically has the form of the Lorenz inhomogeneous wave equation in case it is described in terms of the potentials:

$$-\nabla^*(\nabla A + S) = \mu J - \nabla^* S \quad \text{or} \tag{50}$$
$$\Box A = \mu J \quad \text{or} \tag{51}$$
$$-\nabla^* \left(-S + \vec{i} \cdot \left[\vec{B} - \tfrac{1}{c}\vec{E}\right]\right) = \mu J \tag{52}$$

Therefore, solutions of the Maxwell equation are also solutions of the inhomogeneous wave equation. The transformed Maxwell equation also contains scalar field S and is a generalisation of the original Maxwell equation. Rewriting this equation in seperate scalar and vector equations results into the *generalised* Maxwell equations:

$$\frac{1}{c^2}\frac{\partial^2 \Phi}{\partial t^2} - \vec{\nabla}^2 \Phi = c^2 \mu \rho = \frac{\rho}{\epsilon} \tag{53}$$

$$\frac{1}{c^2}\frac{\partial^2 \vec{A}}{\partial t^2} - \vec{\nabla}^2 \vec{A} = \mu \vec{J} \tag{54}$$

$$\vec{\nabla} \cdot \vec{E} + \frac{\partial S}{\partial t} = c^2 \mu \rho = \frac{\rho}{\epsilon} \tag{55}$$

$$\vec{\nabla} \cdot \vec{B} = 0 \tag{56}$$

$$\frac{\partial \vec{B}}{\partial t} + \vec{\nabla} \times \vec{E} = 0 \tag{57}$$

$$\vec{\nabla} \times \vec{B} - \frac{1}{c^2}\frac{\partial \vec{E}}{\partial t} - \vec{\nabla}S = \mu \vec{J} \tag{58}$$

Analogous to the addition of the displacement current, which allowed Maxwell to derive the homogeneous field wave equations, the addition of the scalar field related charge and current terms allow for the derivation of the inhomogeneous potential wave equations without the Lorenz gauge. Because the potential gauge freedom is lost, one should regard the expression S as a real and physical field. It is doubtful if potential gauge freedom exists, which means that $\nabla \cdot \vec{A}$ has an arbitrary value, considering the fact that S can be deduced from the technical specifications of electrodynamic equipment. We assume it is possible that S can be measured, just like the electric or magnetic field. Next, it is shown that a natural current gauge must satisfy the condition $\Box S_g = 0$. First, the gauge transform of the partial derivative of current is described by:

$$\nabla(\mu J) \quad \rightarrow \quad \nabla(\mu J') = \nabla(\mu J) - \nabla\nabla^* S_g = \nabla(\mu J) + \Box S_g \tag{59}$$

Since $\square S_g$ is strictly a scalar, only the scalar part of $\nabla(\mu J)$ is transformed:

$$\mu(\frac{\partial \rho'}{\partial t} + \vec{\nabla} \cdot \vec{J}') = \mu(\frac{\partial \rho}{\partial t} + \vec{\nabla} \cdot \vec{J}) + \square S_g \tag{60}$$

Because charge is conserved, the scalar part of $\nabla(\mu J)$ and of $\nabla(\mu J')$ is zero:

$$\mu(\frac{\partial \rho}{\partial t} + \vec{\nabla} \cdot \vec{J}) = \mu(\frac{\partial \rho'}{\partial t} + \vec{\nabla} \cdot \vec{J}') = 0 \tag{61}$$

And therefore $\square S_g = 0$. Hence, a natural regauge is possible only if $\square S = \square S_g = 0$. The current gauge condition has a straightforward and physical interpretation: the conservation of charge. If an electrodynamical system has a natural tendency to satisfy the Lorenz inhomogeneous wave equation, then this can be described by a conditional current regauge, such that charge conservation is not violated. By assuming that the scalar is zero everywhere (S=0) the original Maxwell equations are found. The condition S=0 is not a free-to-choose gauge condition, but it simply is a physical condition. A physical interpretation of the current gauge transform is the following.

Within the framework of quantum electrodynamics, charge polarizations are a possibility in vacuum, and this means that the charge density and current density in the vacuum can fluctuate. A current gauge transformation is the classical equivalent of this concept. A regauge of the biquaternion current can be regarded as the change in vacuum charge-current density, such that the potentials associated with both real and "virtual" charge-current, are solutions of the Lorenz inhomogeneous wave equation. During a current regauge the microscopic "virtual" charge-current fluctuations are forced by the presence of a real charge-current to form an orderly pattern on a larger scale. The current regauge is therefore a physical process, instead of a pure mathematical transformation, and we suggest that the entropy of the current regauge process is negative.

4.1 Vacuum field wave solutions

Assuming the current gauge condition is true, the partial derivatives of the left and right hand side of the Maxwell equation are gauge invariant. After current regauging, this equation is:

$$\nabla \square A = -\nabla \nabla \nabla^* A = -\nabla \nabla^* \nabla A = \square \left(-S + \vec{i} \cdot \left[-\frac{i}{c}\vec{E} + \vec{B} \right] \right) = \mu \nabla J \tag{62}$$

and reformulated into seperate scalar and vector wave equations:

$$\frac{1}{c^2}\frac{\partial^2 S}{\partial t^2} - \vec{\nabla}^2 S = \mu\left(\frac{\partial \rho}{\partial t} + \vec{\nabla} \cdot \vec{J}\right) = 0 \tag{63}$$

$$\frac{1}{c^2}\frac{\partial^2 \vec{E}}{\partial t^2} - \vec{\nabla}^2 \vec{E} = \mu\left(-\frac{\partial \vec{J}}{\partial t} - c^2\vec{\nabla}\rho\right) \tag{64}$$

$$\frac{1}{c^2}\frac{\partial^2 \vec{B}}{\partial t^2} - \vec{\nabla}^2 \vec{B} = \mu\left(\vec{\nabla} \times \vec{J}\right) \tag{65}$$

In case J=0 the latter equations become the homogeneous field wave equations. It is well known that, for S=0 and J=0, a transversal electromagnetic wave is a natural solution of the Maxwell equations. Now suppose that $\vec{B} = 0$ and J=0. From the generalised Maxwell equations the following wave equations can be deduced:

$$\frac{1}{c^2}\frac{\partial^2 S}{\partial t^2} - \vec{\nabla}^2 S = 0 \tag{66}$$

$$\frac{1}{c^2}\frac{\partial^2 \vec{E}}{\partial t^2} - \vec{\nabla}^2 \vec{E} = 0 \qquad (67)$$

that have the solutions:

$$\vec{E} = \vec{E}_0 e^{i(\vec{k}\cdot\vec{r}-\omega t)} \qquad (68)$$

$$S = S_0 e^{i(\vec{k}\cdot\vec{r}-\omega t)} \qquad (69)$$

$$\frac{\omega}{c^2}\vec{E} = \vec{k}S \qquad (70)$$

$$\vec{k}\cdot\vec{E} = \omega S \qquad (71)$$

This is a longitudinal electroscalar wave (LES wave). The possible existence of such a wave might be the subject of experimentation, and therefore our theory is testable. Associated with a LES wave is an energy flow density vector $\vec{E}S$. In the next sections it is shown that this energy flow density vector is part of an extended energy theorem. A more general set of vacuum wave equations can be found via the following equation

$$\vec{\nabla}S + \frac{k}{c^2}\frac{\partial \vec{E}}{\partial t} = 0 \qquad (72)$$

After combining the extended Maxwell equations with this extra field equation, we find the following vacuum wave equations

$$\vec{\nabla}\cdot\vec{\nabla}S - \frac{k}{c^2}\frac{\partial^2 S}{\partial t^2} = 0 \qquad (73)$$

$$\vec{\nabla}\vec{\nabla}\cdot\vec{E} - \frac{k}{c^2}\frac{\partial^2 \vec{E}}{\partial t^2} = 0 \qquad (74)$$

$$\vec{\nabla}\times\vec{\nabla}\times\vec{B} + \frac{(1-k)}{c^2}\frac{\partial^2 \vec{B}}{\partial t^2} = 0 \qquad (75)$$

$$\vec{\nabla}\times\vec{\nabla}\times\vec{E} + \frac{(1-k)}{c^2}\frac{\partial^2 \vec{E}}{\partial t^2} = 0 \qquad (76)$$

$$\qquad (77)$$

These equations are valid wave equations only if $k \in [0,1]$ and have superluminal solutions for $k \in <0,1>$. The longitudinal part of the electric field interacts with the S field, forming a LES wave with speed $\frac{c}{\sqrt{k}}$. The transversal part of the electric field interacts with the B field, forming a TEM wave with speed $\frac{c}{\sqrt{1-k}}$. For $k = \frac{1}{2}$ both waves have speed $c\sqrt{2}$. For $k=0$ the TEM wave solution is the usual luminal TEM wave with speed c, and the LES wave speed is infinite. This means there might be a simultaneous coexistance of retarded and immediate action at-a-distance. Vice verse, for $k=1$ the speed of the LES wave is c, and the speed of the TEM wave is infinite, and this might be the classical equivalent of tunneling photons. Such photons are accompanied by LES waves that have phase velocity of c, according to this theory. If one of the wave types has infinite speed then the other wave type has speed c, and this adds new meaning to the value c. In [15] a superluminal microwave with average speed of 4.7c was measured by G. Nimtz. This value is close to $\sqrt{22}c$, and this suggests a value of $k = \frac{21}{22}$ and a LES wave speed of $\sqrt{\frac{22}{21}}c \approx c$. G.F. Ignatiev [16] measured a signal speed of 1.12c, and this value is close to $\sqrt{\frac{5}{4}}c$. This suggest a value of $k = \frac{1}{5}$ and a LES wave with speed $\sqrt{5}c \approx 2.24c$.

4.2 The generalised Lorentz force

A generalised Lorentz force definition is expressed by the following equation:

$$F = J\nabla A \tag{78}$$

In expression $J\nabla A$ the scalar field S is no longer cancelled, while in $J(\nabla A + S)$ in classical Electrodynamics the S field is cancelled. This is similar to the generalisation of the Maxwell equation. In the next section it is shown that this generalisation of the Lorentz force is also based on the current regauge transformation. Rewriting the latter biquaternion equation into separate scalar and vector equations:

$$\vec{J} \cdot \vec{B} = 0 \tag{79}$$

$$\vec{J} \cdot \vec{E} - \rho c^2 S = P \tag{80}$$

$$\rho\vec{E} + \vec{J} \times \vec{B} - \vec{J}S = \vec{F} \tag{81}$$

$$\vec{B} - \frac{\vec{v}}{c^2} \times \vec{E} = 0 \tag{82}$$

The first equation shows two power flow terms: $\vec{J} \cdot \vec{E}$ and $\rho c^2 S$. The first term is the electrical energy flow that is usually associated with a current $\vec{J}$ in a wire, and that compensates for the dissipated energy. The second power flow term is new and has to be associated with a 'static' charge. It is like the Zero Point Energy exchange between a charge and the vacuum. The energy flows inwards and outwards with respect to the volume of the charge. The fourth equation defines the vector Lorentz force and it contains an extra term $\vec{J}S$. This term is similar to a radiation reaction force, but it is independent of the acceleration of the charge.

It is also interesting to examine the special case of $S\vec{v} = \vec{E}$. For this case the Lorentz force quaternion reduces to:

$$\rho(v^2 - c^2)S = P \tag{83}$$

$$\vec{0} = \vec{F} \tag{84}$$

This means that a charge can exchange energy with an external electromagnetic potential due to the S-field, despite of the absence of an electromagnetic force. The speed of the particle is constant in this situation. If S and $\vec{E}$ are also wave solutions then the relation $S\vec{v} = \vec{E}$ shows a longitudinal electroscalar wave with speed $\vec{v}$. The energy exchange diminishes with $v^2 S$ and becomes zero when v approaches c. In [14] an alternative description of the De Brogly wave is given, based on a new principle of *intrinsic* (belonging to the particle) potential energy of electromagnetic origin. The wave nature of a particle is in essence a periodic transformation of kinetic energy into intrinsic potential energy, and vice verse. The intrinsic potential energy of a particle might as well be electroscalar in nature. In other words: the LES wave with $S\vec{v} = \vec{E}$ might be a new description of the De Brogly wave if we consider the S field and the E field as intrinsic to the particle.

4.3 Generalised energy and momentum theorems

Within the generalised Lorentz force equation $J\nabla A = F$ we can substitute for J its definition in terms of potentials, $\frac{1}{\mu}\Box A$. Then we get:

$$\Box A \nabla A = \mu F \tag{85}$$

When we evaluate the imaginary scalar part of this equation in terms of fields and sources, we get the following energy equation:

$$\mu(\vec{E} \cdot \vec{J} - \rho c^2 S) = -\frac{\partial}{2\partial t}\left[S^2 + \frac{1}{c^2}E^2 + B^2\right] - \vec{\nabla} \cdot (\vec{E} \times \vec{B} + \vec{E}S) \tag{86}$$

349

The term $\vec{E}S$ represents an extra power flux vector that can be associated with the longitudinal electroscalar wave. The Poynting vector $\vec{E} \times \vec{B}$ is usually associated with the transversal electromagnetic wave. The energy term S^2 is the energy stored in the field S. The real vector part of the biquaternion equation is:

$$\mu\left(\rho\vec{E} + \vec{J} \times \vec{B} - \vec{J}S\right) = \left[\frac{1}{c^2}\left((\vec{\nabla} \cdot \vec{E})\vec{E} + (\vec{\nabla} \times \vec{E}) \times \vec{E}\right) + (\vec{\nabla} \times \vec{B}) \times \vec{B}\right] +$$
$$\left[S\vec{\nabla}S - \vec{\nabla} \times (S\vec{B})\right] + \frac{1}{c^2}\frac{\partial(\vec{E}S - \vec{E} \times \vec{B})}{\partial t} \qquad (87)$$

This equation is the extended momentum theorem in the generalised Maxwell theory. Usually, Maxwell's stress tensor represents the terms between the first pair of square brackets. It can be generalised, such that it represents also the terms between the second pair of square brackets. The power flow terms of both TEM waves and LES waves are present also in this equation.

These equations can be derived also by applying the current regauge to the original energy and momentum theorems:

$$\vec{J} \cdot \vec{E} \quad \rightarrow \quad \left(\vec{J} + \frac{1}{\mu}\vec{\nabla}S\right) \cdot \vec{E} =$$
$$\vec{J} \cdot \vec{E} + \frac{1}{\mu}(\vec{\nabla}S) \cdot \vec{E} =$$
$$\vec{J} \cdot \vec{E} + \frac{1}{\mu}\vec{\nabla} \cdot (S\vec{E}) - S\vec{\nabla} \cdot \vec{E} =$$
$$\vec{J} \cdot \vec{E} + \frac{1}{\mu}\left(\vec{\nabla} \cdot (S\vec{E}) - S\vec{\nabla} \cdot \vec{E}\right) =$$
$$\vec{J} \cdot \vec{E} + \frac{1}{\mu}\left(\vec{\nabla} \cdot (S\vec{E}) - S\left(\frac{\rho}{\epsilon} - \frac{\partial S}{\partial t}\right)\right) =$$
$$\left(\vec{J} \cdot \vec{E} - Sc^2\rho\right) + \frac{1}{\mu}\left(\vec{\nabla} \cdot (S\vec{E}) + \frac{\partial(S^2)}{2\partial t}\right) \qquad (88)$$

$$\rho\vec{E} + \vec{J} \times \vec{B} \quad \rightarrow \quad \left(\rho - \epsilon\frac{\partial S}{\partial t}\right)\vec{E} + \left(\vec{J} + \frac{1}{\mu}\vec{\nabla}S\right) \times \vec{B} =$$
$$\left(\rho\vec{E} + \vec{J} \times \vec{B}\right) - \epsilon\frac{\partial S}{\partial t}\vec{E} + \frac{1}{\mu}\vec{\nabla}S \times \vec{B} =$$
$$\left(\rho\vec{E} + \vec{J} \times \vec{B}\right) - \epsilon\frac{\partial(\vec{E}S)}{\partial t} + \frac{1}{\mu}\vec{\nabla} \times (\vec{B}S) + \frac{S}{\mu}\left(\frac{\partial\vec{E}}{c^2\partial t} - \vec{\nabla} \times \vec{B}\right) =$$
$$\left(\rho\vec{E} + \vec{J} \times \vec{B}\right) - \epsilon\frac{\partial(\vec{E}S)}{\partial t} + \frac{1}{\mu}\vec{\nabla} \times (\vec{B}S) + S\left(-\vec{J} - \frac{1}{\mu}\vec{\nabla}S\right) =$$
$$\left(\rho\vec{E} + \vec{J} \times \vec{B} - S\vec{J}\right) + \frac{1}{\mu}\left(-\frac{\partial(\vec{E}S)}{c^2\partial t} + \vec{\nabla} \times (\vec{B}S) - S\vec{\nabla}S\right) \qquad (89)$$

This shows that the current regauge not only changes the biquaternion Lorentz force, but also introduces extra terms with respect to the field energy, radiation energy flow and the stress tensor.

4.4   The source of scalar field S

Including the extra scalar field in the Maxwell equations rises the question how to induce this field. Keeping in mind that the potentials always satisfy the Lorenz inhomogeneous wave equations after the current regauge, and that the retarded potentials are solutions of these wave equations, it is necessary to verify if a scalar field can be derived from the retarded potentials:

$$\Phi(\vec{x}, t) = \frac{1}{4\pi\epsilon_0} \int_V \frac{\rho(\vec{x}', t'_{ret})}{|\vec{x} - \vec{x}'|} d^3x' \tag{90}$$

$$\vec{A}(\vec{x}, t) = \frac{\mu_0}{4\pi} \int_V \frac{\vec{j}(\vec{x}', t'_{ret})}{|\vec{x} - \vec{x}'|} d^3x' \tag{91}$$

$$t'_{ret} = t - \frac{|\vec{x} - \vec{x}'|}{c} \tag{92}$$

This can be done by Fourier analysis. The Fourier transformed retarded potentials are defined by:

$$\Phi(\vec{x}, \omega) = \frac{1}{4\pi\epsilon_0} \int_V \rho(\vec{x}', \omega) \frac{e^{ik|\vec{x}-\vec{x}'|}}{|\vec{x} - \vec{x}'|} d^3x' \tag{93}$$

$$\vec{A}(\vec{x}, \omega) = \frac{\mu_0}{4\pi} \int_V \vec{j}(\vec{x}', \omega) \frac{e^{ik|\vec{x}-\vec{x}'|}}{|\vec{x} - \vec{x}'|} d^3x' \tag{94}$$

with

$$\rho(\vec{x}', \omega) = \frac{1}{2\pi} \int_{-\infty}^{\infty} \rho(\vec{x}', t) e^{i\omega t} dt \tag{95}$$

$$\vec{j}(\vec{x}', \omega) = \frac{1}{2\pi} \int_{-\infty}^{\infty} \vec{j}(\vec{x}', t) e^{i\omega t} dt \tag{96}$$

The Fourier transformed scalar field is defined as

$$\begin{aligned}
S(\vec{x}, \omega) &= \frac{1}{c^2} \frac{\partial \Phi(\vec{x}, \omega)}{\partial t} + \vec{\nabla} \cdot \vec{A}(\vec{x}, \omega) \\
&= -i\omega \frac{1}{4\pi c^2 \epsilon_0} \int_V \rho(\vec{x}', \omega) \frac{e^{ik|\vec{x}-\vec{x}'|}}{|\vec{x} - \vec{x}'|} d^3x' + \vec{\nabla} \cdot \left(\frac{\mu_0}{4\pi} \int_V \vec{j}(\vec{x}', \omega) \frac{e^{ik|\vec{x}-\vec{x}'|}}{|\vec{x} - \vec{x}'|} d^3x' \right) \\
&= -i\omega \frac{\mu_0}{4\pi} \int_V \rho(\vec{x}', \omega) \frac{e^{ik|\vec{x}-\vec{x}'|}}{|\vec{x} - \vec{x}'|} d^3x' + \frac{\mu_0}{4\pi} \int_V \vec{j}(\vec{x}', \omega) \cdot \nabla \left(\frac{e^{ik|\vec{x}-\vec{x}'|}}{|\vec{x} - \vec{x}'|} \right) d^3x'
\end{aligned} \tag{97}$$

By using the Fourier transform of the continuity equation

$$\vec{\nabla}' \cdot \vec{j}(\vec{x}', \omega) - i\omega\rho(\vec{x}', \omega) = 0 \tag{98}$$

it is possible to evaluate further the first integral:

$$\begin{aligned}
-i\omega \frac{\mu_0}{4\pi} \int_V \rho(\vec{x}', \omega) \frac{e^{ik|\vec{x}-\vec{x}'|}}{|\vec{x} - \vec{x}'|} d^3x' &= -\frac{\mu_0}{4\pi} \int_V \left(\vec{\nabla}' \cdot \vec{j}(\vec{x}', \omega) \right) \frac{e^{ik|\vec{x}-\vec{x}'|}}{|\vec{x} - \vec{x}'|} d^3x' \\
&= -\frac{\mu_0}{4\pi} \int_V \vec{\nabla}' \cdot \left(\vec{j}(\vec{x}', \omega) \frac{e^{ik|\vec{x}-\vec{x}'|}}{|\vec{x} - \vec{x}'|} \right) d^3x' \\
&\quad + \frac{\vec{\mu_0}}{4\pi} \int_V \vec{j}(\vec{x}', \omega) \cdot \nabla' \left(\frac{e^{ik|\vec{x}-\vec{x}'|}}{|\vec{x} - \vec{x}'|} \right) d^3x' \\
&= \frac{\vec{\mu_0}}{4\pi} \int_V \vec{j}(\vec{x}', \omega) \cdot \nabla' \left(\frac{e^{ik|\vec{x}-\vec{x}'|}}{|\vec{x} - \vec{x}'|} \right) d^3x' \tag{99}
\end{aligned}$$

In the last step Gauss's theorem was applied: the first integral term vanishes if $\vec{j}(\vec{x}', \omega)$ is assumed to be limited and tends to zero at large distances. The final expression for the scalar field in frequency domain is:

$$
\begin{aligned}
S(\vec{x}, \omega) &= \frac{\vec{\mu_0}}{4\pi} \int_V \vec{j}(\vec{x}', \omega) \cdot \nabla' \left(\frac{e^{ik|\vec{x}-\vec{x}'|}}{|\vec{x}-\vec{x}'|} \right) d^3x' + \frac{\vec{\mu_0}}{4\pi} \int_V \vec{j}(\vec{x}', \omega) \cdot \nabla \left(\frac{e^{ik|\vec{x}-\vec{x}'|}}{|\vec{x}-\vec{x}'|} \right) d^3x' \\
&= 0
\end{aligned}
\tag{100}
$$

The inverse Fourier transform of this expression gives us the expression for the S field in time domain: $S(\vec{x}, t) = 0$. In a similar way one can derive the same result for the advanced potentials. Also in case that S is derived from the Liénard-Wiechert potentials, which are a special case retarded potentials, one finds that S=0. However, the retarded and advanced potentials are not the most general solutions of the Lorenz inhomogeneous wave equations. An interesting special case for getting a non zero S is by setting $\vec{E} = \vec{0}$ and $\vec{B} = \vec{0}$. In this case the field equations are

$$
\frac{\partial S}{\partial t} = \frac{\rho}{\epsilon}
\tag{101}
$$

$$
-\vec{\nabla} S = \mu \vec{J}
\tag{102}
$$

Adding the gradient of the first equation to the time differential of the second equation gives

$$
0 = \frac{1}{\epsilon} \vec{\nabla} \rho + \mu \frac{\partial \vec{J}}{\partial t}
\tag{103}
$$

The divergence of this equation (and using the continuity of charge equation) finally gives

$$
0 = \vec{\nabla} \cdot \vec{\nabla} \rho - \frac{1}{c^2} \frac{\partial^2 \rho}{\partial t^2}
\tag{104}
$$

which is a charge density wave. Also in this case the scalar field is a wave, but directly associated with a charge density wave. In classical electrodynamics there cannot be sources without the presence of an electric or magnetic field. In this theory it is possible to have sources that only induce a scalar field. This example shows that the scalar field can be sourced directly by a dynamic charge density distribution.

5   Conclusions

It is possible to describe classical electrodynamics in the form of two biquaternion equations. This form is very useful in order to generalise electrodynamics. Generalising the Maxwell equation by introducing an extra scalar field is comparable with Maxwell's introduction of the displacement current that allowed for the derivation of the homogeneous field wave equations. This theory predicts the existance of longitudinal electroscalar waves in vacuum. Such a wave might be used to transmit and receive signals. The power density vector of LES waves is $\vec{E}S$, thus energetic and wireless signals might be transmittable in LES wave form and received at a distance.

6   Acknowledgement

The authors are grateful to V. Onoochin, H. Puthoff and M. Ibison for valuable discussions and scientific comments.

References

[1] W.R. Hamilton, *On a new Species of Imaginary Quantities connected with a theory of Quaternions*. Proceedings of the Royal Irish Academy 2 (13 November 1843) 424-434

[2] J.C. Maxwell, *A Treatise on Electricity & Magnetism*, , (1893) Dover Publications, New York ISBN 0-486-60636-8 (Vol. 1) & 0-486-60637-6 (Vol. 2)

[3] D. Sweetser and G. Sandri, *Maxwells vision: Electromagnetism with Hameltons quaternions*, http://www.emis.de/proceedings/QSMP99/sweeters.pdf

[4] A. Waser, *Quaternions in electrodynamics*, http://www.aw-verlag.ch/Documents/QuaternionsInElectrodynamicsEN02.pdf

[5] J.W. Marshall, *Quaternions as 4-Vectors*, American Journal of Physics 24 (1956) 515-522

[6] R. Fueter, Comm. Math. Helv., 7, 307, (1934-35), 8, 371, (1936-37)

[7] Y. Yong-Gwan, *On the Nature of Relativistic Phenomena*, Apeiron 6 Nr.3-4 (July-Oct.1999)

[8] L.V. Lorenz, *Über die Identität der Schwingungen des Lichts mit den elektrischen Strömen*, Poggendorfs Annalen der Physik, Bd. 131, 1867.

[9] B. Riemann, *Schwere, Elektrizität und Magnetismus* (1861),K. Hattendorf, Ed. Hannover, Germany: Carl Rümpler, 1880.

[10] A. Liénard, L' Éclairage Électique 165 (1898); E. Wiechert, Ann. Phys. 4 676 (1901).

[11] L.D. Landau and E.M. Lifshitz, Teoria Polia (Nauka, Moscow, 1973) [English translation: *The Classical Theory of Fields* (Pergamon, Oxford, 1975)].

[12] A.E. Chubykalo and S.J. Vlaev, *Double (implicit and explicit) dependence of the electromagnetic field of an accelerated charge on time: Mathematical and physical analysis of the problem* , http://xxx.lanl.gov/abs/physics/9803037

[13] D.G. Boulware, *Radiation from a Uniformly Accelerated Charge* , Annals of Physics 124, 169 (1980).

[14] W.A. Hofer, *Beyond Uncertainty: the internal structure of electrons and photons* , http://xxx.lanl.gov/abs/quant-ph/9611009, TU Wien, November 1996

[15] G. Nimtz, *Superluminal signal velocity*, http://xxx.lanl.gov/abs/physics/9812053

[16] G. F. Ignatiev and V. A. Leus, *On a superluminal transmission at the phase velocities*, http://www.truth.myweb.nl/Art-p16.pdf

19 WARDENCLYFFE

How does Tesla's World Wireless tower work?
Gary Peterson[1]

An electric current flowing through a conductor carries electrical energy. Earth is an electrical conductor, nearly spherical in shape, insulated in space. It possesses an electric charge relative to the upper atmosphere beginning at about 31 miles or 50 kilometers elevation. When a second body, directly adjacent to Earth, is charged and discharged in rapid succession this causes an equivalent variation of its electrostatic charge resulting in the passage of electric current through the ground. The Tesla coil magnifying transmitter is an electrical machine specifically designed to create as large a displacement as possible of Earth's natural electric charge. It does this by alternately charging and discharging the oscillator's elevated terminal capacitance at a specific frequency, periodically altering the electrostatic charge of the earth, and consequently, with sufficient power, the pressure over its entire surface.

> A connection to earth, either directly or through a condenser is essential. – "The Disturbing Influence of Solar Radiation on the Wireless Transmission of Energy," *Electrical Review and Western Electrician*, July 6, 1912.

The placement of a grounded Tesla coil receiver tuned to the same frequency as the transmitter at another point on the surface results in the flow of electric current through the earth between the two, "while an equivalent electric displacement occurs in the atmosphere." The electrical displacement takes place predominantly by electrical conduction through the oceans, and metallic ore bodies and similar subsurface structures. The electrical displacement may also be by means of electrostatic induction through the more dielectric regions such as quartz deposits and other non-conducting minerals.

> There is no appreciable radiation and the receiver is energized through the earth while an equivalent electric displacement occurs in the atmosphere. – "The Disturbing Influence of Solar Radiation on the Wireless Transmission of Energy, " *Electrical Review and Western Electrician*, July 6, 1912.

This current can be used at the receiver to drive an electrical load, which in the case of an individual World Wireless Telecommunications System receiver is a sensitive device using only a small amount of energy. This energy transfer technique is suitable for wireless broadband telecommunications in conjunction with low level energy harvesting

---

[1] Mr. Peterson is an independent researcher specializing in the Tesla wireless system. He is the owner of 21st Century Books, www.tfcbooks.com, publishing critical material about Nikola Tesla and his work.

for battery charging, and possibly for transmission of electrical power in industrial quantities.

What was the intended purpose of the Long Island Wardenclyffe tower and the additional towers to be built around the world?

The Wardenclyffe Tower facility was a commercial venture designed for trans-Atlantic wireless telecommunications, broadcasting and for proof-of-concept demonstrations of global wireless power transmission. These applications were mentioned by Tesla in 1923 in the Wardenclyffe Foreclosure Proceedings.

> Well, the primary purpose of the tower, your Honor, was to telephone, to send the human voice and likeness around the globe through the instrumentality of the earth. That was my discovery that I announced in 1893, and now all the wireless plants are doing that. There is no other system being used. And the idea was to reproduce this apparatus and then connect it just with a central station and telephone office, so that you may pick up your telephone and if you wanted to talk to a telephone subscriber in Australia you would simply call up that plant and the plant would connect immediately with that subscriber, no matter where in the world, and you could talk to him. And I had contemplated to have press messages, stock quotations, pictures for the press and these reproductions of signatures, checks and everything transmitted from there throughout the world. The tower was so designed that I could apply to it any amount of power and I was planning to give a demonstration in the transmission of power which I have so perfected that power can be transmitted clear across the globe with a loss of not more than five per cent, and that plant was to serve as a practical demonstration. And then I was going to interest people in a larger project and the Niagara people had given me 10,000-horse power. – Nikola Tesla On His Work With Alternating Currents and Their Application to Wireless Telegraphy, Telephony, and Transmission of Power, Leland I. Anderson, Editor, 21st Century Books, 1992.

Earlier Tesla had stated,

> It is intended to give practical demonstrations of these principles with the plant illustrated. As soon as completed, it will be possible for a business man in New York to dictate instructions, and have them instantly appear in type at his office in London or elsewhere. He will be able to call up, from his desk, and talk to any telephone subscriber on the globe, without any change whatever in the existing equipment. An inexpensive instrument, not bigger than a watch, will enable its bearer to hear anywhere, on sea or land, music or song, the speech of a political leader, the address of an eminent man of science, or the sermon of an eloquent clergyman, delivered in some other place, however distant. In the same manner any picture, character, drawing, or print can be transferred from one to another place. Millions of such instruments can be operated from but one plant of this kind. More important than all of this, however, will be the

transmission of power, without wires, which will be shown on a scale large enough to carry conviction. These few indications will be sufficient to show that the wireless art offers greater possibilities than any invention or discovery heretofore made, and if the conditions are favorable, we can expect with certitude that in the next few years wonders will be wrought by its application. – "The Future of the Wireless Art," Wireless Telegraphy & Telephony, 1908, pp. 67-71 Walter W. Massie & Charles R. Underhill.

Tesla intended to use a 200 kilowatt Westinghouse generator for the Wardenclyffe transmitter power supply. Factoring in an estimated 100 H.P. or 75 kilowatt terrestrial transmission-line loss, this leaves around 125 kW to work with. When demonstrating wireless power transfer the plant could have run the equivalent of about 100 modern-day toaster ovens. Stated another way, the facility would have been capable of powering the heaters, lighting and appliances of a little more than a single all-electric residential house when demonstrating wireless power transmission.

The Wardenclyffe facility was erected in 1901 at Wardenclyffe-On-Sound (now Shoreham) on Long Island, New York. Built entirely of wood, except for spar junction plates and the 55-ton skeleton spheroid at the top, the tower was designed so that every piece of framing could be taken out at any time and replaced if necessary. It was to be the first of many wireless installations constructed near population centers around the world. If plans had moved forward without interruption, the Long Island prototype would have been followed by a second plant built in the British Isles, on the west coast of Scotland near Glasgow. Each facility would include a magnifying transmitter of a design loosely based upon the apparatus assembled at the Colorado Springs Experimental Station in 1899.

> The plant in Colorado was merely designed in the same sense as a naval constructor designs first a small model to ascertain all, the quantities before he embarks on the construction of a big vessel. I had already planned most of the details of the commercial plant, subsequently put up at Long Island, except that at that time the location was not settled upon. The Colorado plant I have used in determining the construction of the various parts, and the experiments which were carried on there were for the practical purpose of enabling me to design the transmitters and receivers which I was to employ in the large commercial plant subsequently erected. – Nikola Tesla On His Work With Alternating Currents and Their Application to Wireless Telegraphy, Telephony, and Transmission of Power, Leland I. Anderson, Editor, 21st Century Books, 1992, p. 170.

On the next page is a wiring diagram of the preferred wireless magnifying transmitter design developed by Tesla at the Colorado Springs Experimental Station in 1899. The oscillator's primary LC circuit is not shown.

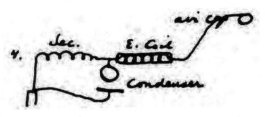

The schematic is redrawn below with standardized labeling of various components.

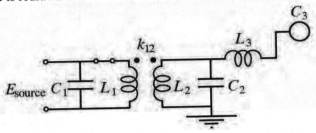

Primary capacitor C1 and primary inductor L1, and secondary capacitor C2 and secondary inductor L2 comprise the primary and secondary tuned LC circuits of what is called the master oscillator transformer. Extra coil inductor L3 and elevated terminal capacitance C3 comprise the transmitting element that is connected through the tuned secondary lumped LC driver circuit to a robust ground terminal connection.

How did Tesla envision energy transmission by means of his system?

Between 1897 and 1902 Tesla applied for ten patents directly related to his wireless system. They describe two different propagation modes that he claimed can be excited by means of his wireless system apparatus. The first involves electrical conduction through the upper atmosphere with a return circuit through Earth. The second depends upon the transmission of sufficient alternating current energy into Earth so it responds by vibrating electrically. In 1932 journalist John J. O'Neill conducted an interview with Tesla in which he talked about these two different methods of using his wireless system apparatus for the transmission of electrical energy.

> I also asked him if he is still at work on the project which he inaugurated in the '90's of transmitting power wirelessly anywhere on earth. He is at work on it, he said, and it could be put into operation. . . . He at that time announced two principles which could be used in this project. In one the ionizing of the upper air would make it as good a conductor of electricity as a metal. In the other the power is transmitted by creating "standing waves" in the earth by charging the earth with a giant electrical oscillator that would make the earth vibrate electrically in the same way a bell vibrates mechanically when it is struck with a hammer. . . . – "Tesla Cosmic Ray Motor May Transmit Power 'Round' Earth," Brooklyn Eagle, July 10, 1932.

Terrestrial Transmission Line with Atmospheric Return Method

Present day earth return AC electrical power transmission systems rely on current flowing through the earth plus a single wire insulated from the earth to complete the circuit. In emergencies high-voltage direct current power transmission systems can also operate in the 'single wire with earth return' mode. Elimination of the raised insulated wire, and transmission of high-potential alternating current through the earth with an atmospheric return circuit is the basis of the atmospheric conduction method of wireless electrical power transmission.

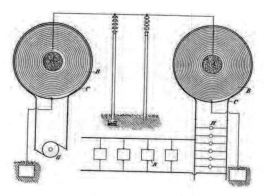

A pair of Tesla coils can be used to transmit and receive electrical energy using the earth as a conductor. The one-wire with ground return electrical power transmission system shown here relies on current flowing through a single elevated wire plus current flowing through the earth to complete the circuit.

The process is essentially the same as transmitting electricity by conduction through a wire. The earth itself is one of the conducting media involved in this implementation of "the disturbed charge of ground and air system." The other medium is the atmosphere above approximately 3 miles (4.8 km) elevation. While not an ohmic conductor, in this region of the troposphere and upwards, the density or pressure is sufficiently reduced to so that, according to Tesla's theory, the atmosphere's insulating properties can be easily impaired thus allowing an electric current to flow. Theory further asserts that the conducting region is developed through the process of atmospheric ionization in which the affected portions thereof are changed to plasma. The presence of the magnetic fields developed by each plant's helical resonator suggests that an embedded magnetic field and flux linkage might also be involved, with flux linkage with Earth's natural magnetic field being a possibility. The atmosphere below 3 miles is also a propagating medium for a portion of the above-ground circuit. Being an insulating medium, electrostatic induction would be involved rather than true electrical conduction.

The earth is 4,000 miles radius. Around this conducting earth is an atmosphere. The earth is a conductor; the atmosphere above is a conductor, only there is a little stratum between the conducting atmosphere and the conducting earth which is insulating. Now, you realize right away that if you set up differences of potential at one point, say, you will create in the media corresponding fluctuations of potential. But, since the distance from the earth's surface to the conducting atmosphere is minute, as compared with the distance of the receiver at 4,000 miles, say, you can readily see that the energy cannot travel along this curve and get there, but will be immediately transformed into conduction currents, and these currents will travel like currents over a wire with a return. The energy will be recovered in the circuit, not by a beam that passes along this curve and is reflected and absorbed, . . . but it will travel by conduction and will be recovered in this way. – <u>Nikola Tesla On His Work With Alternating Currents and Their Application to Wireless Telegraphy, Telephony, and Transmission of Power</u>, Leland I. Anderson, Editor, 21st Century Books, 1992, pp. 129-130.

Tesla believed the practical construction limitation imposed upon the height of the elevated terminals could be overcome by charging them to a sufficiently high electrical potential. With a highly energetic transmitter, as was intended at Wardenclyffe, the elevated terminal would be charged to the point where the atmosphere around and above the facility would become strongly ionized. This would lead to a flow of true conduction currents between the two terminals by a path up to and through the upper atmosphere, and back down to the other facility. He believed the ionization of the atmosphere above a plant's elevated terminal might be facilitated by the use of a vertical ionizing beam of ultraviolet radiation to form what might be called a high-voltage plasma transmission line. Upon further investigation he found this idea not to be practical.

I mastered the technique of high potentials sufficiently for enabling me to construct and operate, in 1899, a wireless transmitter developing up to twenty million volts. Sometime before I contemplated the possibility of transmitting such high tension currents over a narrow beam of radiant energy ionizing the air and rendering it, in measure, conductive. After preliminary laboratory experiments, I made tests on a large scale with the transmitter referred to and a beam of ultra-violet rays of great energy in an attempt to conduct the current to the high rarefied strata of the air and thus create an auroral such as might be utilized for illumination, especially of oceans at night. I found that there was some virtue in the principal but the results did not justify the hope of important practical applications. . . . – "The New Art of Projecting Concentrated Nondispersive Energy Through Natural Media System of Particle Acceleration for Use in National Defense," circa May 16, 1935. (Reprinted as part of <u>Nikola Tesla's Teleforce & Telegeodynamics Proposals</u>, edited by Leland I. Anderson, 21st Century Books, 1998, pp. 11-33.]

Terrestrial Transmission Line Zenneck Surface Wave Method

Given the apparent impracticality of atmospheric conduction method, Tesla directed his attention to the second technique that involves Earth's conductivity and the excitation of a non-radiating guided wave propagation mode known as the Zenneck surface wave.

The Zenneck wave was first identified in 1907, when Jonathan Zenneck described an electromagnetic wave that travels over a flat surface bounding two homogeneous media of different conductivity and dielectric constants, which exists as an exact solution to Maxwell's equations. [J. Zenneck, "Über die Fortpflanzung ebener elektromagnetischer Wellen längs einer ebenen Leiterfläche und ihre Beziehung zur drahtlosen Telegraphie" ("On the propagation of plane electromagnetic waves along a planar conductor surface and its relation to wireless telegraphy"), *Ann. Physik* [4] 23, 846 (1907).] The Zenneck wave has a phase velocity greater than that of light and its field strength falls off exponentially at a rate of $e^{-\alpha d}/\sqrt{d}$ in the direction of propagation along the interface, where α is a frequency-dependent attenuation constant. As the wavelength is increased the propagation attenuation decreases and the fields extend over a greater distance. The field intensity of the wave is at a maximum at the bounding surface, has a small attenuation in the direction along the interface, and high attenuation with height above the surface. In 1909 Arnold Sommerfeld performed a theoretical analysis of the propagation of radio waves around the earth, solving for the problem of a vertical dipole over a finitely conducting homogeneous ground.[2] He divided the expression for the resulting field into "space wave" and "surface wave" components. The surface wave part had nearly identical properties to the unique plane surface wave solution to Maxwell's equations that had been identified by Jonathan Zenneck two years previously. The field amplitudes varied inversely as the square root of the horizontal distance from the source and decayed exponentially with height above the interface. [Sommerfeld, Arnold N., "Uber die Ausbreitung der Wellen in der drahtlosen Telegraphie," *Annalen der Physik*, March 16, 1909 (Vol. 28, No. 4), pp. 665-736.] For a more detailed discussion of Zenneck surface waves, reference is made to U.S. Patent Application Publication Nos. 2014/0252886 and 2016/0072300, which were filed by J.F. Corum and K.L. Corum and are the property of Texzon Technologies, LLC.

Tesla was clear on the point that his wireless system did not depend upon the production of electromagnetic radiation or radio waves, also known as the Hertz wave:

> There is no radiation in this case. You see, the apparatus which I devised was an apparatus enabling one to produce tremendous differences of potential and currents in an antenna circuit. These requirements must be fulfilled, whether you transmit by currents of conduction, or whether you transmit by electromagnetic [space] waves. You want high potential currents, you want a great amount of vibratory energy; but you can graduate this vibratory energy. By proper design and choice of wave lengths, you can arrange it so that you get, for instance, 5 percent in these electromagnetic waves and 95 percent in the current that goes through the earth. That is what I am doing. Or, you can get, as these radio men, 95 percent in the energy of electromagnetic waves and only 5 percent in the energy of the current. . . . The apparatus is suitable for one or the other method. I am not producing radiation in my system; I am suppressing electromagnetic waves. But, on the other hand, my apparatus can be used effectively with electromagnetic waves. The apparatus has nothing to do with this new method except that it is the only means to practice it. So that in my system, you should free yourself of the idea that there is radiation, that energy is radiated. It is not radiated; it is conserved.

---

[2] See the chapter "Bell Labs and the Radio Surface Wave Propagation Experiment" by James and Kenneth Corum elsewhere in this book, for a history of radio-wave propagation theory development as it relates to the physical reality of the Zenneck surface wave. – Ed. Note

He goes on to say,

> It is just like this: I have invented a knife. The knife can cut with the sharp edge. I tell the man who applies my invention, you must cut with the sharp edge. I know perfectly well you can cut butter with the blunt edge, but my knife is not intended for this. You must not make the antenna give off 90 percent in electromagnetic and 10 percent in current waves, because the electromagnetic waves are lost by the time you are a few arcs around the planet, while the current travels to the uttermost distance of the globe and can be recovered. This view, by the way, is now confirmed. Note, for instance, the mathematical treatise of Sommerfeld, who shows that my theory is correct, that I was right in my explanations of the phenomena. – Nikola Tesla On His Work With Alternating Currents and Their Application to Wireless Telegraphy, Telephony, and Transmission of Power, Leland I. Anderson, Editor, 21st Century Books, 1992.

A basic difference between those transmitters designed to excite radio space waves and those designed to excite Zenneck surface waves is in the geometry of the launching structure. In the case of a simple monopole radio wave antenna for example, the vertical height would be close to ¼ of a wavelength of the operating frequency. The launching structure of a Tesla coil transmitter designed to operate at the same frequency would have a vertical height that is only a small fraction of a ¼ wavelength.

> Tesla's claim that "his" system is different from "Hertz's" is based on the fact that at low frequencies, and with small antenna in terms of wavelength, radiation of Hertzian type electromagnetic wave is small. "Tesla's waves," if we are allowed to use such a name, are in fact surface waves in modern terminology (as known, this type of waves are significant in the range of long waves) or the Earth cavity waves, known better as ELF (extremely low frequency) waves. In "pure Hertzian" wave (in Tesla's terminology) there is no induced current in the Earth, except on reflection region which is not essential for the discussion. In contrast to the latter, guided surface or ELF waves do not exist without current in the Earth crust. Having this in mind, we can conclude that there is a truth in Tesla's statements about specific behavior of low frequency, guided to the Earth [surface] waves. – Marinčić, Aleksandar (1990). "Research of Nikola Tesla in Long Island Laboratory," Energy and Development at the International Scientific Conference in Honor of the 130th Anniversary of the Birth of Nikola Tesla". *The Tesla Journal, An International Review of the Sciences and the Humanities* (Tesla Memorial Society, Inc.) (Numbers 6 & 7): 25-28.

Earth Resonance

With a sufficiently powerful magnifying transmitter operating at a frequency at or below 25 kHz, Tesla believed its emissions would reach to the opposite side of the earth and be reflected back, thus creating stationary terrestrial waves. He explained that Earth has self-inductance and also capacitance, the latter which is relative to the conducting region of the upper atmosphere. This natural electrical system behaves as a resonant LC circuit when it is electrically excited at a harmonic of the 11.78 Hz fundamental earth resonance frequency. The operating frequencies used at Wardenclyffe ranged from 1,000 Hz to 100 kHz, and those up to about 25 kHz were found "to be most economical."

> At present it may be sufficient, for the guidance of experts, to state that the waste of energy is proportional to the product of the square of the electric density induced by the transmitter at the earth's surface and the frequency of the currents. Expressed in this manner it may not appear of very great practical significance. But remembering that the surface density increases with the frequency it may also be stated that the loss is proportional to the cube of the frequency. With waves 300 meters in length [1 MHz] economic transmission of energy is out of the question, the loss being too great. When using wave-lengths of 6,000 meters [50 kHz] it is still noticeable though not a serious drawback. With wave-lengths of 12,000 meters [25 kHz] it becomes quite insignificant and on this fortunate fact rests the future of wireless transmission of energy. – "The Disturbing Influence of Solar Radiation On the Wireless Transmission of Energy," *Electrical Review and Western Electrician*, July 6, 1912.

In describing his work he speaks of the Earth resonance principle over and over again. Following an experiment at Colorado Springs in which sufficient energy was transmitted through the ground to cause sparking on the oscillator's lightning arresters, he wrote,

> This is certainly extraordinary for it shows more and more clearly that the earth behaves simply as an ordinary conductor and that it will be possible, with powerful apparatus, to produce the stationary waves which I have already observed in the displays of atmospheric electricity. – Marincic, Aleksandar (1978). Nikola Tesla Colorado Springs Notes 1899–1900. Nolit. p. 103; July 24, 1899.

In the patent *Art of Transmitting Electrical Energy Through the Natural Mediums* he described some of the necessary requirements to achieving earth resonance.

> The powerful electrical oscillations in the system E C D being communicated to the ground cause corresponding vibrations to be propagated to distant parts of the globe, whence they are reflected and by interference with the outgoing vibrations produce stationary waves the crests and hollows of which lie in parallel circles relatively to which the ground–plate E may be considered to be the pole. Stated otherwise, the terrestrial conductor is thrown into resonance with the oscillations impressed upon it just like a wire.

> Three requirements seem to be essential to the establishment of the resonating condition.

> First. The earth's diameter passing through the pole should be an odd multiple of the quarter wave length—that is, of the ratio between the velocity of light— and four times the frequency of the currents.

Second. It is necessary to employ oscillations in which the rate of radiation of energy into space in the form of hertzian or electromagnetic waves is very small . . . say smaller then twenty thousand per second, though shorter waves might be practicable. The lowest frequency would appear to be six per second, in which case there will be but one node, at or near the ground-plate.

Third. Irrespective of frequency the wave or wave-train should continue for a certain interval of time, estimated to be not less then one-twelfth or probably 0.08484 of a second [11.78 Hz] and which is taken in passing to and returning from the region diametrically opposite the pole. – Art of Transmitting Electrical Energy Through the Natural Mediums, May 16, 1900, U.S. Patent No. 787,412, Apr. 18, 1905.

In response to objections raised by U.S. Patent Examiner G.C. Dean regarding these three requirements, Tesla's attorneys responded,

These three requirements, as stated are in agreement with his numerous experimental observations. . . . We would point out that the specification does not deal with theories, but with facts which applicant has experimentally observed and demonstrated again and again, and in the commercial exploitation of which he is engaged. – Corum, J. F.; Corum, K. L.; Daum, J. F. X. (1987). "Spherical Transmission Lines and Global Propagation, An Analysis of Tesla's Experimentally Determined Propagation Model." CPG Communications. p. 3n.

In the article *The Problem of Increasing Human Energy* he writes of an observation made at Colorado Springs convincing himself that Earth can be electrically resonated.

That communication without wires to any point of the globe is practicable with such apparatus would need no demonstration, but through a discovery which I made I obtained absolute certitude. Popularly explained, it is exactly this: When we raise the voice and hear an echo in reply, we know that the sound of the voice must have reached a distant wall, or boundary, and must have been reflected from the same. Exactly as the sound, so an electrical wave is reflected, and the same evidence which is afforded by an echo is offered by an electrical phenomenon known as a "stationary" wave — that is, a wave with fixed nodal and ventral regions. Instead of sending sound-vibrations toward a distant wall, I have sent electrical vibrations toward the remote boundaries of the earth, and instead of the wall the earth has replied. In place of an echo I have obtained a stationary electrical wave, a wave reflected from afar. [THE PROBLEM OF INCREASING HUMAN ENERGY, Century Magazine, June 1900]

In 1905 he wrote of the challenge he encountered in developing an oscillator capable of exciting Earth resonance.

This problem was rendered extremely difficult, owing to the immense dimensions of the planet . . . but by gradual and continuous improvements of a generator of electrical oscillations . . . I finally succeeded in reaching electrical movements or rates of delivery of electrical energy not only approximately, but, as shown in comparative tests and measurements, actually surpassing those of lightning discharges . . . By the use of such a generator of stationary waves and receiving apparatus properly placed and adjusted in any other locality, however remote, it is practicable to transmit intelligible signals, or to control or actuate at will any one apparatus for many other important and valuable purposes. – "Tesla's Reply to Edison," English Mechanic and World of Science, July 14, 1905, p. 515.

And, in 1908 he summarized this achievement and describe some of the ramifications.

When the earth is struck mechanically, as is the case in some powerful terrestrial upheaval, it vibrates like a bell, its period being measured in hours. When it is struck electrically, the charge oscillates, approximately, twelve times a second. By impressing upon it current waves of certain lengths, definitely related to its diameter, the globe is thrown into resonant vibration like a wire, stationary waves forming, the nodal and ventral regions of which can be located with mathematical precision. Owing to this fact and the spheroidal shape of the earth, numerous geodetically and other data, very accurate and of the greatest scientific and practical value, can be readily secured. Through the observation of these astonishing phenomena we shall soon be able to determine the exact diameter of the planet, its configuration and volume, the extent of its elevations and depressions, and to measure, with great precision and with nothing more than an electrical device, all terrestrial distances.

In the densest fog or darkness of night, without a compass or other instruments of orientation, or a timepiece, it will be possible to guide a vessel along the shortest or orthorhombic path, to instantly read the latitude and longitude, the hour, the distance from any point, and the true speed and direction of movement. By proper use of such disturbances a wave may be made to travel over the earth's surface with any velocity desired, and an electrical effect produced at any spot which can be selected at will and the geographical position of which can be closely ascertained from simple rules of trigonometry.

This mode of conveying electrical energy to a distance is not 'wireless' in the popular sense, but a transmission through a conductor, and one which is incomparably more perfect than any artificial one. All impediments of conduction arise from confinement of the electric and magnetic fluxes to narrow channels. The globe is free of such cramping and hinderment. It is an ideal conductor because of its immensity, isolation in space, and geometrical form. Its singleness is only an apparent limitation, for by impressing upon it numerous non-interfering vibrations, the flow of energy may be directed through any number of paths which, though bodily connected, are yet perfectly distinct and separate like ever so many cables. Any apparatus, then, which can be operated through one or more wires, at distances obviously limited, can likewise be worked without artificial conductors, and with the same facility and precision, at distances without limit other than that imposed by the physical dimensions of the globe. – "The Future of the Wireless Art," <u>Wireless Telegraphy & Telephony</u>, Walter W. Massie & Charles R. Underhill, 1908, pp. 67-71.

What is the function of the tunnels below the tower and its mushroom-shaped dome?

The Ground Terminal Electrode

Electrical energy can be transmitted through inhomogeneous Earth with low loss because the net resistance between Earth antipodes is less than one ohm, as analytically demonstrated by James Corum in *Spherical Transmission Lines and Global Propagation, An Analysis of Tesla's Experimentally Determined Propagation Model*. The resistance of Earth is negligible due to its immense cross-sectional area and relative shortness as compared to its diameter. The greatest losses are apt to occur at the points where the transmitting-receiving plants and dedicated receiving stations are connected with the ground. The key to good performance is a robust ground connection. This is why Tesla stated:

> You see the underground work is one of the most expensive parts of the tower. In this system that I have invented it is necessary for the machine to get a grip of the Earth, otherwise it cannot shake the Earth. It has to have a grip on the Earth so that the whole of this globe can quiver, and to do that it is necessary to carry out a very expensive construction. – <u>Nikola Tesla On His Work With Alternating Currents and Their Application to Wireless Telegraphy, Telephony, and Transmission of Power</u>, Leland I. Anderson, Editor, 21st Century Books, 1992.

And again,

> However surprising, it is a fact that a sphere of the size of a little marble offers a greater impediment to the passage of a current than the whole earth. Every experiment, then, which can be performed with such a small sphere, can likewise be carried out, and much more perfectly, with the immense globe on which we live. This is not merely a theory, but a truth established in numerous and carefully conducted experiments. – "The Future of the Wireless Art," <u>Wireless Telegraphy & Telephony</u>, 1908, pp. 67-71 Walter W. Massie & Charles R. Underhill.

On Long Island, Tesla continued to refine the magnifying transmitter with a focus on designing the underground portion of the tower that served as a massive ground terminal electrode connecting the transmitter to Earth. Describing one of his critical scientific observations he stated,

In many instances when working with an oscillatory system transmitting its vibrations upon the ground I observed that the effects produced were stronger when the oscillating system was connected to earth through a condenser. This may have been due to some secondary causes but theoretically there are reasons for expecting such a result. – <u>Nikola Tesla - From Colorado to Long Island: Research Notes - Colorado Springs 1899-1900 - New York 1900-1901</u>, Nikola Tesla Museum, Belgrade.

As a further improvement, the master oscillator secondary shunt capacitor C2 was combined with the ground terminal electrode.

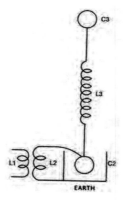

Shown here is a magnifying transmitter design dated May 19, 1901, placing it at Wardenclyffe. Included are, L1, L2, C2, L3, C3. Notice how the master oscillator secondary capacitor C2 and the ground terminal electrode are combined as a single unit. – Credit: <u>Nikola Tesla - From Colorado to Long Island: Research Notes - Colorado Springs 1899-1900 - New York 1900-1901</u>, Nikola Tesla Museum, Belgrade. Long Island, N.Y.

Some additional ground terminal construction details were provided by Mr. Tesla during the 1923 foreclosure proceedings:

A. *Yes. You see the underground work is one of the most expensive parts of the tower. In this system that I have invented it is necessary for the machine to get a grip of the earth; otherwise it cannot shake the earth. It has to have a grip on the earth so that the whole of this globe can quiver, and to do that it is necessary to carry out a very expensive construction. I had in fact invented special machines. But I want to say this underground work belongs to the tower.*

By Mr. Hawkins:

Q. Anything that was there, tell us about.

A. *There was, as your Honor states, a big shaft about ten by twelve feet goes down about one hundred and twenty feet and this was first covered with timber and the inside with steel and in the center of this there was a winding stairs going down and in the center of the stairs there was a big shaft again through which the current was to pass, and this shaft was so figured in order to tell exactly where the nodal point is, so that I could calculate every point of* distance. *For instance I could calculate exactly the size of the earth or the diameter of the earth and measure it exactly within four feet with that machine.*

Q. And that was a necessary appurtenance to your tower?
A. *Absolutely necessary. And then the real expensive work was to connect that central part with the earth, and there I had special machines rigged up which would push the iron pipe, one length after another, and I pushed these iron pipes, I think sixteen of them, three hundred feet, and then the current through these pipes takes hold of the earth. Now that was a very expensive part of the work, but it does not show on the tower, but it belongs to the tower.*

Based upon available drawings and descriptions of the underground work the following simplified schematic may be a fair representation of the Long Island facility.

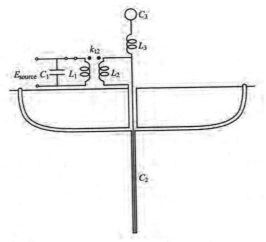

Shown are the earth terminal electrode assembly and also the legendary Wardenclyffe tunnels. The cylindrical subterranean chamber with its insulated central electrode, comprising capacitor **C2**, would have been oil-filled. According to George Scherff the brick-lined passages starting at the bottom of the 120-foot deep vertical and curving up to the surface are for ground water drainage. They serve to keep the earth around the tower dry, thus reducing capacitive coupling of the elevated terminal with the surrounding terrestrial ground plane and mitigating as much as possible the partial short circuit. A sump pump would have been placed into continuous operation for the water drainage system to be effective.

> Mr. Scherff, the private secretary of the inventor, told an inquirer that the companionway led to a small drainage passage built for the purpose of keeping the ground about the tower dry. – "Cloudborn Electric Wavelets To Encircle the Globe," *New York Times*, March 27, 1904.

In the well were four stone-lined tunnels, each of which gradually rose back to the surface. Large enough for a man to crawl through, they emerged like isolated, igloo-shaped brick ovens three hundred feet from the base of the tower. – Seifer, Marc, <u>Wizard : The Life and Times of Nikola Tesla</u>, Chapter 33, p. 291.

The Elevated Terminal Electrode

It appears that Tesla discovered some problems with the Colorado Springs magnifying transmitter design. And, when he began operational testing of the Wardenclyffe plant in July 1903 he was not satisfied with its performance. Experiments with the 1899 through 1901 transmitter configuration led him to write his financial underwriter J.P. Morgan on November 5, 1903,

> Dear Mr. Morgan:-
>
>   The enclosed bears out my statement made to you over a year and a half ago. The old plant has never worked beyond a few hundred miles. Apart of imperfections of the apparatus design there were four defects, each of which was fatal to success. It does not seem probable that the new plant will do much better, for these faults were of a widely different nature and difficult to discover.
>   As to the remedies, I have protected myself in applications filed 1900-1902, still in the office.
>
>   Yours faithfully,
>
>   N. Tesla

N. TESLA.
APPARATUS FOR TRANSMITTING ELECTRICAL ENERGY.
APPLICATION FILED JAN. 18, 1902. RENEWED MAY 4, 1907.

1,119,732.   Patented Dec. 1, 1914.

The "old plant" probably refers to the Colorado Springs Experimental Station. As for the "remedies" protected in applications filed between 1900 and 1902, and "still in the office," the only patented invention meeting these criteria is "Apparatus for Transmitting Electrical Energy," No. 1,119,732, issued on December 1, 1914.

Shown are the closely coupled primary C and secondary A of the master oscillator transformer. The lower end of the tertiary or helix extra coil resonator is connected to the upper end of the relatively low impedance secondary. The lower end of the transformer secondary is connected to a substantial ground terminal set in the bulk composite material that is Earth as described above. The upper end of the extra coil is connected to the elevated terminal insulated in space on non-conducting supports.

Comparing the Colorado Springs apparatus with the Wardenclyffe design, one obvious difference is in the geometry of the elevated terminal electrode, which is now greatly enlarged and covered with closely spaced hemispherical attachments. This modification allowed the air terminal to be charged to a much higher potential than was previously possible, due to the prevention of streamers.

The design of an even more advanced elevated terminal than the 1902 version was published in 1935 in, *The New Art of Projecting Concentrated Non-dispersive Energy Through Natural Media – System of Particle Acceleration for Use in National Defense.*

FIG.3
NEW TERMINAL FOR EXCEEDINGLY HIGH POTENTIALS
CONSISTING OF SPHERICAL FRAME WITH ATTACHMENTS

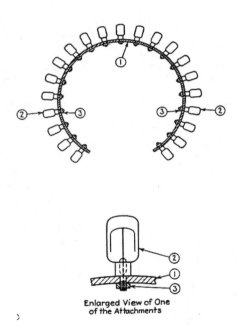

Enlarged View of One
of the Attachments

As will appear from the inspection of the drawing, the spherical frame of the terminal is equipped with devices, one of which is shown in the enlarged view below and comprises a bulb 2, of glass or other insulating material and an electrode of thin sheet suitable rounded. The latter is joined by a supporting wire to a metallic socket adapted for fastening to the frame 1, by means of nut 3. The bulb is exhausted to the very highest vacuum obtainable and the electrode can be charged to an immense density. Thus, it is made possible to raise the potential of the terminal to any value desired, so to speak, without limit, and the usual losses are avoided. I am confident that as much as one hundred million volts will be reached with such a transmitter providing a tool on inestimable value for practical purposes as well as scientific research. – "The New Art of Projecting Concentrated Non-dispersive Energy Through Natural Media System of Particle Acceleration for Use in National Defense," circa May 16, 1935. (Reprinted as part of <u>Nikola Tesla's Teleforce & Telegeodynamics Proposals,</u> edited by Leland I. Anderson, 21st Century Books, 1998, pp. 11-33.)

This is the basic earth-resonance transmitter configuration as incorporated into the initial Wardenclyffe design. To the right is a Tesla coil receiver. [Leland Anderson, "Rare Notes from Tesla on Wardenclyffe," *Electric Spacecraft*, Issue 26, Apr/May/Jun 1997.]

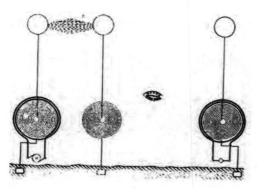

Tesla described this configuration in a 1906 patent.

I shall typically illustrate the manner of applying my discovery by describing one of the specific uses of the same - namely, the transmission of intelligible signals or messages between distant points. . . . For the present it will be sufficient to state that the planet behaves like a perfectly smooth or polished conductor of inappreciable resistance with capacity and self-induction uniformly distributed along the axis of symmetry of wave propagation and transmitting slow electrical oscillations without sensible distortion and attenuation. . . . The specific plan of producing the stationary waves, here-in described, might be departed from. For example, <u>the circuit which impresses the powerful oscillations upon the earth might be connected to the latter at two points.</u> . . . In collecting the energy of these disturbances in any terrestrial region at a distance from their source, for any purpose and, more especially, in appreciable amounts, the most economical results will be generally secured by the employment of my synchronized receiving transformer. – "Art of Transmitting Electrical Energy Through the Natural Mediums," April 17, 1906, Canadian Patent No. 142,352, August 13, 1912.

These words harken back to a statement made in 1893.

Now, it is quite certain that at any point within a certain radius of the source S a properly adjusted self-induction and capacity device can be set in action by

resonance. But not only can this be done, but another source S1 . . . similar to S
. . . can be set to work in synchronism with the latter, and the vibration thus
intensified and spread over a large area. . . . – *On Light and Other High
Frequency Phenomena*, Franklin Institute, Philadelphia and National Electric
Light Association, St. Louis, 1893.

The following drawing shows a similar earth-resonance transmitter. In this case the two
side-by-side synchronized resonator structures are energized by separate power plants.
The drawing is from Modern Mechanix, July 1934, "Radio Power will Revolutionize the
World" as told by Tesla to Alfred Albelli. The illustration is captioned, "We are on the
threshold of a gigantic revolution, based on the commercialization of the wireless
transmission of power."

While the distant future of Tesla World Wireless system technology may be industrial
power transmission, one immediate application is the harvesting of electrical energy
combination with broadband wireless telecommunications. Great possibilities may be
easy achieved. Here are some suggested services that might become available:

- Enhanced emergency alert and response resources, including global 911 services
  and passive transponder-type locator beacons;
- Emergency battery trickle charging for mobile devices;
- Global common-carrier wireless telecom services with no 'dead' zones;
- Global multi-channel broadcasting (digital radio and television);

- Global broadband wireless Internet backbone to compliment middle-mile and last-mile fiber-optic infrastructure;
- Global synchronization of precision timekeepers;
- Enhanced Deep Space Network.

All of Tesla's achievements in the area of wireless transmission have been formally validated through replication by others, with the exception of the earth resonance principle. Given the fact that, to date, all of his other patented inventions in this area have been demonstrated as fully functional and reproducible, one cannot help but look forward to the day when empirical evidence is presented that fills in this one missing piece of the puzzle. Closing with Tesla's own words,

> I gave to the world a wireless system of potentialities far beyond anything before conceived. I made explicit and repeated statements that I contemplated transmission, absolutely unlimited as to terrestrial distance and amount of energy. But, altho I have overcome all obstacles which seemed in the beginning unsurmountable and found elegant solutions of all the problems which confronted me, yet, even at this very day, the majority of experts are still blind to the possibilities which are within easy attainment. – "The True Wireless," *Electrical Experimenter*, May 1919.

20   Bell Labs and the Radio Surface Wave Propagation Experiment[*]

K.L. Corum[†] and J.F. Corum, Ph.D.[†]

TEXZON TECHNOLOGIES, 202 I-35 N. Suite C, Red Oak, Texas 75154

Dedicated to the memory of Professor R. G. Kouyoumjian

(Ohio State EE Dept.)

"With which type are wireless telegraphy waves to be identified? Are they like Hertzian waves [radiation] or are they like [guided] waves on wires? . . . The main task ... is to ... settle the question: space waves or surface waves?"

Arnold Sommerfeld
(March 1909)[40]

"After a half-day conference among T.C. Fry, W.H. Wise, K.A. Norton, and C.R. Burrows was unsuccessful in finding the source of any error in either paper [Sommerfeld's 1909 or Weyl's 1919], Dr. Fry suggested the experimental approach. This was made in 1935."

C. R. Burrows
(May 1967)[1]

Abstract

In this article we recount the story of one of the least cited but broadly influential physics experiments conducted during the 20th century: Charles Burrows' key 1936 experiment to delineate between Sommerfeld's 1909 theory and Weyl's 1919 theory of radiowave propagation. Burrows (Bell Labs) used a simple vertical doublet antenna disposed over a deep freshwater lake at a frequency of 150 MHz. Burrows (and others) declared that the 'crucial' experiment vindicated Weyl's theory over Sommerfeld's, and Kenneth Norton (FCC) used it to justify the famous "Sommerfeld sign error" myth that held sway in wave propagation theory for almost 70 years.

---

[*]   Originally entitled: "Bell Labs and the 'Crucial' 1936 Seneca Lake Experiment". The turn of phrase is due to C. R. Burrows (1962) and J. R. Wait (1964, 1971).

[†]   The authors are with TEXZON Technologies, LLC, 202 I-35 N., Suite C, Red Oak, TX 75154 and can be reached at info@texzont.com. The Texzon technology described herein is Patent Pending.

1.0 Introduction

Electric field strength measurements today are conducted regularly in the broadcast industry and seldom produce extraordinary or unanticipated results. However, as late as the mid-1930s theoretical analyses led to two dramatically diverse propagation predictions for conventional field strength. A critical experiment was conducted at Seneca Lake, New York, by Dr. Charles R. Burrows, of Bell Laboratories, and the results used to justify the assertion of a sign error in Sommerfeld's 1909 analysis of a vertical doublet above a lossy interface, and set the FCC engineering standards used for the past 75 years. Surprisingly, in spite of its being vital to our national prosperity and one of the more "crucial" experiments of 20th century physics (radiowave propagation is to be compared only with the printing press in terms of its impact upon civilization down through the last century!), mention of the Seneca Lake story is virtually absent in most significant electromagnetics textbooks. (Stratton's being a notable exception. [2]) By unraveling the background circumstances of this historical experiment, this article provides an introductory vehicle for grasping the arcane science of classic radiowave propagation (in the presence of lossy media) and the magnificent contributions of the great investigators that brought us to this point.

2.0 Background

The radiowave propagation account starts, sensibly, with the pioneering efforts of Heinrich Hertz (1857-1894) during his 1887-1890 experimental and analytical investigations [3], [4] the publication of which, according to one AIEE Vice-President, *"... caused a thrill as had scarcely ever been experienced before."* [5] In spite of the fact that we are separated from Hertz, his apparatus, and his experiments by well over a century, the shadow of their enchantment continues to hover over our technical world.

Early in 1895, the notion that these Hertzian waves could be used to send messages wirelessly fired the imagination of Guglielmo Marconi (1874-1937), who, in his 1909 Nobel lecture wrote,

"After a few preliminary experiments with Hertzian waves I became very soon convinced, that if these waves or similar waves could be reliably transmitted and received over considerable distances a new system of communication would become available possessing enormous advantages ..." [6]

There exists a certain romance associated with the history of radiowave propagation. Recall that it was December 12, 1901, a raw, blustery Thursday afternoon with gail-force winds under slate-gray skies on Signal Hill near St. John's, Newfoundland when Marconi, assisted by George Kemp and William Paget, reported successfully receiving a transatlantic Morse signal at 12:30, 1:30 and 2:30 PM, Newfoundland time. The letter "S", was transmitted from Poldhu, Cornwall, England, 2170 miled distant. Marconi and Kemp testified hearing the signal, which was intercepted by an end-fed long-wire (a kite antenna, wind-lofted with 400 feet of tether) operated against a ground plate. [7]

Figure 1.   Signal Hill, St. Johns, Newfoundland, December 1901. Marconi on extreme left, assistant William Paget on far right. Three local men (William Holwell – holding the tether, James Hoham, and Peter Edstrom assisting).

Since Marconi's investigations had evolved from initial experiments with Hertzian waves, it was sensible for contemporaneous engineers and physicists to seek a phenomenological explanation within the framework of Hertz-wave propagation. The critical point was the transmission around the bulge of the earth. (Questions concerning signal-to-noise and detector sensitivity are still debated even today.) According to British IEE President, John Ratcliffe, the question of the day was, *"How did the signal surmount the 160 km-high wall of sea water between Cornwall and Newfoundland?"* [8] In 1903 Lord Rayleigh (1904 Nobel Laureate), himself, would write,

> *"… nothing of this sort is observed in the case of light.*
> *The relation of wavelength to diameter of an object is about the same as when visible light impinges on a sphere one inch in diameter. So far as I am aware, no creeping of light into the dark hemisphere through any sensible angle is observed under these conditions even though the sphere is highly polished."* [9],[10]

In his IRE golden anniversary review of "Radiowave Propagation up to World War I", IRE Director, Charles Burrows, states that speculation as to the propagation mechanism of radio waves around the curvature of the earth fell into three categories: (1) ionospheric propagation; (2) guided propagation; and (3) diffraction into the earth's shadow. [11]

In retrospect, the insightful, but relatively informal, 1902 conjectures of Arthur Kennelly [12] (1861-1939) and Oliver Heaviside [13] (1850-1925) concerning an ionized mirror (the term "ionosphere" was actually coined by Sir Robert Watson-Watt [14],[15] in 1926) provided the subsequently acceptable explanation. [16] And, the radiowave *diffraction* problem required not only the analyses of Watson [17],[18] (the perfectly conducting sphere, 1918) but also that of Vvedensky, [19-21] Burrows, [22] and van der Pol and Bremmer, [23-28] (the lossy dielectric sphere).

However, even as late as 1909 Marconi [6], himself, asserted that he was *not* using Hertzian waves, and speculated that the critical propagation mechanism had something to do with the ground connection of both the transmitter and receiver. [29] [Incidentally, it was Andre Blondel who, in 1898, first pointed out that,

"... the action of the earth in a grounded transmitter could be replaced by that of an image of the transmitter, i.e., that a grounded transmitter can properly be conceived as one-half of a Hertz linear oscillator." [30]

Today, in antenna engineering, the notion of an image plane is taken as almost self-evident.]

Balthazar van der Pol (1889-1959), who made numerous contributions to radio science, mathematics, and electro-technology, and was known at Philips Research Labs as "the grand old man of radio", asserted that the problem of radiowave propagation at that time was *of foremost interest to "almost every nation"*. He stated that, "Zenneck was one of the first known scientists to study the interesting problem of electromagnetic wave propagation over the surface of the earth." [31] Who was Jonathan Zenneck? Zenneck (1871-1959) had performed post-doctoral experiments under Karl Ferdinand Braun at Strasbourg [32] (prior to Braun's sharing the 1909 Nobel Prize with Marconi). In 1907, while Professor of Physics at Munich's Technischen Hochschule, Zenneck determined formal analytical expressions which rigorously demonstrate that a certain surface wave solution actually satisfies Maxwell's equations *exactly*. [33], [34] (See Appendix A.4, below.) [In his 1915 text on wireless telegraphy [30], Zenneck, himself, pointed out that it was actually Andre Blondel [35] in Paris and Ernst Lecher [36] at Vienna who first intuitively *speculated* that radio waves were probably *surface guided waves* (like ocean surface waves) and not Hertzian waves (radiation).] One may assert with conviction that it is at this turning point, with the field analysis of Zenneck, the fascinating science of radiowave propagation formally moved into the 20th century and began to gain intellectual stature. In fact van der Pol stated that Zenneck was one of the first to study the influence of the earth on radiowave propagation, and his analysis demonstrated for the first time the difference that the earth's boundary conditions have on propagation. (See Appendix A.4.)

The dissimilarity between radiation and guided waves has to do with the distinction between the *continuous* and the *discrete* components of the wave equation's eigenvalue spectrum. (See Section 5.0, below.) [Incidentally, Jonathan Zenneck was the first Fellow of the IRE (1914). He later received the IRE Medal of Honor (1928) and he served as IRE Vice President in 1933.] As will be seen a little later, the Zenneck *surface wave* (we will be using the term "surface wave" in the sense defined by Sommerfeld, see below) arises analytically from a distinct *pole* yielding a *discrete* component in the plane wave spectrum. [37] Its field strength decays exponentially with distance (much like propagation in a lossy waveguide) and resembles radial transmission line propagation, as opposed to classical Hertzian radiation, which propagates spherically, possesses a *continuum* of eigenvalues, falls off geometrically, and results from branch-cut integrals. Zenneck's *surface wave* is, physically and mathematically, not the same as the Norton *ground wave* now so familiar from radio broadcasting. We use the terms *"ground wave"* and *"surface wave"* to identify two *distinctly different* physical phenomena. They are as markedly dissimilar as "Hertzian waves" (radiation) are from transmission-line modes (guided waves). These two propagation mechanisms arise from the excitation of different types of eigenvalue spectra (continuum or discrete) on the complex plane. As distinguished above, the former waves fall off *geometrically*, while the latter decrease (attenuate) *exponentially*. [The authors wish to express their gratitude to the late Francis

J. "Hans" Zucker (1922-1999) of AFCRL for pointing this out to us during an after-dinner conversation on the topic, late one evening several decades ago at a Los Alamos UWB Radar conference.]

Dr. Zenneck did not specify any physical source distribution for producing these waves, he simply obtained a rigorous analytical set of fields that, when plugged back into Maxwell's equations, satisfies them *exactly* in the space above, below, and along the boundary interface between vacuum and a lossy medium. [The situation is analogous to Kerr's prominent metric tensor in General Relativity, which satisfies the Einstein field equations *exactly* and which describes certain source attributes (such as the angular momentum of a rotating mass), but for which no explicit physical source mass distribution is specified. [38]] On the other hand, in rectangular coordinates, Hill and Wait determined a non-physical source with an aperture consisting of an infinite sheet of vertically decaying (exponentially) horizontally-directed magnetic current, "...*which excites a pure Zenneck wave with no radiation field.*" [39] These Zenneck waves have the property that, once launched, they propagate along the surface of the earth, *without radiation*, much like a waveguide mode with the earth as the guiding surface.

Then, in a masterful 1909 technical article, Munich Physics Professor Arnold Sommerfeld (1858-1951) powerfully conceived and skillfully executed a brilliant sequence of penetrating steps leading to a mathematically "elegant" solution of the wave equation for the fields produced by an incremental Hertzian antenna in the presence of a flat lossy ground. [40-44] His compelling, almost breathtaking, analysis passes to the conventional Hertzian radiation solution when either the antenna elevation becomes infinite or the soil becomes a perfect conductor. What is intriguing though, is that a surface wave similar to that obtained hypothetically by Zenneck in 1907 appeared to emerge in Sommerfeld's 1909 paper, this time from a specific incremental source geometry. A year later Sommerfeld published flat earth propagation predictions. [45]

In order to grasp the context of Sommerfeld's 1909 *wireless* paper, it is of value to back up and consider the "wire-wave" problem that he solved a decade earlier, in which he exercised many of the same analytical thoughts that he used in the 1909 analysis. (Both are boundary value solutions of the wave equation.) The issue in the 1899 paper was what Sommerfeld called *"Hertz's experiment that failed"*. Hertz's problem of the "wire-wave propagation" question was first successfully treated analytically in this 1899 paper by Sommerfeld, a decade after Hertz's experiments, as a boundary value solution of Maxwell's equations.

> "As is well known, the experiments of Heinrich Hertz dealt with 'surface waves' progressing along wires as well as with 'space waves' propagated freely through the air. Hertz expected their velocity to also be equal to c, but could confirm this result neither experimentally nor theoretically. The reason for *his experimental failure* was the influence of the walls of the laboratory; the reason for *his theoretical failure* was an excessive idealization of the problem. He treated the wire as infinitely thin and hence could not set up electromagnetic boundary conditions. This was first accomplished by [Sommerfeld] [46] … The experimental difficulties were overcome by Ernst Lecher by using a two wire line." [47]

[The history behind the Lecher line, the distinction between common-mode and differential-mode propagation, and the introduction of the notion of RF distributed element circuits, can be can be traced by consulting the original papers of Heaviside [48], Lecher [49], Tesla [50-52], Lodge [53], Mie (who used part of the original apparatus of Hertz at Karlsruhe) [54], Hondros [55], Sommerfeld [47, pp. 199-211], and Terman [56].]

The compelling motivation driving Sommerfeld's 1909 wireless investigation is now clear, and revolves around the distinction between radiation and surface-guided waves. (Incidentally, Sommerfeld's use of the term "surface wave" is defined in Volume 6 of his Lectures on Theoretical Physics [41, pp. 254-255] and is different from Schelkunoff's classification. [57]) The mystery that Sommerfeld wanted to resolve was explicitly stated in his 1909 paper as follows:

> *"With which type are wireless telegraphy waves to be identified? Are they like Hertzian waves* [radiation] *or are they like* [guided] *electrodynamic waves on wires? . . . The main task of the present investigation is ... to settle the question: space waves or surface waves?"* [40], [58]

Forty years later, in 1949, Sommerfeld would write,

> *"It was the main point of the Author's work of 1909 to show that these fields* [Zenneck's surface waves] *are automatically contained in the wave complex, which, according to our theory, is radiated from a dipole antenna. This fact has, of course, not been changed. What has changed is the weight which we have attached to it. At the time it seemed conceivable to express the overcoming of the earth's curvature by radio signals with the help of the character of the <u>surface waves</u>; however we know now that this is due to the ionosphere."* [41, p. 255]

While the "160 km wall of water" question was resolved ionospherically, the issue of wireless surface wave propagation was not.

As a human sidelight on the propagation history, we note that Sommerfeld owned an alpine ski lodge to which students (such as Wolfgang Pauli, Werner Heisenberg, Hans Bethe, Peter Debye, Isidor Rabi, Linus Pauling, Alfred Landé, and Ernst Guillemin) and colleagues were often invited for discussions of physics. Interestingly, Sommerfeld and Zenneck owned adjacent alpine dwellings and discussed these wave propagation problems while vacationing. In fact, in his 1907 paper, Zenneck thanks Sommerfeld. And, in his 1909 paper, Sommerfeld thanks his "colleague" Professor Zenneck "for special expert advice on technical aspects of the problem". (The first German edition of Zenneck's text on wireless telegraphy had just been published the previous year, in 1908.) Also of interest is the fact that in 1915 Karl Braun (who shared the 1909 Nobel Prize with Marconi) and Jonathan Zenneck were in New York together to join Nikola Tesla in providing expert testimony on behalf of the Atlantic Communication Company (Telefunken), which had retained noted American patent council Frederick P. Fish (former President of AT&T) as defense attorney in a lawsuit initiated by American Marconi (which became RCA in 1919). At the time of the 1915 IRE banquet at Luchow's (a legendary German restaurant in NYC – See Figure 2), Zenneck met Nikola Tesla (1856-1943). Tesla had served two years (1892-1894) as Vice-President of the AIEE [59], [60] and his published lectures had introduced Zenneck to RF technology 15 years earlier [61]. The two became warmhearted friends. (This was just prior to Zenneck's being detained at an Ellis Island facility (and later at Fort Oglethorpe, Georgia) as a prisoner of war during WWI.) A year later, in 1916, Tesla, whose propagation experiments appeared to contradict the Hertz wave theory, confided to his own attorney,

> "Note, for instance, the mathematical treatise of <u>Sommerfeld</u>,[40,43] who shows that my theory is correct, that I was right in my explanations of the phenomena, and that the profession was completely mislead." [62]

And,

> "The effect at a distance is due to the current energy that flows through the surface layers of the earth. That has already been mathematically shown, really, *by Sommerfeld*. [40] He agrees on this theory; but as far as I am concerned, that is positively demonstrated." [62, p. 133]

[Be aware that Tesla is speaking of the electrodynamic propagating currents associated with what we now call surface waves, *not* the simple ohmic (diffusion [2, pp. 278-279]) ground current mechanism proposed by Karl Steinheil (1801-1870) or James Bowman Lindsay (1799-1862) in the mid-19th century.]

Figure 2. Gathering of the Titans. The second IRE banquet, held at Luchow's German restaurant in NYC on April 24, 1915. Standing (L-R) George W. Pierce (4th VP 1915, President 1918, 1919), Karl Ferdinand Braun (Nobel Laureate, 1909), John Stone Stone (3rd VP 1914, President 1915), Jonathan Zenneck (1st IRE Fellow, VP 1933), Lee de Forest (President 1930), Nikola Tesla (VP AIEE 1892, 1893), Fritz Lowenstein (1st VP IRE, 1912), Alfred Goldsmith (Editor, Proc. IRE, President 1928). Other notables include David Sarnoff (Secretary IRE 1915, 1916, 1917; later VP and President of RCA), far end of left side of table); E.F.W. Alexanderson (1921 President of IRE), 5th from far end of table on right. [62, p. ii]

Tesla's assistant at the spectacular Colorado Springs propagation experiments in 1899, and later at his Wardenclyffe, Long Island laboratory (which, incidentally, still exists and was purchased in 2013 by Friends of Science East and is being renovated as an historic site and museum of science), was Fritz Lowenstein (1874-1922) [63], who was also present at the April 1915 IRE banquet. In his assessment of Sommerfeld's theory, presented in a December 1, 1915 lecture before the IRE (published in the June 1916 issue of the Proceedings of the IRE), Lowenstein said,

> "The analysis … when transmitting over poorly conducting ground, as given by Sommerfeld, is of great value and interest. … The fundamental question remains, however, whether we may designate the method now used as a transmission by Hertzian waves similar to those emitted by the Hertzian oscillator of the early Marconi apparatus, or as true conduction along the ground as was proposed by Tesla in 1893. Sommerfeld's work … led him to decompose the total wave action into a *space wave* and a *surface wave*. Thus the belief has been established in the minds of many radio engineers that transmission of the energy was carried by two distinct and different phenomena." [64]

Recall that Fritz Lowenstein was a pioneer-expert in the area of radio-remote control and guided weaponry. [65], [66] He was the inventor of the grid-biased Class-A amplifier (which he sold to AT&T) [67-70] and the shaped-plate variable capacitor [71-73]. As IRE member #14 he was elected as the first Vice-President of the IRE (now the IEEE) when it was formed in 1912. [74] Zenneck, himself, was in the audience at Lowenstein's 1915 lecture and added his supporting comments. This was the general view of the state of radiowave propagation in the WWI era.

Attention must also be directed to the electromagnetic wave propagation studies of another distinguished world-class mathematical physicist: Hermann Weyl (1885-1955), professor of mathematics and colleague of Albert Einstein's at the ETH (Eidgenössche Technische Hochschule – Federal Institute of Technology) at Zürich. In 1918 Weyl had written the first paper attempting to combine general relativity with electromagnetism (actually the first of the relativistic unified field theories that followed in the 1920-1930s) [75], and published the first edition of his textbook <u>Raum, Zeit Materie</u> (Space, Time, Matter) also in 1918. In 1919, he turned his attention, briefly, to the Sommerfeld radiowave problem, and developed a propagation solution as an angular spectrum of plane waves [76-78], complimenting Sommerfeld's 1909 solution. Weyl's approach was to expand a time-harmonic spherical wave into a bundle of inhomogeneous plane waves incident on the earth at various complex angles. [2, pp. 577-587], [79], [80] While his solution involves complicated conformal mapping in transform space, it is valid at large distances where it can be evaluated asymptotically in a straightforward manner. However, while Weyl's solution can be interpreted much like Sommerfeld's, as the superposition of a space wave plus a ground wave, the physical nature of *Weyl's ground wave* (which has more in common with a *radiation field*) is distinctly different from the nature of the *surface wave* of Zenneck and Sommerfeld (which arises from a discrete pole, and propagates more like a guided wave) [2 p. 584], [81], [82]. The Weyl solution fails to capture this Sommerfeld-Zenneck pole and leaves only what later became known as the *ground wave* of Norton, Burrows, Niessen and van der Pol.

As a final point of introduction, we note that the 1909-1910 propagation studies by Arnold Sommerfeld were distilled by Bruno Rolf (1885-1934), and later published by the IRE in 1930 as a set of propagation prediction curves for engineers. [83] However, it was soon determined that these curves, based on Sommerfeld's 1909 analysis, rigorously formal though that was, *did not* accurately predict the *observed* ground wave emitted from radio transmitting antennas at close range. [84], [85-88] And, here our saga begins...

3.0 The 1935 Meeting At Bell Labs

The spot-light now turns from these giants in the early development of radio science to focus on Bell Labs and the second wave of radio savants in the mid-20th century. Kenneth Alva Norton (at the FCC) and Charles R. Burrows (at Bell Labs) recognized that the predictions of the 1909 Sommerfeld formula did not agree with either reported physical observations or the novel, but equally accurate, 1919 solution by Hermann Weyl. This was a serious issue. National prosperity rested on its correct resolution. International agencies, governments and engineers had to know. A small conference was held at Bell Labs to determine which theory was correct. The attendees at the conference (including **T.C. Fry**, **W.H. Wise**, **K.A. Norton**, and **C.R. Burrows**) were faced with a dilemma in that they could find no error in either Sommerfeld's or Weyl's analysis. According to Burrows, it was ***Dr. Fry**[*] who suggested <u>performing an experiment</u> to decide whether Sommerfeld or Weyl was correct. Concerning Fry's suggestion, Burrows tells us that, *"This was made in 1935."* [1, see footnote 6 on p. 408] In his IRE 50th anniversary review, Burrows told the story as follows,

> "Burrows discovered that these two formulations [Sommerfeld's and Weyl's], in fact, *differed by exactly the surface wave of Sommerfeld*, and, in collaboration with his associates, L.E. Hunt and A. Decino, conducted *the crucial experiment* which showed that Weyl's formulation was correct and that the surface wave of Sommerfeld did not exist." [11]

According to this 1962 account by Burrows it was *Rice and Niessen* that *actually* discovered the mathematical error in Sommerfeld's work ... and that was done *after* the crucial experiment had been performed:

> *"Later* Rice and Niessen independently found the source of the error in Sommerfeld's work – the incorrect choice of the square root of a complex quantity in an intricate mathematical derivation." [11]

This account is at variance with the conventionally accepted story that it was Kenneth Norton who first asserted that there was a sign error in Sommerfeld's 1909 paper. The National Bureau of Standards (now NIST) asserted that, "Norton was led to his discovery of the error from anomalies of measurement values compared with computed values in

---

[*] An industrial mathematician, Dr. Thornton C. Fry (Ph.D. Math, Physics and Astronomy; U. Wisconsin, 1920) is best known for his text, <u>Probability and Its Engineering Applications</u>, which was one of the volumes in the famous Bell Labs series. After a lengthy career at Bell Labs (1916-1961), Dr. Fry helped inaugurate the National Center for Atmospheric Research, NCAR, (1961-1968).

the observations of field intensities made by him and Kirby at an earlier time." [89], [85] James Wait, who heard the story personally from Kenneth Norton wrote,

"It appears that the *first* explicit statement of the error in Sommerfeld's 1909 paper was published by K.A. Norton (1935) in a letter to Nature." [90], [78]

(A retrospective appraisal of Norton's analysis has been outlined by Wait. [91]) In that June 8, 1935 letter to Nature, Norton wrote,

"The purpose of this letter is to point out an *error in sign* in Prof. A. Sommerfeld's original paper (1909) on the attenuation of radio waves. This error in sign has recently been reflected in Bruno Rolf's graphs [83] of the Sommerfeld formula … Correcting this error in sign, I have found …" [92]

Burrows, in a letter to Nature a year later (August 15, 1936) even appears to recognize Norton as the discoverer (which would seem to be at variance with his 1962 and 1967 accounts),

"In the mathematical development of the problem of radio propagation over plane earth, Sommerfeld expressed his solution in the form of three terms, one of which he identified with the surface wave of Zenneck. Curves calculated from Sommerfeld's formula have been given by Rolf. Weyl, approaching the problem in a different manner, obtained a solution which did not explicitly contain this term. A *formula given by Norton* [in his 1935 Nature letter] gives the values agreeing with Weyl. It appears that Weyl was of the opinion that his result was numerically equivalent to that of Sommerfeld. The purpose of this letter is to point out that this is *not* true, that the evaluation of Sommerfeld's formula by Rolf differs from *the formulae of Weyl and Norton* by exactly the 'surface wave' component, and to give the results of *a recent experiment* showing the *Weyl-Norton* values to be the correct ones, which raises a question as to whether such waves do or do not physically exist." [93]

Several months later, in an October 1936 IRE paper (published *after* the Seneca Lake experiment), Norton cites the above Nature letter by Burrows,

"Some recent experimental results obtained by C.R. Burrows and described in a letter to Nature [August 15, 1936] substantiate the theoretical ground-wave formulas and graphs [in Norton's present paper] … These measurements were made on a frequency of 150,000 kilocycles over fresh water so that $[\varepsilon_r] \gg x$. [94][*] In his last sentence, Mr. Burrows states, 'It seems evident that a revision of the Sommerfeld-Rolf curves is required for propagation over all types of ground for which the dielectric constant cannot be neglected.' Revised data of this kind are given in Figure 1 [a plot of the ground wave attenuation factor F(p) vs. p as a function of b] and Table I [a tabulation of the calculated values] of this paper and should be adequate in all cases such that the heights of the transmitting and

---

[*] Norton's quantity $x = \sigma/(\omega\varepsilon_o) = 1.8 \times 10^{18} \sigma_{emu}/f_{kc} = 1.8 \times 10^7 \sigma_{MKS}/f_{kHz} = 60\lambda\sigma_{MKS}$ is the ratio of conduction current to free-space displacement current. [Note that: $\sigma_{MKS} = \sigma_{emu} \times 10^{11} = 4\pi\varepsilon_o\sigma_{esu} = \sigma_{esu}/(9 \times 10^9)$.]

receiving antennas are a small fraction of a wavelength above the surface of the earth." [95]

A slightly modified version of the cited Figure [in the form of F(p)/p vs. p] subsequently became incorporated into the FCC Rules and Regulations [96]. (See Fig. A.1 below.) In January 1937, Burrows published a set of curves to facilitate the calculation of radio propagation over plane earth in the BSTJ (Bell System Technical Journal), [97] with an "Addendum", added in October 1937 responding to a note from K. A. Norton. [98]

Most of the information that we can glean about the Seneca Lake experiment was published in four sources. First, Burrows' August 15, 1936 letter to the journal Nature,

"The purpose of this letter is to point out that ... the evaluation of Sommerfeld's formula by Rolf differs from the formulae of *Weyl and Norton* by exactly the 'surface wave' component and to give the results of a recent experiment showing the Weyl-Norton values to be the correct ones. [93, p. 284]

Second, the Radio Club of America report of his talk delivered on February 11, 1937,
"Mr. Burrows stated [that] resort to experiment was indicated as being desirable in order to decide which of these two [formulas] is correct. In making such an experiment it is highly desirable to make transmission tests under conditions where the received field strength predicted by these two formulas differ greatly. This occurs for propagation over a perfect dielectric and ... fresh water is the nearest approach to a perfect dielectric available in sufficient volume and area ..." [99]

Third, a February 1937 paper by Burrows published in the Proceedings of the IRE,
"Since *Norton has derived his results from a formula of van der Pol and Niessen*, their formulas presumably agree with those of Weyl. ... As a result of the realization that the mathematics contained an ambiguity, the writer [Burrows] on September 23, 1933, attempted to decide the question experimentally by measurements at Budd Lake, New Jersey, employing ultra-short waves. The results indicated that the water was too shallow to meet the requirements of the experiment, since the transmission resembled that over land instead of over deep fresh water. At that time an experiment over deep fresh water was planned which has recently been successfully carried out. The *results prove conclusively* that simple antennas do not generate a Sommerfeld surface wave. This is in agreement with recent theoretical work by Wise [100-103], [104], and Rice [105]." [106]

Fourth, there was also a short note in the *Bell Labs Record* (not the BSTJ).

"Only since the development of ultra-short wave radio has it been possible for Laboratories' engineers to perform a *crucial* experiment which would settle the question as to which result was correct. The decision, *which has since been confirmed theoretically by S. O. Rice*, was found to be in favor of Weyl's formula, which does not contain any term corresponding to the surface wave. ... Calculations from the two conflicting formulas indicate that at a distance of one kilometer over Seneca Lake the received field strength, on a wavelength of two meters, should be 44 dB greater for a surface wave than without it. ... no such surface wave was present. ... Taken together with Rice's recent review of the work of Sommerfeld and Weyl, which has brought the two in agreement and established the fact that the prediction of a surface wave was due to a mathematical error, these tests *prove conclusively* that simple antennas do not generate a surface wave and that this time-honored concept must be given up, at least in the sense that radio engineers have customarily used it." [107]

The bizarre story of the "Sommerfeld error and its discovery" was further elaborated in a footnote by Burrows in 1967, and differs somewhat from that commonly accepted. [See Wait, below.] According to Burrows' 1967 account, it was Steven Rice (later of Ricean noise fame) who actually discovered the famous Sommerfeld sign error, not Kenneth Norton, and the discovery was made *after* the Seneca Lake experiment.

"*After the experiment*, S.O. Rice found that Sommerfeld took the wrong square root of a complex quantity introducing the surface wave which does not belong in the formulation. Publication of Rice's paper was withheld by the Bell Telephone Laboratories on the basis that an isolated error by a leading physicist had been discussed enough" [1 (see footnote 6 on p. 408)]

[Rice did publish a contemporaneous 1937 paper discussing Sommerfeld's dipole radiation over the earth's surface, but with mathematical analyses omitting any discussion of a sign error in the Sommerfeld solution. [105]] No mention of Norton (or Niessen) being the discoverer here. This strange 1967 assertion by Burrows would seem to be in contradiction to the 1935 note by Kenneth Norton, who had written,
"The purpose of this letter is to point out an *error* in sign in Prof. Sommerfeld's 1909 paper." [92]

The commonly-held opinion was stated by James Wait, who, as mentioned above, had been a colleague of Kenneth Norton's. In 1971 Dr. Wait wrote,
"The first explicit statement of the *error* in Sommerfeld's 1909 paper was published by K.A. Norton in a 1935 letter in Nature."

Collin, in his 2004 assessment, points out,

"In 1935 Norton published a short paper in which he asserted that Sommerfeld had made an error in sign in one of his formulas. Unfortunately, Norton did not provide any specific details as to [78]

So, ... was it Steven Rice or Kenneth Norton (or even Karel Niessen) who actually "found" the so-called Sommerfeld error?
which of Sommerfeld's expressions had the sign error, or what had gone wrong in Sommerfeld's analysis." [108]

Figure 3. The main players in the radio surface-wave controversy.

The "correction" of this "error" has had a remarkable impact on 20th century radio coverage predictions. Burrows, who tells the story, was present at the 1935 conference meeting along with Rice, Fry, and Norton. And, Burrows even refers to Norton's 1935 letter in Nature in both his own 1936 letter to Nature [93] and in his 1937 IRE paper [106]. But, how are those remarks by Burrows and his subsequent footnote comments in 1967 to be harmonized?

The sign error controversy is further clouded by the 1930's German language papers of van der Pol and Niessen.[*] [109-114] And then, we have the late Professor Collin's 2004 remarks,

"In 1937 Niessen published a paper [115] in which he also claimed that Sommerfeld had made a sign error in his 1909 paper. According to Niessen, the sign error came about because Sommerfeld did not take the value of the angle of the square root of a complex number using the convention that this should always be taken to be between 0 and 2π. ... *Niessen's argument was not a valid one.* ... this explanation was widely accepted and has been propagated throughout the technical literature from that time forward. ... *What both Norton and Niessen had observed was that by a simple change in sign* – in the square root of a parameter called the numerical distance – they could provide a ***quick fix*** to Sommerfeld's 1909 solution *that would bring his solution into conformity with that of later workers.* ... From a mathematical perspective, a change in sign of Sommerfeld's closed-form expression for his solution is *not allowed*, and Norton's and Niessen's assertions are *not* acceptable." [108]

[The sign change "imposed" by Norton/Niessen (changing the sign of -j$\sqrt{p}$ in Sommerfeld's solution to that given in Equation (A.20) in Appendix A, below) [116] artificially removed the Zenneck-Sommerfeld *surface wave* (a transmission line mode) while leaving the Norton *ground wave* (a radiation mode). Realize that many people today loosely use the term "radio surface wave" when they actually mean the Norton "ground wave", not the Zenneck surface wave!]

Actually, Sommerfeld never admitted to making such an error, and recently Prof. R.E. Collin has made the striking assertion that there was, in fact, no sign error!

"The sign error that has been claimed in the technical literature for more than 65 years ***is a myth***. ... in spite of the long-held belief, Sommerfeld did ***not*** make a sign error in his 1909 paper. ... There is no sign error . . . ***The famous sign error is a myth***." [108]

---

[*] Dr. Karel Niessen, a Dutch theoretical physicist (Ph.D. at Utrecht, 1922), did post-doctorial studies under Arnold Sommerfeld (1925-1926). He is best known for the Pauli-Niessen model of the Hydrogen molecule, which demonstrated the inadequacy of the Bohr-Sommerfeld framework for quantum mechanics, and which ultimately paved the way for Heisenberg's matrix mechanics and Schrodinger's wave mechanics. Niessen spent most of his career as an industrial theoretical physicist at Phillips Electronics in Holland.

So, what is going on? Which account by Burrows is the correct story, the one in the 1936-1937 time frame or the one in the 1962-1964 time frame? (Or both?) Can they be harmonized?

When Sommerfeld (who used $e^{-i\omega t}$ time variation) wrote his 1909 solution, he effectively expressed the field attenuation factor (rewritten for $e^{+j\omega t}$ time variation [116]) as

$$F(w) = 1 + j\sqrt{\pi w}\, e^{-w}\, erfc(-j\sqrt{w})\,. \tag{1}$$

(When the source and observer are at ground level, $w \rightarrow p$.) It appears that Niessen and Norton discovered that by re-expressing this as (see Equation (A.19) in Appendix A below, also written for $e^{+j\omega t}$ time variation)

$$F(w) = 1 - j\sqrt{\pi w}\, e^{-w}\, erfc(j\sqrt{w}) \tag{2}$$

the predicted results would conform to the analytical consequences of Weyl's 1919 solution [117] and the experimental results obtained by the Bell Labs team led by Charles Burrows. A *ground wave* now emerges *without* the $1/\sqrt{r}$ behavior of a *surface wave*. This was what Collin called the *"quick fix"*, above. However, apparently it was actually Steven Rice[*] who determined that Sommerfeld's term (which had a negative sign in the lower limit)

$$\frac{2}{\sqrt{\pi}} \int_{-j\sqrt{w}}^{\infty} e^{-u^2}\, du \qquad \text{"ought" to be} \qquad \frac{2}{\sqrt{\pi}} \int_{j\sqrt{w}}^{\infty} e^{-u^2}\, du\,. \tag{3a,b}$$

(with the positive lower limit). By doing this and performing the subsequent integration on a different Riemann sheet in complex space, the *surface wave* of Zenneck disappears leaving, instead, only the Norton *ground wave*. [95, pp. 1382-1387]

4.0 The "Crucial" Seneca Lake Experiment

Be all that as it may, the critical experimental resolution suggested by Dr. Fry was actually performed during the summer of 1936. [99] In 1964 (and in 1971) Dr. Wait described the event as follows,

"Burrows felt that the problem could finally be resolved by performing a *crucial* experiment. Accordingly, he carried out measurements on Seneca Lake in New York State. The experimental data agreed with Weyl's expressions for the field strength as a function of distance." [78],[90]

---

[*] Steven O. Rice: BSEE and honorary Sc.D. (1929, 1961, Oregon State). Following grad work at Cal. Tech, Rice worked at Bell Labs from 1930-1972. His Ph.D. dissertation was rejected by Columbia. An IEEE Fellow, he is noted for his contributions to FM, communication theory and the statistics of random noise.

In a more recent review of the issue, Wait wrote,

> "Another leading researcher in the USA, working at the Bell Laboratories, was *Charles Burrows.*[*] He carefully measured the field strength at 150 MHz, for a distance range of 2000 meters, over a deep, calm lake near Seneca, in upper New York state.[94] He showed, using data on the electrical properties of the lake water (independently obtained), that the observed field strength vs. distance was in *full conformity with Norton*, but differed from Rolf." [118]

Subsequent interpretation is illustrated by Terman's Radio Engineers' Handbook remark:

Figure 4. Experimental set-up used by Burrows in 1936 on Seneca Lake in New York. It may be argued *forcefully* that this was one of the *most significant experiments in all of 20th century physics*. (Yet, the entire topic is absent in modern graduate-level texts on electrodynamics. Neither the controversy, nor its major participants are even recognized in recent books on the tory of wireless"! [119])

---

[*] Dr. Charles R. Burrows (Ph.D. in Physics from Columbia, Fellow of both the American Physical Society and the IEEE, and served as an IEEE Director) was a member of the Research Staff at Bell Labs from 1924-1945. During WWII he was responsible for *all* of the propagation research affecting the war effort. Later, he became Director of the School of Electrical Engineering at Cornell University, where he further developed research in radio wave propagation, radio astronomy, and radar. There is a full-page bio of Dr. Burrows on p. 1104 of the September 1956 issue of the Proceedings of the IRE. He concluded his distinguished radio science career as a corporate leader and engineering consultant.

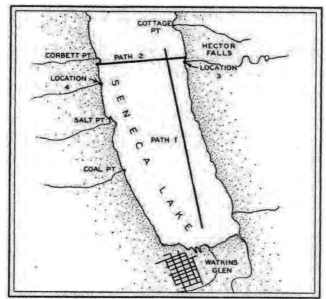

Figure 5. The region of Seneca Lake where the Burrows experiment was conducted.

"Work by C.R. Burrows gave *experimental verification* that … a wave of the Zenneck type is *not* produced by a vertical radiator, as was indicated by Sommerfeld's 1909 paper." [73]

In his authoritative monograph, Professsor Alfredo Banos wrote,

"Burrows made careful measurements which show that the results of Weyl for radio propagation over a plane earth are entirely consistent with the experimental work, but they differ from those of Sommerfeld by the surface wave term [arising from the residue of the pole in Figure 8 below]. … Burrows concludes that his *experimental results* prove conclusively that simple antennas do not generate a Sommerfeld surface wave…" [79, p. 155]

Norton published a set of flat earth radio propagation curves in 1936, [95] as did Burrows in 1937. [120] The final Norton curves give normalized ground wave field strength vs. distance for various conductivities, fixed permittivity ($\varepsilon_r = 15$), and a selection of frequencies. [Norton's plots for 540-1640 kHz, ultimately ended up as the 20 famous charts in the FCC Rules and Regulations. (See FCC Rules and Regulations, § 73.184, Graphs 1-19A, "Ground Wave Field Intensity vs. Distance".)]

Burrows reported that the original Budd Lake experiment was performed on September 23, 1933. This puzzling date predates the 1935 conference and the suggestion for an experimental resolution made by Fry appearing in Burrows' 1967 article. (See below.) The Budd Lake experiment gave results "resembling propagation over land instead of fresh water."[106] While Budd Lake in northern New Jersey is shallow and very boggy (it has an average depth of only 7-12 feet), Seneca Lake is 38 miles long, with a mean depth of 291 feet, and a maximum depth of 838 feet. (The 1/e depth at Seneca Lake, at 150 MHz with $\varepsilon_r = 82.1$ and $\sigma = 0.045$, is approximately 1 meter.) No wonder Burrows repeated the experiment at Seneca Lake. Incidentally, it appears that two distinguished Soviet physicists, Leonid Mandelshtam and N.D. Papalexi repeated this experiment *over land* with essentially the same results as Burrows, but we have been unable to obtain a translation of their publication. [121] Kirby and Norton [85] had earlier measured field strength over lossy soil at large distances, and, in retrospect, their results, as did Feldman's, concur with the absence of a Zenneck surface wave. (Feldman wrote, "We have found no case in which this term can be said to predominate experimentally." [86]).

5.0 Theoretical Predictions

A concise synopsis introducing the various solutions to the problem of an antenna above a lossy medium is presented in Appendix A, below. Using previously developed series solution representations, Burrows found that when he wrote down the solution for Norton [which matched Weyl (1919) asymptotically] and the solution for Sommerfeld (1909), the difference was just the surface wave term of Sommerfeld (which Sommerfeld had identified as Zenneck's surface wave).

Following the analyses of Norton [92], [95, p. 1369], [122], Norton-Weyl broadcast field strength predictions were determined from a simplified engineering procedure. The unattenuated field (essentially the RMS value of the vertically polarized Hertz field at $\theta = \pi/2$ on a perfectly conducting plane), Equation (A.23),

$$E_{u,RMS}(r_k) = \frac{173.2 \sqrt{P_{kW} \, G_T}}{r_k} \quad \text{mV/m} \tag{4}$$

and the (Norton) augmented van der Pol-Shuleikin attenuation function, Equation (A.25),

$$A = |F(p)| = \frac{2 + 0.3\,p}{2 + p + 0.6\,p^2} - \sqrt{\frac{p}{2}}\, e^{-0.625p}\, \sin(b) \tag{5}$$

are then multiplied as Equation (A.24)

$$E_N = A \times E_{u,RMS} \quad \text{mV/m} \quad . \tag{6}$$

where the symbols are defined in Appendix A.

In his analyses Burrows [120, pp. 49-50 and 70-71], [106, p. 225], [100, pp. 36-38] noticed that, numerically, the Norton/Weyl solution was equal to the Sommerfeld solution minus the Zenneck surface wave term. Zenneck's (exact) solution is expressed in cylindrical coordinates by Equation (A.35)

$$E_{2z} = C\left(\frac{-\gamma}{\omega\varepsilon_o}\right) e^{-u_2 z}\, H_0^{(2)}(-j\gamma\,\rho). \tag{7}$$

So, following Burrows' ansatz, the Sommerfeld solution is calculated as the RMS sum of Equation (6) plus Equation (7).

Similar expressions prompted Burrows to then calculate and plot the predicted fields in the region of interest. Equation (4) predicts the unattenuated field, Equation (6) is the Norton ground wave prediction (which blends smoothly with Weyl's solution out in the far zone), Equation (7) predicts the Zenneck surface wave alone, and according to Burrows, the sum of Equations (6) plus (7) gives Sommerfeld's 1909 prediction. These are illustrated in Figure 6.

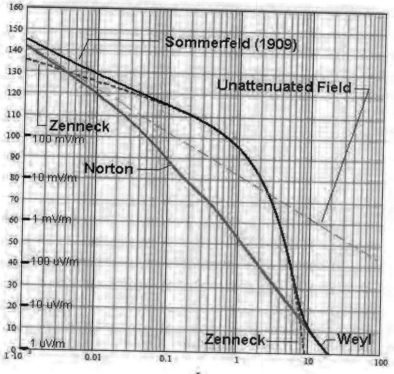

Figure 6. Theoretically predicted electric field strengths (E in dBμV/m) vs. range (r in km) for ground wave propagation on Seneca Lake. ($\varepsilon_r = 82.1$, $\sigma = 0.045$ mhos/m, f = 150 MHz) Note how Sommerfeld's solution transitions to Weyl's solution out beyond where the exponentially decaying surface wave contribution is negligible. As the influence of the Zenneck surface wave term diminishes, Sommerfeld's 1909 solution morphs smoothly back onto the Norton-Weyl curve.

Norton's procedure extracts the surface wave component from Sommerfeld's 1909 solution. Because of a saddle-point integration path, Weyl's solution also excludes the Sommerfeld surface wave singularity. The predicted and measured fields in Burrows' heroic Seneca Lake experiment are shown in Fig. 7. [106],[99],[107],[123, Fig. 6]

[As a side-note (the voice of experience), it is really difficult for theoreticians and armchair experimentalist reading this to actually grasp the unmentioned obstacles and demanding challenges confronted in performing these sensitive experiments under such unconventional circumstances. Yet, Burrows and his assistants managed to make this critical experiment happen with adequate (and admirable) precision . . . enough so that Dr. Fry and the attendees of that 1935 Bell Labs meeting, and virtually everyone concerned with RF propagation since then (including signatory governments of international radio treaties) accepted his results as confirmation of Weyl's theory over Sommerfeld's.]

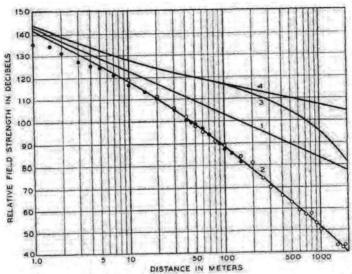

Fig. 7. Burrows' measured Seneca Lake data over a range of 1 m to 2 km. Curve 1 is the lossless inverse field. Curve 2 is the van der Pol-Norton *ground wave* over a lossy medium (Seneca Lake) with measured data. Curve 3 is the Sommerfeld-Zenneck *surface wave* over a lossy medium (Seneca Lake). Curve 4 is the lossless dielectric surface wave. Curve 1 (the "inverse field") varies as 1/r, and curve 4 varies as 1/√r. Burrows used a dipole antenna at 150 MHz. [106]

Burrows gives the measured relative permittivity of Seneca Lake water as $\varepsilon_r = 82.1$ and the measured conductivity as $\sigma = 0.045$ mhos/m, which is surprisingly greater than what is usual for a freshwater lake. There are several large salt mines under Seneca Lake. Near Watkins Glen, one is operated by Cargill, and another is operated by US Salt. Perhaps these may account for Burrows' sizeable values for σ. [Burrows gives $\sigma = 4.05 \times 10^8$ *electrostatic units* in his February 1937 paper, [106, p. 224] which, according to his January 1937 publication, [120, p. 74] may be converted as $\sigma_{MKS} = \sigma_{esu} \times 1/(9 \times 10^9)$.] The *complex* Brewster angle at a frequency of 150 MHz would be 76.1 - j2.3 degrees, measured from the surface normal.

There are two additional details to be noted from Figure 7. First, as the conductivity increases, the *Norton ground wave* (a radiation field with E_z and H_φ in-phase, carrying real power - watts), curve 2, morphs into curve 1, the unattenuated inverse field (which varies as 1/r). Secondly, as the conductivity decreases, the Sommerfeld-Zenneck wave (a surface wave with E_z and H_φ in complex phase relation) increases until it matches the lossless, perfect dielectric field, which falls off as the $1/\sqrt{r}$ curve shown as curve 4. Kahan and Eckart comment,

> "The discrepancy between the experimental points and curve 3, which is a plot of Sommerfeld's formula, is so great that there can be no doubt as to the incorrectness of the latter . . . the Sommerfeld curves predict a field strength about 100 times that measured by Burrows." [123]

With this 40 dB observational difference, Burrows now seemed to have decisive *experimental* evidence to resolve the analytical dilemma faced by T.C. Fry, W.H. Wise, *K.A. Norton*, and C.R. Burrows in their 1935 conference at Bell Labs. Recall that it was T.C. Fry, an industrial mathematician, who proposed performing an experiment to resolve whether the Sommerfeld or the Weyl formulation was correct. Burrows' resolution, [120] based on his 150 MHz data, was plainly that,

> "These *experiments* on the propagation of 2-meter waves over Seneca Lake have shown that the surface wave component of Sommerfeld is *not* set up by simple antennas... The results prove *conclusively* that simple antennas *do not* generate a Sommerfeld surface wave." [93], [99], [106], [107]

His Bell Labs colleague, W. H. Wise[*] (who had previously studied the analytical origin of Sommerfeld's surface wave [84], [101-103]) stated the experimental results this way,

> "Burrows presents experimental data supporting the correctness of the Weyl-Norton values and raises the question as to whether a surface wave really is set up by a radio antenna. A vertical current dipole *does not* generate a wave which at great distances behaves like Zenneck's plane surface wave." [100]

However, commenting on Burrows interpretation of these results, Professor Banos wrote:
> "Burrows concludes that his experimental results prove *conclusively* that simple antennas do not generate a Sommerfeld surface wave and, in further support for his conclusion, he cites the work of Wise (1937) and Rice (1937), who had obtained asymptotic expansions showing that that the [surface wave] term in Sommerfeld's resolution is cancelled ...   That is, Wise and Rice showed that, *asymptotically*, the complete result does not exhibit the electromagnetic surface wave of Sommerfeld, completely in accord with our own findings ... But it is important to observe that this does not mean, as one might erroneously read by implication, that the surface wave term does not belong in Sommerfeld's (correct) formula in the first place. Thus Burrows (1937) had no right to invoke the results of Wise and Rice ... The expansions of Wise and Rice merely demonstrate that the

---

[*] Dr. W. Howard Wise, (Ph.D. Cal. Tech, 1926), joined the R&D Department of AT&T in 1926.

[surface wave] term in Sommerfeld's resolution altogether disappears *in the asymptotic range...*" [79, p. 155]

(Also see the comments of Ott. [104]) In 1941 Julius Adams Stratton, professor of physics at MIT (later President of MIT) wrote,

"Burrows has pointed out that numerically the transmission formulas based on Sommerfeld's results differ from those of Weyl's by just the surface wave term, and he has made careful *measurements* that support the results of Weyl." [2, p. 585]

So, what is going on? We appear to have a mathematically correct theory (Sommerfeld's) which predicts what *is not* observed, and an erroneous theory (Sommerfeld's with the Norton/Niessen mythical error 'correction' removing the surface wave contribution) predicting what *is* observed. [Yogi Berra is reported to have said, "In theory there's no difference between theory and practice... In practice, there is."] What happened?

6.0 Pole-Residue Excited Waves

Theoretical analyses notwithstanding, the conclusion to be drawn from the experiment on Seneca Lake in New York would seem to be quite simple: a vertical *current* element launches the Norton *ground wave* (a radiation field proportional to the square root of the power) but, consistent with the Weyl-Norton theory, it does *not* launch the Sommerfeld-Zenneck *surface wave*.

However, Burrows' experiment does not address Norton's 1937 comment:

"The final establishment of Sommerfeld's view that the surface wave is similar to a guided wave on a wire must await further theoretical and experimental studies." [58]

[Historically, this was later observed and developed by Georg Goubau, [124], [125] who, interestingly, had obtained his doctorate under Jonathan Zenneck at the Munich Technical Institute in 1931. At the close of WWII, Dr. Goubau was brought to the US and served as a technical consultant at Fort Monmouth, NJ, as part of "Operation Paperclip". It was there, where he developed the Goubau line and invented that celebrated broadband *electrically small antenna* that bears his name.]

Recall that it was Sommerfeld's guided wave query that motivated his 1909 analysis. There is an analytical distinction to be drawn, which is inherent in the wave equation and may clarify some of these issues. It goes back to Sommerfeld's lectures and was most clearly described by Friedman.

The wave equation is a linear operator: how is it that there exist two different solutions at all? What is happening is the effect of boundary conditions. It has been asserted that in the Sommerfeld integral the branch cuts are contributing radiation fields and the residue from the pole is contributing the surface wave. The wave equation, itself, contains the distinction between the nature of radiation and guided waves.

In empty space (and in the presence of an infinite, smooth, perfectly conducting plane) the wave equation is a differential operator whose eigenfunctions possess a *continuous* spectrum of eigenvalues on the complex wave-number plane.

(This is called the radiation field, and those propagating fields are called "Hertzian waves".) However, in the presence of a conducting boundary [126] the wave equation plus boundary conditions lead to a spectral representation composed of a continuous spatial spectrum *plus* a sum of *discrete* spatial spectra [127], [81, pp. 474-477]. The discrete spectrum of guided wave modes and the radiation field (which has a continuous eigenvalue spectrum) satisfy the orthoganality integral relation. [81, p. 483], [170], [171] ("...the Zenneck surface wave must be orthogonal to the radiation field of any source distribution." [39])

(1) The *continuous* part of the eigenvalue spectrum (corresponding to *branch-cut integrals*) produces space waves (radiation).

(2) The *discrete* spectra (and corresponding *residue sum* arising from the poles enclosed by the contour of integration) result in traveling waves that are exponentially damped in the direction transverse to the propagation. (These surface waves are "guided transmission line modes", i.e. - *"non-Hertzian waves"*.)

(Weyl's path integration harvests only a continuous spectrum of eigenvalues.) The analytical particulars associated with these intriguing topics are treated in elegant detail in the lectures of Sommerfeld [41, pp. 188, 196, and 240] and in the classic text by Friedman, [82, pp. 214, 283-286, 290, 298-300] as well as many other advanced texts.

To illustrate the simple distinction between radiated and guided fields, consider Fig. 8, which compares propagation curves on a log-dB plot for RF links of length r(km) composed of either a waveguide transmission line or two separated horn antennas.

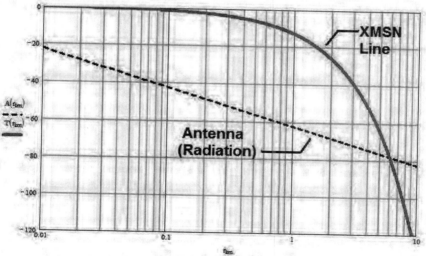

Figure 8. Amplitude (20 dB/div) vs Range (km) comparison for free-space radiation link (dashed) and waveguide propagation (solid). Recall that radiation fields fall off geometrically, as the spreading factor 1/r. Transmission line fields decay exponentially as $e^{-\alpha x}$ for longitudinal guides and $e^{-\alpha\rho}/\sqrt{\rho}$ for radial

waveguides. [Take f = 1 GHz, G_T = G_R = 15 dB, and $\alpha_{waveguide}$ = 13 dB/km. Employ the classic Friis formula: $A(r_{km})$ = G_T + G_R – 92.44 – 20 log (r_{km}) – 20 log (f_{GHz}). Also note that: $T(\ell)$ = 20 log ($e^{-\alpha\ell}$).]

In this case, the radiation field (dashed) is for horn antenna radiation, and the transmission line mode (solid) is the field strength along a conventional waveguide or a low-loss coaxial cable having the same input power and operating frequency. Notice that the waveguide link is stronger than the antenna link out to 6 km, where their responses cross over. After that, the antenna link wins. (*When path length doubles, cable losses double in dB but antenna link signals decrease by only 6 dB. As distance is increased, antennas eventually have lower losses than any cable.*" [128] Eventually, radiation spreading loss is less than transmission line attenuation.)

The shapes of the curves for radiation and for transmission line propagation are different. The radiation field's curve (dashed) falls off geometrically (1/r) and is a straight line on a log-dB scale. The transmission line curve (solid), on the other hand, has the characteristic exponential decay of $e^{-\alpha r}$ and exhibits a distinctive knee, very similar to Zenneck wave propagation. [The major dissimilarity between the cable transmission line curve in Figure 8 (which varies with distance as $e^{-\alpha\ell}$) and the Zenneck component in Figure 6 is that the latter varies more like a *radial transmission line* (radially varying as $e^{-\alpha\rho}/\sqrt{\rho}$) and this gives its curve a tilt before the exponential decay predominates. See Figure B.1, below.] Antennas want to give you 1/r radiation fields. But, if they are too small electrically, the radiation component of their fields is choked down while the reactance predominates; they become poor radiators and leave only large reactive near fields varying as $1/r^2$ and $1/r^3$. The inverse-distance fields exhibit geometrical spreading loss. Wave *guiding* structures, on the other hand, want to give you $e^{-\alpha r}$ fields (exponential propagation loss). Apparently this distinction occurred to Norton, who wrote,

"It will be noted that … our [ground] wave is *not* attenuated exponentially at large distances as was the Zenneck wave; this is due to the fact that the Zenneck wave was a plane wave guided along a plane imperfectly conducting surface, while our [ground] wave originates in an antenna and its 'attenuation factor' varies with distance first exponentially and finally, at large distances, inversely with the distance." [58]

This fundamental distinction between guided waves and radiation fields is further illuminated when we note that, in empty space, the wave equation is a differential operator whose eigenfunctions possess a *continuous spectrum of eigenvalues* on the complex wave-number plane. (This is the radiation field, and the propagating fields are called "Hertzian waves".) However, in the presence of a conducting boundary the wave equation plus boundary conditions lead to Sommerfeld's integral and an eigenfunction spectral representation composed of the continuous spectrum *plus* a sum of discrete *spatial* spectra, corresponding to $-2\pi j \times \Sigma$ (residues). (The negative sign arises for $e^{+j\omega t}$ time-variation when traversing the contour in the clockwise direction.) [81, p. 490], [129], [116, p. 610]

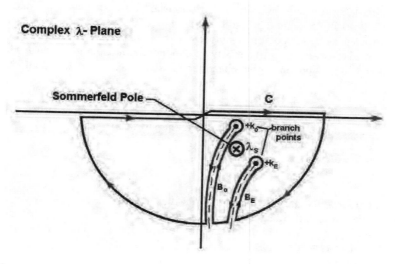

Figure 9.   **The singularities of Sommerfeld's 1909 solution. [41, p. 251], [130], [108, p. 66)] (See Appendix A below.) The surface wave pole, branch-points, branch cuts, and integration contour are for $e^{+j\omega t}$ time variation.**

(Discrete propagating and degenerate modes in waveguides are characterized by a string of complex poles on the λ-plane.) The analytical particulars associated with these topics are treated in elegant detail in the classic text by Friedman [82] and in advanced texts [41, pp. 182-200, 236-289, 295-296], [81, pp. 453, 474-477], [79 pp. 149-158], [131, pp. 261, 511], [133, pp. 377-385], [116, pp. 448-476] and papers. [108], [37] Although this distinction is common knowledge among EEs, relatively few consider the inherent analytical basis for this aspect of the wave equation's Green function. As can now be appreciated, this wave-number domain distinction is just as deep-rooted as the more casual space-domain distinction graphically illustrated in Figure 8. "The radiation field has a continuous eigenvalue spectrum." [81, p. 484] Stated another way,

"The *branch-cut integrals* represent the radiation field with a continuous eigenvalue spectrum, while *residues at the enclosed poles* are the surface-wave modes." [81, p. 491]

Physically, the <u>continuous</u> part of the eigenvalue spectrum (corresponding to *branch-cut integrals*) produces space waves (radiation, including Norton's ground wave). The <u>discrete</u> spectra (and corresponding *residue sum arising from enclosed poles*) result in traveling waves ("waveguide modes" [129, pp. 133, 139-140]) that are exponentially damped in the direction transverse to propagation. (These *surface waves* are like "transmission line modes", i.e. - *"non-Hertzian waves"*.)

In 1955 a special symposium on Normal Mode Theory (including surface waves) was convened at the Navy Electronics Laboratory in California under the chairmanship of Sergi Schelkunoff. Professor Banos led the discussion on surface waves, but Schelkunoff read a lengthy description published by Zenneck to the audience [30, Section 139, pp. 248-255, and the first footnote on p. 250], which misinterprets the nature of surface waves as being governed only by a $r^{-1/2}$ spatial dependence throughout the radiation zone. However, the surface wave's exponential attenuation is stronger than the geometrical $r^{-1/2}$ decay beyond some characteristic distance. Banos called attention to the fact that, even if the Zenneck wave is excited,

> "... *at some intermediate range* the residue (or surface wave term) has an $r^{-1/2}$ dependence and hence is the dominant part of the solution. However, for large r the branch-cut integrals asymptotically cancel the residue term." [134]

The $r^{-1/2}$ factor is only part of the surface wave expression. As Dr. Wait points out, because of its *exponential* decrease with range, the Zenneck-wave contribution to the solution is negligible out beyond some characteristic distance, and the $r^{-1/2}$ dependence does *not* dominate as $r \rightarrow \infty$.

> "... it is apparently in the intermediate range (where the numerical distance $|p| \approx 1$) where the Zenneck wave is most significant. For larger distances, the ultimate exponential decay of the Zenneck wave takes it out of contention. However, it should be noted that the *intermediate range* can be quite large for frequencies in the vicinity of 1 MHz ..." [135, pp. 71-72]

For propagation distances beyond this "knee", or 1/e value, one is left with only the Norton ground wave from the antenna [the Weyl solution], which dominates along the earth as the ground wave radiation field so familiar from AM broadcasting.

However, at ELF, *if it could be excited* the surface wave component can reach the antipode before you get to the "knee" of the residue's contribution to the propagation curve, and global wave interference patterns (stationary waves) become a possibility for coherent sources. At higher frequencies, as in Figure 8 where the exponential attenuation of a transmission line mode for distances ρ out beyond the cross-over point, the advantage of guided waves such as the Zenneck wave falls off from view. Collin, however, did not miss this feature and, prior to subsequently asserting the fallacy of Norton's "Sommerfeld error" in 2004, he observed in his 1985 text,

> "... the residue term is not the dominant contribution from the [Sommerfeld] integral. When the integral is evaluated asymptotically for large ρ it turns out that the surface wave term is cancelled and what is left over is a much more rapidly attenuating surface field, which is sometimes called the *Norton* [ground] *wave*. This [Norton] field is *not a true surface* wave but nevertheless is generally referred to as 'the [Norton ground wave]' field." [133, pp. 381-382]

Apparently, in Burrows' Seneca Lake experiment (as with virtually all conventional radio antennas today) the current source did not physically excite the *discrete* eigenvalue spectrum of the wave equation but rather only the continuum part of the eigenvalue spectrum, resulting instead in the so-called Norton ground wave: a radially outward propagating wave for which E_z and H_φ are in-phase, and the energy is lost forever like radiation. But, why was this so? There is one further feature of the Sommerfeld solution that brings the resolution of the Zenneck wave controversy into focus.

7.0 The Noether-Fock-Ott Resolution

The significance of Sommerfeld's silence concerning the "sign error" asserted by Norton, Niessen, and others now becomes apparent. In spite of the fact that Sommerfeld's 1926 paper gives results consistent with Weyl's 1919 approach, there is no mention made in it of a 1909 "sign error". Wait, who personally knew Kenneth Norton, has said (on multiple occasions) that Sommerfeld never acknowledged making a sign error in the lower limit of an ensuing integration decomposition. See Equation (A.20), below. [118], [90], [117, p. 249] On the other hand, Paul Epstein (a former student of Sommerfeld's) asserted that,

> *"Finally in 1935, Sommerfeld himself* [130, p. 932] *conceded that the surface wave has no reality. Referring to Fritz Noether,* [136, p. 167] *he attributed this to an inaccuracy in the evaluation of his general solution."* [137]

[We think that Epstein's remark could be misunderstood. No, Sommerfeld did not say that a surface wave "has no reality". Yes, (in 1935) Sommerfeld did grant that he made *approximations* in the evaluation of his closed-form integral solution that were inadequate in regions that are not asymptotically large. And, he explains why, *"... these approximations were inadaquate. The reason for this is that in Figure* (9 above), *the singular points of P and Q_1, i.e. $\lambda = \lambda_s$ and $\lambda = k_o$, practically coalesce, ..."* [130, p. 932]
　　　　The context of this 1935 textbook remark by Sommerfeld is as follows. After treating surface waves on a conducting plane, leading to Zenneck waves, in a previous chapter (Chapter XX1, §1 and §2), Sommerfeld then exploited a correlated similarity between those waves and those emerging from his antenna field analysis (Chapter XXIII, §1, section 5, p. 932). Here is our translation of the passage in question.
　　　　"Through this correlation we have provided a demonstration that the plane waves studied previously are contained in the wave complex which emanates from a linear transmitter. In the original publication by the author on the theory of dipole radiation (1909), great significance was placed upon this because it appeared to be in accord with the earlier developments of Zenneck (1907) and Uller (1903). However, this conformance is only formal. Actually, the P component of the wave complex is accompanied by the two components Q_1 and Q_2, (i.e., $\Pi_o = P + Q_1 + Q_2$), and *cannot* be separated (isolated) from them. The latter components were characterized as space waves on the basis of certain computational approximations, which should be individually characteristic of propagation in the upper and lower media. However, these approximations were inadaquate. (On this point, friendly communication from F. Noether and V. Fock is acknowledged.) The reason for this is that in Figure (9) [above], the singular points, $\lambda = \lambda_s$ and $\lambda = k_o$, of P and Q_1, respectively, practically coalesce, i.e. for large |k|. (By comparison, the Q_2 component can always be neglected.) Therefore, as has been said, wave types P and Q_1 cannot be separated from one another. There should be no conditions under which a pure surface wave (type

P) develops and represents the main component of the wave complex [from a radiating antenna]. As a result, we *cannot* explain the overcoming of the earth's curvature by the phenomenon of of surface waves. On the contrary, it must be due to other circumstances; refer to §4 [p. 963, 'Wireless Telegraphy Around the Earth']." [130, p. 932]

Sommerfeld was not admitting a sign error or the nonexistence of surface waves, but rather was emphasizing, as he did later in his 1949 book chapter (titled "Problems of Radio") that, *"...we cannot explain the overcoming of the earth's curvature by the phenomenon of surface waves."* [41, p. 255] More explicitly, Zenneck surface waves were not the physical mechanism for overcoming the earth's curvature in Marconi's 1901 transcontinental radiowave signals: the phenomenon responsible was skywave propagation and ionospheric reflection.

There is another interesting sidelight in the history of the Sommerfeld "sign error". Recall the temporal order of the evolving events. The Budd Lake, New Jersey experiment was in September 1933. Sommerfeld published his chapter on Wireless Telegraphy in the Frank-von Mises book in 1935. The correspondence to Sommerfeld from Fritz Noether (University of Breslau and later Tomsk Polytechnic Institute) and Vladimir Fock (Leningrad University), pointing out the restriction emerging from the *virtual co-location of the Zenneck pole and the branch-point, λ_s and k_o,* respectively, and the inadequacy of the approximations that Sommerfeld had used is acknowledged by Sommerfeld in his chapter in the 1935 book: *Differential und Integralgleichungen in der Mechanik und Physik, Vol. II,* by Frank and von Mises. [130, p. 939] (Noether's book chapter was subsequently published in 1931 in the book *Funktionentheoretie ünd Anwendung in der Technik, Vol. II,* which was translated and published by MIT in 1942. [136]) Noether published the following comment.

> "In his 1909 paper, Sommerfeld attempts to provide reasons for it [the separation of Q_1 and P] by means of a development of series into the function under the integral sign (p. 703) [the integral expression for Q_1]. But it is in the neighborhood of the pole P that this development has to become divergent; the series development, therefore, does not in a sufficient degree take account of the influence of the pole upon this part of the integral, which part is directly opposite to the previously mentioned residue." [136, p. 180]

[Aside: Fritz Noether (1884-1941), younger brother of Emmy Noether, while Professor of Mathematics at the Tomsk Polytechnic Institute in the Soviet Union was arrested as a German spy by the NKVD during the 'Great Purge'. While in prison he was sentenced to death and shot. Sadly, 47 years too late, in 1988 the Supreme Court of the Soviet Union declared him innocent of any crimes.]

The thorny issue is the evaluation of the integral in Equation (A.7) or Equation (A.8) in Appendix A.1 below, i.e. - the presence of the Sommerfeld-Zenneck pole so close to the saddle-point integration path. Specific integration details were published by Noether [130, p. 939], [136, pp. 177-183] and by Heinrich Ott (1894-1962). [138], [139] Ott was also a former student of Sommerfeld's at Munich. The detailed saddle-point analysis has recently been addressed by Collin [108], and again by Sarkar et al [160], [161], and the interested reader is referred to their publications. The bottom line is that the Zenneck-Sommerfeld pole contribution is absent in the final complete field expression for the integral evaluation. Sommerfeld's 1909 integral was correct, but the particular evaluation published is restricted to only large values of the numerical distance ... out where his residue contribution would actually be negligible.

The conference at Bell Labs was in 1935: Norton announced the "sign error" discovery in Nature in June 1935. Burrows' Seneca Lake measurements were announced in Nature in August 1936. The publications by Ott detailing the application of the saddle-point method of integration, which avoids the Zenneck pole integration approximation made by Sommerfeld (resulting in Zenneck-free propagation), were in 1942 and 1943. Against this framework, consider the Norton-Sommerfeld correspondence described by Professor Wait as follows.

"The late Kenneth Norton, an eminent radio engineer, exchanged numerous letters with Sommerfeld. An example from Sommerfeld to Norton [September 5, 1937] was written while the former was on holiday at Zirmerhof in the Austrian Tyrol. Sommerfeld never agreed an error or miscue had ever been made. Two later letters [both dated September 20, 1937] from Norton indicated his views on the separation of surface waves and space waves. Sommerfeld acknowledged Norton's communications, and suggested he compare his results with those of van der Pol and Niessen (1930). [109] Of course, there was consistency here, because Norton based his papers (1935, 1936, 1937a, 1937b) on van der Pol and Niessen!" [118]

Elsewhere, Wait also wrote that, *"It is evident in the exchange of correspondence between Norton and Sommerfeld that the error was never acknowledged."* [90] We speculate that Sommerfeld understood that he had not made any *sign error*, but rather the complicated issue was what Fritz Noether and Vladimir Fock had called to his attention: the approximation he had made in performing the final integration, while formally appropriate for asymptotic field-strength predictions (out where the exponential dissipation has attenuated the effect of the pole) it was inadequate for practical calculations at reasonable ranges (i.e., relatively small "numerical distances"). The approximations made by Sommerfeld in 1909 *limit the range of usefulness to very large distances.* One suspects that these on-going (1931-1943) investigations of the consequences of the close proximity of λ_s and k_o, and the analytical developments applying saddle-point integration, were, in Sommerfeld's opinion, just too involved to broach and explain to the American radio engineer through the post office. Sommerfeld doesn't even present the analytical details in his 1949 textbook, but rather refers the reader directly to Ott's publications with the words,

"This has been carried out most completely by Ott. However, we have to forego the presentation of his results in order not to get lost in the details of the problem." [41, p. 256]

The closed form expressions given in Equations (A.1) and (A.4) were rigorous and correct. And, specialized formal publication was a better venue to explain such issues. In 2004, Collin pointed out that there was no sign error, the Sommerfeld solution is okay and even his integral approximations are fine, ... they just restrict application of the resulting expressions to only regions far out (i.e., asymptotically at large numerical distances). And, out in this region the solutions of Weyl (whose integration does not capture the discrete pole) and Sommerfeld give identical results. Collin remarked,

"... both Norton's and Niessen's manipulations of Sommerfeld's solution and claiming that an error in sign had been made has no merit. Sommerfeld's first solution is given by his asymptotic series plus the Zenneck surface wave. His second solution is given by a power series, which is consistent with his first solution. ... There are inherent *limitations* in Sommerfeld's [post-integration] solution, but they are not caused by a sign error." [108]

In spite of the fact that the Seneca Lake experiment did not satisfy the asymptotically large numerical distance *restriction* on field strength determination, it was unjustifiably used to substantiate the Niessen-Norton-Rice assertion of a "Sommerfeld sign error".

8.0 Summary

While this article has ranged broadly over radiowave history, its focus was primarily concentrated on the significance of the legendary Seneca Lake experiment. We have also attempted to provide the enquiring reader with specific and detailed citations to follow the pathway through the historic wave propagation labyrinth. The purpose of the paper was three-fold, and can be summarized as:

1. To provide a general introduction to the salient features of RF waves propagating over lossy earth.

2. To point out a discrepancy in the conflicting assertions made by those who were participants about *who actually discovered Sommerfeld's mythical sign 'error'*. (The authors have been unable to resolve that issue and harmonize published assertions.)

3. But, mostly to call attention to the significance of the critical (Wait calls it "crucial") Bell Labs experiment performed at Seneca Lake, NY in 1936.

It can even be asserted that the Seneca Lake investigation was one of the most *far-reaching* experiments in all of 20th century physics! (This audacious assertion is in line with Balthazaar van der Pol's comment above, that it had a substantial impact on "almost every nation" on the planet!) Surely, its intricate technical analyses and consequent widespread influence on contemporary civilization *compels* prominent consideration, not only in modern engineering curricula and textbooks on electromagnetics, but also in broader discussions exploring the critical influence of physics and the practice of engineering (and radio science in particular) upon society and culture during the past century.

APPENDIX – A
ANALYTICAL CONSIDERATIONS
"It is of great advantage to the student of any subject to read the original memoirs on that subject…" Maxwell's Treatise, 1st edition, 1873

A.1. The Sommerfeld Solution

Sommerfeld solved the antenna-excited wave equation subject to boundary conditions that include the presence of a lossy earth. [40], [41, pp. 236-289, 295-296] His detailed analyses, though the most ambitious and intricate in the early history of radio science, are also among the most elegant in the chronicles of electromagnetism. [41, p. 249, Eq. 9], [2, pp. 573-587], [79, p. 150], [131, pp. 506-512 and 554-559], [116, pp. 447-476 and 609-620], [117, p. 249] His analytical expressions have been geometrically interpreted as a direct wave, a ground reflected wave, plus a ground wave propagating along parallel to the earth. His heroic analysis and "general solution" for a small vertical doublet oscillator gives the z-component of the Hertz potential (generalizing Hertz's free-space solution for incremental dipoles) to now satisfy the boundary conditions for the presence of a smooth, homogeneous, lossy ground. For $z > 0$ and $e^{-i\omega t}$ time variation, it corresponds to [116, pp. 451-452]

$$\Pi_o (z > 0) = \frac{e^{ik_o R_o}}{R_o} + \int_0^\infty f_o(\lambda) J_o(\lambda \rho) e^{-\sqrt{\lambda^2 - k_o^2}\,(z + h_T)} \lambda\, d\lambda$$

(A.1)

(which is essentially Sommerfeld's 1909 Eq. (11a) when the doublet antenna's height above ground, h_T, is nonzero). The parameter λ is Sommerfeld's radial propagation constant (not wavelength), the source-observer distance is $R_o = \sqrt{\rho^2 + (z - h_T)^2}$, and a quantity

$$f_o(\lambda) = \frac{1}{\sqrt{\lambda^2 - k_o^2}} \frac{k_E^2 \sqrt{\lambda^2 - k_o^2} - k_o^2 \sqrt{\lambda^2 - k_E^2}}{k_E^2 \sqrt{\lambda^2 - k_o^2} + k_o^2 \sqrt{\lambda^2 - k_E^2}}$$

(A.2)

emerges as a result of satisfying the boundary conditions imposed by the earth. With the stentorian tenor of an 'ordinarius professor', Sommerfeld (1909) flatly states, "By introducing $f_o(\lambda)$ into Equation (A.1), *our general solution is completely specified.*"

By employing the Sommerfeld integral identity [133], [118, p. 9]

$$\frac{e^{ik_o R}}{R} = \int_0^\infty \frac{J_o(\lambda \rho) e^{-\sqrt{\lambda^2 - k_o^2}\,|z - h_T|}}{\sqrt{\lambda^2 - k_o^2}} \lambda\, d\lambda \ ,$$

(A.3)

the components of Equation (A.1) may be combined and, when $h_T \to 0$, the expression reduces to [2, p. 577], [132]

$$\Pi_o(z > 0) = 2k_E^2 \int_0^\infty \frac{J_0(\lambda\rho)}{k_E^2\sqrt{\lambda^2 - k_o^2} + k_o^2\sqrt{\lambda^2 - k_E^2}} e^{-\sqrt{\lambda^2 - k_o^2}\,z}\,\lambda\,d\lambda\,, \qquad (A.4)$$

which is Sommerfeld's 1926 Eq. (9a). In 1926 Sommerfeld declared, "... this is our solution *in rigorous, closed form.*"

Returning to Equation (A.2), note that $f_o(\lambda)$ may be written either as

$$f_o(\lambda) = \frac{1}{\sqrt{\lambda^2 - k_o^2}} \frac{k_E^2\sqrt{\lambda^2 - k_o^2} + k_o^2\sqrt{\lambda^2 - k_o^2} - 2k_o^2\sqrt{\lambda^2 - k_E^2}}{k_E^2\sqrt{\lambda^2 - k_o^2} + k_o^2\sqrt{\lambda^2 - k_E^2}} \qquad (A.5)$$

or as

$$f_o(\lambda) = \frac{1}{\sqrt{\lambda^2 - k_o^2}} \frac{2k_E^2\sqrt{\lambda^2 - k_o^2} - k_E^2\sqrt{\lambda^2 - k_o^2} - k_o^2\sqrt{\lambda^2 - k_E^2}}{k_E^2\sqrt{\lambda^2 - k_o^2} + k_o^2\sqrt{\lambda^2 - k_E^2}}. \qquad (A.6)$$

Taking the distances from the elevated source, and its geometrical image, to the observation point as $R_{o,E} = \sqrt{\rho^2 + (z \mp h_T)^2}$, Sommerfeld's integral identity then permits one to write (for $z > 0$) either

$$\Pi_o = \frac{e^{ik_o R_o}}{R_o} + \frac{e^{ik_e R_E}}{R_E} - 2k_o^2 \int_0^\infty \frac{\sqrt{\lambda^2 - k_E^2}\,J_0(\lambda\rho)\,e^{-\sqrt{\lambda^2 - k_o^2}\,(z + h_T)}}{\sqrt{\lambda^2 - k_o^2}\left(k_E^2\sqrt{\lambda^2 - k_o^2} + k_o^2\sqrt{\lambda^2 - k_E^2}\right)}\,\lambda\,d\lambda \quad (A.7)$$

Or

$$\Pi_o = \frac{e^{ik_o R_o}}{R_o} - \frac{e^{ik_e R_E}}{R_E} + 2k_E^2 \int_0^\infty \frac{J_0(\lambda\rho)}{k_E^2\sqrt{\lambda^2 - k_o^2} + k_o^2\sqrt{\lambda^2 - k_E^2}} e^{-\sqrt{\lambda^2 - k_o^2}\,(z + h_T)}\,\lambda\,d\lambda \quad (A.8)$$

A solution has been expressed in contemporary terms by Collin, for $e^{+j\omega t}$ time variation [133] (Collin also did it for $e^{-i\omega t}$ time variation [108, see Sections 2 and 3]), in the form:

$$\Psi_z \sim \frac{e^{-jk_o R_o}}{R_o} - \frac{e^{-jk_o R_E}}{R_E} + 2n^2\,I \qquad (A.9)$$

where

$$I = \int_{\lambda = -\infty}^\infty \frac{\lambda H_0^{(2)}(\lambda\rho)\,e^{-\gamma_o(z + h_T)}}{2(n^2\gamma_o + \gamma_E)}\,d\lambda \qquad (A.10)$$

and $\gamma_{o,E} = \left(\lambda^2 - k_{o,E}^2\right)^{\frac{1}{2}}$ are for air and earth, respectively. We take $k_o = \omega\sqrt{\mu_o\varepsilon_o}$ and $k_E = \sqrt{\omega^2\mu\varepsilon - j\omega\mu\sigma}$. Consequently, $n = \dfrac{k_E}{k_o} = \sqrt{\varepsilon_r - j\sigma/(\omega\varepsilon_o)}$ is the complex index of refraction. The integrand in Equation (A.10) has a singularity (historically called the Sommerfeld-Zenneck pole), which is determined as a zero of the denominator. Collin calls attention to this pole and physically interprets it as follows:

"Note that there is a pole when[*]

$$n^2 \gamma_o + \gamma_E = 0 \tag{A.11}$$

which is the Zenneck surface wave pole. This occurs for

$$\lambda = \lambda_s = \frac{n}{\sqrt{n^2 + 1}} k_o = k_o \sqrt{\frac{\varepsilon_r - jx}{1 + \varepsilon_r - jx}} \tag{A.12}$$

... In principle, the contour of integration, which is the real axis in the λ-plane, can be deformed so that it encircles the pole. If this is done the residue at the pole gives the following *discrete-mode* solution

$$\Psi_s = \frac{\pi k_E}{n^4 - 1} \frac{n}{\sqrt{n^2 + 1}} \; H_0^{(2)}\!\left(\frac{k_E \rho}{\sqrt{n^2 + 1}} \right) e^{-jk_o(z + h_T)/\sqrt{n^2 + 1}} \tag{A.13}$$

which is called the Zenneck surface wave. When the horizontal distance ρ is large and $z = 0$, we find, by using the *asymptotic* expression for the Hankel function, that

$$\Psi_s \sim C \frac{e^{-jk_E \rho/\sqrt{n^2 + 1}}}{\sqrt{\rho}} \tag{A.14}$$

where $k_E = nk_o$ and C is a suitable constant. This field solution decays with distance like $\rho^{-1/2}$ instead of ρ^{-1} for free space propagation. The surface wave is guided by the interface, and its field is confined to the region close to the surface: this is why it decays like a cylindrical wave and not a spherical wave.

However, the residue term is *not the dominant contribution* from the integral. When the integral is evaluated asymptotically *for large ρ* it turns out that the surface wave term is cancelled and what is left over is a much more rapidly attenuating surface field, which is sometimes called the *Norton [ground] wave*. This field is *not* a true surface wave ..." [133, p. 381-382]

There are two critical observation to be made. First, the Zenneck wave contribution is not significant at great ranges (out where the separated contour integration approximation would actually be suitable), but, under certain surface impedance conditions it can be a

---

[*] Actually, the pole, λ_s, is found from the *rationalized* algebraic form of the denominator equation:

$$(n^2 + 1) \lambda^2 - n^2 k_o^2 = 0 .$$

major player at intermediate ranges and, as Wait suggests, "… is most effective near its 1/e range." [135] [Note that the presence of finite ground conductivity in $x = \sigma/\omega\varepsilon_0$ will compel exponential attenuation of the cylindrically propagating Zenneck mode. Because the Zenneck surface wave falls off as $e^{-\alpha r}/\sqrt{r}$, depending on the constitutive parameters the *exponential* knee of the Zenneck wave may have been long past (see Figure B.1 in Appendix B, below) and its contribution be negligible out in the radiation zone … even though under certain circumstances at closer distances the amplitude of the Zenneck surface wave may be several hundred times greater than the Norton ground wave (radiation) field. Being a *discrete-mode* guided radial transmission line field (arising from discrete poles), terrestrial surface wave energy decreases principally by attenuation, not radiation, without violating Sommerfeld's conventional (continuum-mode) radiation condition! The geometrical spreading loss is different for a radial transmission line than for free-space radiation.]

Secondly, as Sommerfled explained in the 1935 quote given in Section 6.0 above, "…*in Figure (8), the singular points, $\lambda = \lambda_s$ and $\lambda = k_o$, of P and Q_1, respectively* [P is the *surface* wave and Q_1 is the primary *space* wave], *practically coalesce … Therefore, as has been said, wave types P and Q_1 cannot be separated from one another.*" So, the above decomposition made by Collin in this 1985 antenna textbook quote is generally *unsuitable* for a simple radiating antenna (for example, one *not* over a stratified or a corrugated medium). However, it would be appropriate for a probe in a surface-wave radial waveguide having an *inductively reactive* surface impedance (e.g., corrugated or stratified), such as is considered in chapter 11 of Collin's 1960 textbook on guided waves. [81, pp. 453-508]

A.2. The Hertz Solutions

There are two particular cases worth special consideration:

(1) Sommerfeld (1909) showed (see his p. 687) that, for an unbounded homogeneous medium (i.e., $k_E = k_o$), the surface wave pole vanishes, and Equation (A.1) for $z > 0$ plus the solution for $z < 0$ reduce to

$$\Pi_o\,(z > 0) \;=\; \Pi_E(z < 0) \;=\; \frac{e^{ik_o R}}{R} \tag{A.15}$$

(where $R = \sqrt{\rho^2 + z^2}$), which is equivalent to Hertz's 1889 free-space radiation solution. [3], [131, pp. 477-481]

(2) Sommerfeld (see his p. 688) also showed that, for a perfect conductor or infinitely great dielectric constant in the lower medium ($k_E \rightarrow \infty$), his solution reduces to

$$\Pi_o\,(z > 0) \;=\; 2\,\frac{e^{ik_o R}}{R} \tag{A.16}$$

in air and $\Pi_E\,(z < 0) = 0$ in the lower medium. This is as expected over a ground plane for Hertz's original solution. (Usually the *modal analysis* for the simple Hertz problem is carried out in spherical coordinates [140], [141] instead of cylindrical coordinates. [131, pp. 477-481], [141, pp. 105-129, 133-161], [142])

A.3. The Norton Fields[92],[95],[122],[2, pp. 584, 573-587]

The integration in Equation (A.1) or (A.4) was the crux of the controversy and involved various approximations and several thorny issues (related to branch cuts and Riemann sheets). With the transmitting antenna on the ground $h_T = 0$, and the receiver near ground level $z = 0$, the integration (for $e^{+j\omega t}$ time variation) leads to

$$\Psi_z = \frac{I_o \ell}{2\pi j \omega \varepsilon_o} \frac{e^{-jk_o \rho}}{\rho} F(p) \tag{A.17}$$

with [133, p. 382], [2, p. 586]

$$p = -j(k_o - \lambda_s)\rho \approx -j\frac{k_o \rho}{2n^2}\left(1 - \frac{1}{n^2}\right) = |p|e^{-jb} \tag{A.18}$$

where the root λ_s is the radial wave-number of the surface wave, and (for Norton)

$$F(p) = 1 - j\sqrt{\pi p}\, e^{-p}\, erfc(j\sqrt{p}) \tag{A.19}$$

$$erfc(j\sqrt{p}) = 1 - erf(j\sqrt{p}) = \frac{2}{\sqrt{\pi}} \int_{j\sqrt{p}}^{\infty} e^{-u^2}\, du \quad . \tag{A.20}$$

The sign in Equation (A.19) and in the latter integral's lower limit follows the historical "sign error" choice introduced by Norton [58, p. 1195], Niessen [115], and others [90], [78, p. 167], [117, p. 254], [116, p. 609]. The complex quantity p is called Sommerfeld's "numerical distance" and b is the all-important phase of the numerical distance. (The sign of the exponent b in Equation (A.18) has been chosen in accordance with Fig. (A.1). [133, p. 382]) In his 1926 paper, Sommerfeld explains that, physically, the complex "numerical distance" is to be understood and calculated as the product of the range times the wave-number *difference* between the space-wave and the surface-wave traversed during the same time interval. [42, p. 1140], [41, p. 256] The (Norton) ground wave field then follows from the Hertz potential as [143, pp. 130, 202], [116, p. 315], [81, p. 174]

$$E_z = \left(k_o^2 + \frac{\partial^2}{\partial z^2}\right)\Psi_z \tag{A.21}$$

giving

$$E_z = \frac{-j\omega\mu_o I_o \ell}{2\pi} \frac{e^{-jk_o\rho}}{\rho} F(p) \quad , \tag{A.22}$$

(which, strictly speaking, is valid for $k_o\rho \gg 1$, i.e. well outside Wheeler's "radian sphere"). It follows from Equation (A.1) that the field strength at an arbitrary receiver above ground is composed of a contribution directly from the physical source, plus the specularly reflected field, plus a "ground wave" contribution. For observers on the boundary interface, the direct and reflected fields cancel, leaving only the *ground wave* described by Equation (A.17). Despite having been "adjusted" by Norton, the quantity F(p) is called the "Sommerfeld attenuation function" and its calculation involves careful attention, especially to the phase, b of p, which is the numerical distance. [144] As Wait has pointed out, F(p) may be regarded as the correction to the field of a vertically polarized dipole on the surface of a perfectly conducting plane when the plane becomes lossy. [90], [91, p. 391]

In 1923 M.V. Shuleikin recast the Sommerfeld solution into a form appropriate for calculation. [145] In 1931 Van der Pol [111] arrived at an empirical formula almost identical to Shuleikin's. Van der Pol represented this by the letter A, and his expression approximates Equation (A.19) when b < 5 degrees. [The exponent b is defined as the phase of p in Equation (A.18).] In 1935 Norton [92], [95] introduced an empirical additive correction factor for A that, using his phase angle convention, extends van der Pol's range to all positive values of b. The result makes it possible to calculate Equation (A.22) by employing the *unattenuated field* [146, pp. 190-191] in the azimuthal plane, which, from a simple Poynting calculation, is given by [147, pp. 23, 74]

$$E_{u,RMS}(r_k) = \frac{173.2\sqrt{P_{kW}\ G_T}}{r_k} \quad \text{mV/m} \tag{A.23}$$

where P_{kW} is the transmitter power in kilowatts, G_T is the gain of the transmitting antenna (in the azimuthal plane) over a perfectly conducting earth [148], and r_k is the range in kilometers. By multiplying this unattenuated field by the empirically determined van der Pol approximation to the attenuation function, the ground wave field strength prediction calculation becomes simply

$$E_N = A \times E_{u,RMS} \quad \text{mV/m}, \tag{A.24}$$

where A = |F(p)| could either be calculated directly or obtained from a set of published graphs. (This is actually known as the Shuleikin-van der Pol equation in the Russian literature. [147, p. 75]) These steps permit one to calculate the electric field strength from the specification of the antenna gain and input power without a knowledge of the actual antenna current distribution.

The van der Pol approximation (and Norton correction) [2], [122] for the ground wave attenuation factor is expressed by [92], [94, pp. 645-649], [133, pp. 384-385], [149]

$$A = |F(p)| = \frac{2 + 0.3\,p}{2 + p + 0.6\,p^2} - \sqrt{\frac{p}{2}}\ e^{-0.625p}\ \sin(b) \tag{A.25}$$

where, for vertical polarization, Norton and the FCC take [96, §73.184, p. 107], [150]

$$p = \frac{\pi}{x}\,\frac{r_k\,1000}{\lambda}\cos(b) \tag{A.26}$$

$$b = 2\,b_2 - b_1 \tag{A.27}$$

$$b_1 = \tan^{-1}\left(\frac{\varepsilon_r - 1}{x}\right) \tag{A.28}$$

$$b_2 \;=\; \tan^{-1}\!\left(\frac{\varepsilon_r}{x}\right) \tag{A.29}$$

$$x \;=\; 60\lambda\sigma \tag{A.30}$$

where now p is the magnitude of the complex "numerical distance" with r_k being the physical range ρ (in cylindrical coordinates) measured in kilometers. (Compare Figure A.1, which is from Terman [73, p. 676] and plots A = |F(p)| vs. |p|, with Norton [95, see his Fig. 1], Burrows [120, see insert between pp. 50-51], Jordan and Balmain [94, p. 647], Collin [130, p. 383], Wait and Frazier [144], [129, p. 42], or with the FCC (which plots |F(p)|/p vs. p). [96, Graph 20] And, this "flat earth" approximation is good out to ranges on the order of $50/f_{MHz}^{1/3}$ miles, which at f = 550 kHz would be 61 miles (98 km), and for Burrows'experiment at f = 150 MHz would be about 9.5 miles (15.3 km). [151, p. 625], [73, p. 677] [As an aside, we note that the above is the flat earth ground wave solution. The effect of earth curvature complicates the issue but efficient methods exist for spherical earth propagation calculations. [147, pp. 100-107], [152]-[158] (For readers merely interested in a good back-of-the-envelope approximation for estimating earth curvature effects on Norton ground wave propagation, the note by Trainotti [159] is recommended.)]

Instead of the Zenneck-Sommerfeld *surface wave*, what emerged above is the Norton *ground wave* (which is not a true surface wave), and the latter is quite different from the former. Beyond the "knee" the Zenneck wave is a much more rapidly attenuating field along the interface – see Figure 6 and the measured data of Figure 7. This is popularly called the "Weyl-Norton" solution because it is Norton's version of the Sommerfeld solution that gives the same results as Weyl's 1919 theory (i.e., a ground wave but no surface wave). The field strength graph is often scaled in dB above a microvolt per meter instead of RMS millivolts per meter. The conversion from mV/m to dBμV/m is simply $E_{dB\mu V/m} = 20\,\log E_{mV/m} + 60$.

Lastly, despite the widespread 20th century assumption of the validity of Norton's assertion concerning the sign error [based on Norton's "quick fix" and the apparent support of Burrows' Seneca Lake experiment], Collin has stated in no uncertain terms that an in-depth examination of Sommerfeld's, Niessen's, and Norton's analyses demonstrates unequivocally that such an assertion is a falsehood, and that Sommerfeld was correct. [108] (Collin, with good reason, actually calls the mid-1930's "sign error" assertion by Norton and Niessen a "myth".) When Sommerfeld performed the 1909 integration indicated in Equation (A.1), he employed certain *series approximations* that led to his Equation (47) and the Hertz potential

$$\Pi(\rho,0) \;\approx\; 2\,\frac{e^{ik\rho}}{\rho}\left[\,1 - i\,\sqrt{\pi p}\;e^{-p} - 2\,\sqrt{p}\;e^{-p}\int_{0}^{\sqrt{p}} e^{u^2}\,du\,\right] \tag{A.31}$$

with the resultant Sommerfeld attenuation function expressed by Equation (1) above. However, these approximations restrict the applicable range of Equation (A.31) to large values of the numerical distance, p, and limit its use to regions in the lower right in Figure A.1 (say, p > 50). Collin explains the consistency with experiment by stating that even though Sommerfeld's 1909 solution *"... is mathematically correct, it is only useful when the 'numerical distance' p is very large. ... For such large values of k_oR, the Zenneck surface wave would have decayed to a negligible value. This is the inherent limitation in Sommerfeld's procedure ... "* [108]

When Sommerfeld performed the 1926 integration indicated in Equation (A.4), he used a different set of approximations and was led to the expression [42, Eq. (16)], [41, p. 257], [130, p. 936], [104]

$$\Pi(\rho,0) \;\approx\; Q_1 + P \;\approx\; 2\,\frac{e^{ik\rho}}{\rho}\left[1 + i\,\sqrt{\pi p}\,e^{-p} - 2\,\sqrt{p}\,e^{-p}\int_0^{\sqrt{p}} e^{u^2}\,du\right] \quad (A.32)$$

Taking note of the properties of the error function, Sommerfeld's 1926 attenuation function (the bracketed term in Equation (A.32)) may be expressed as in Equation (A.19), which gives results consistent with Weyl's 1919 analysis. (The 1909 solution agrees, too, but only when it, as required by the analysis, is restricted to ranges for which p >> 1.) Niessen noted that the difference between Equation (A.31) and Equation (A.32) is the change in sign of the second term (which corresponds to a surface wave). [115] Comparing Equation (A.31), which does not fit Burrows' measured data in the 0-2 km range, with Equation (A.32), which does, Norton, Niessen, and others were led to the hasty (but mistaken) conclusion that Sommerfeld had made a sign error in his 1909 analysis, which, of course, was not the case. The early researchers mistakenly associated Sommerfeld's missuse of a formula outside its range of applicability as a physical flawr that could be "fixed" by changing the sign of the lower limit of Sommerfeld's integration.

As pointed out by Noether, Fock and Ott, the technical issue was a restriction placed on Sommerfeld's final expressions (which had been used by Rolf) resulting from the proximity of the Zenneck pole to the branch-cut in the integral approximation technique employed by Sommerfeld. [For example, Burrows' Seneca Lake data had ε_r = 82 and σ = 0.045 at 150 MHz. Using Equation (A.12) gives the relation λ_s = (0.994 − j 0.00004)k_o , which means that these singular points in Figure 9 are practically on top of each other. At 1 MHz in Central Ohio (ε_r = 15 and σ = 0.008), the same calculation gives λ_s = (0.9999 − j 0.0034)k_o , which again is situated overly close for contour integration by the method of residues without paying a severe price in restrictions on the useful range.] The remedy, as pointed out by Sommerfeld in 1949, [41, p. 286] involves saddle-point integration as first proposed by Noether [136], [130, footnote p. 932], skillfully executed in a detailed series by Ott [138]-[139], and subsequently enhanced by many others.

It is to be strongly emphasized that these compelling issues are not merely the technical curiosities of academics and scientific laboratories. The National prosperity turns on such things. These theories and the data measured by Burrows in Figure 7 not only affected the lives, reputations, and professional careers of the men shown in Figure 2, but also played a major decision-making role in the frequency allocations, demographic capture, and financial fortunes of the major commercial network empires, the various advertisers, and the thousands of 20th century medium frequency AM broadcast stations . . .   not to mention the careers of the famous news-readers, sportscasters, entertainers, radio personalities, recording artists, and politicians of the past 100 years! Today, the same issues involve the propagation and signal reliability associated with wireless remote control and wireless communication devices, as recently highlighted by Sarkar et al [160], [161], as well as regulatory bodies and the financial fortunes of related commercial enterprises. It is self-evident that there is hardly a person on the planet (or an astronaut on the moon [162]) that Sommerfeld's problem does not impact in some significant manner.

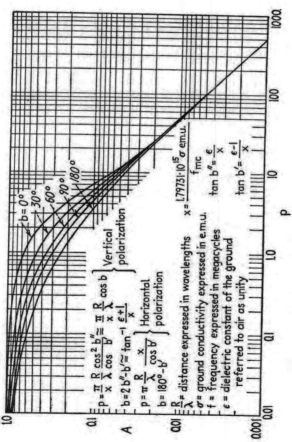

Figure A.1a. Sommerfeld attenuation function A = |F(p)| vs. numerical distance p over a plane earth. Norton [95, Fig. 1], FCC [96], redrawn by Terman [73, p. 676]). For MKS, use Equations (A.26)-(A.30).

Figure A.1b. Normalized Sommerfeld attenuation function vs. numerical distance after Norton. (FCC § 73.184. Graph 20.)

A.4. Zenneck's Solution

Zenneck's 1907 analysis [33] has been carried out in cylindrical coordinates by Barlow and Cullen (1953) [163], and published in book form by Barlow & Brown (1962) [124, pp. 6-12].   Neither they nor Zenneck actually specified a source configuration that can produce these field solutions to Maxwell's equations. [30, pp. 248-263], [164, pp. 729-744] Recently, Kukushkin has stated that the standard (or prevalent) point of view is that *actual* antennas cannot excite such a wave. [165] Using the geometry of Figure A.2 and $e^{+j\omega t}$ time variation, Zenneck's solution for out-going waves can be written as follows:

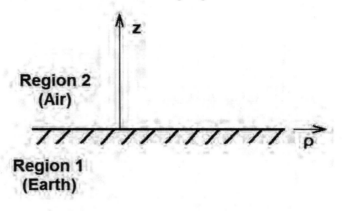

Fig. A.2. Orientation for a radially propagating cylindrical Zenneck wave.

<u>**In Region 2**</u> ($\rho \neq 0$ and $z \geq 0$, Air)

$$H_{2\phi} \; = \; C e^{-u_2 z} \, H_1^{(2)}\left(-j\gamma\,\rho\right) \tag{A.33}$$

$$E_{2\rho} \; = \; C\left(\frac{u_2}{j\omega\varepsilon_o}\right) e^{-u_2 z} \, H_1^{(2)}\left(-j\gamma\,\rho\right) \tag{A.34}$$

$$E_{2z} \; = \; C\left(\frac{-\gamma}{\omega\varepsilon_o}\right) e^{-u_2 z} \, H_0^{(2)}\left(-j\gamma\,\rho\right) \tag{A.35}$$

where:

$$\gamma \; = \; j\sqrt{k_o^2 + u_2^2} \quad \text{and} \quad u_2 \; = \; \frac{-j\,k_o}{\sqrt{1 \, + \, (\varepsilon_r - jx)}} \quad \text{with} \quad x \; \equiv \; \frac{\sigma}{\omega\varepsilon_o} \tag{A.36a,b,c}$$

<u>**In Region 1**</u> ($\rho \neq 0$ and $z \leq 0$, Soil)

$$H_{1\phi} = C e^{u_1 z} H_1^{(2)}(-j\gamma\rho) \tag{A.37}$$

$$E_{1\rho} = C\left(\frac{-u_1}{\sigma_1 + j\omega\varepsilon_1}\right) e^{u_1 z} H_1^{(2)}(-j\gamma\rho) \tag{A.38}$$

$$E_{1z} = C\left(\frac{-j\gamma}{\sigma_1 + j\omega\varepsilon_1}\right) e^{u_1 z} H_0^{(2)}(-j\gamma\rho) \tag{A.39}$$

where in Region 1 (Soil) we have

$$\gamma = j\sqrt{k_1^2 + u_1^2} \quad \text{with} \quad u_1 = -u_2\left(\varepsilon_r - jx\right). \tag{A.40a,b}$$

That this is an *exact* solution can be verified merely by substituting Equations (A.33)-(A.40) back directly into Maxwell's equations. Notice that the lines of magnetic field strength are circles centered on the z-axis, and are normal to the meridians, which are planes of incidence. Except for the E_ρ component, and the fields' exponential decay in the z-direction, the fields bear a slight resemblance to those of an infinite parallel-plate radial transmission line. [127, pp. 464-468], [143, p. 211] However, the electric field vector, while parallel to the meridian planes, is *not* transverse to the direction of propagation of the radial Zenneck wave.

The excitation of Zenneck waves is of some interest, but, as is commonly known, conventional antennas don't launch Zenneck waves. [2, p. 584] Wait noted that it was Goubau [166] who pointed out that "...the Zenneck wave is orthogonal [in the Fourier sense] to the radiation field and therefore it should be possible to excite it." [135, p. 67] One might take a hint from the notion of "mode-matching" in conventional waveguides excitation, supplied by Chu and Barrow long ago. [167] Microwave technologists will recognize that,

"A waveguide is almost always used in conjunction with other types of apparatus that involve coupling of the energy into and out of the waveguide. Devices such as probes and loops for coupling the energy into and out of a waveguide *are mode-sensitive*; that is, their performance is most efficient for the mode for which they are designed to couple, and less so for the other modes. ... For the excitation of a given mode, the probe (or probes) is placed parallel to the electric field in a position where the field exhibited by the mode has its greatest intensity. Where several probes are employed for the excitation, they must *not only be placed in the proper positions within the waveguide, but they must be fed with currents or voltages that have relative phasing of the fields of the mode that is to be excited.*" [168]

In harmony with this strategy, Zenneck's student, Georg Goubau, later wrote,
"... it is, in general, necessary to build up a field which matches that of the surface wave. Efficient excitation is a matter of '*field matching*' and <u>not</u>, as one can find occasionally in the literature, a matter of 'impedance matching'." [169]

Such a stratagem has been performed analytically, by Hill and Wait, using a vertically infinite magnetic sheet-filled aperture possessing a height variation corresponding to the Zenneck wave. They found that, "*...the infinite Zenneck aperture excites a pure Zenneck wave with no radiation field.*" [39], [117, pp. 67-75] As an alternative, consider an infinitely tall wire filament centered at $\rho = 0$, with a time-harmonic vertical electric current distribution of the form $I_o e^{-u_2 z'} \delta(\rho')$. Near the base of the wire the azimuthal magnetic field, $H_\varphi(\rho)$, should match the quasi-static field:

$$H_{2\phi}(\rho,z) \xrightarrow[\substack{\rho \to 0 \\ z \to 0}]{} \frac{I_o}{2\pi\rho} . \tag{A.41}$$

Using the small argument Hankel approximation, [143, pp. 224, 462], [171, p. 33], [172, p. 360, #9.1.9]

$$H_1^{(2)}(x) = J_1(x) - j N_1(x) \xrightarrow[x \to 0]{} \frac{j2}{\pi x} \tag{A.42}$$

the complex excitation amplitude C in Equation (A.33) is then found to be

$$C = \frac{-I_o \gamma}{4} = \frac{-\omega Q_o \gamma}{4} . \tag{A.43}$$

which is equivalent to Goubau's determination in 1951. [166] From the properties of the Hankel function, the asymptotic fields will experience a decrease proportional to $1/\sqrt{\rho}$ due to geometrical spreading in addition to an exponential attenuation (an actual dissipation loss) of the form $e^{-\alpha\rho}$. Further, since the wave velocity $\omega/\beta > c$, the Zenneck wave is a fast wave [90] (as opposed to the Norton ground wave, for which the opposite is true).

One of the more interesting features of the exact solution is that the complex ratio of vertical to horizontal electric field components can be *geometrically* interpreted as belonging to a non-uniform plane wave incident onto the boundary interface at a complex angle determined from:

$$\tan\theta_i = \frac{E_z}{E_\rho} \xrightarrow[\rho \to \infty]{} \frac{-\gamma}{u_2} = \sqrt{\varepsilon_r - j\frac{\sigma}{\omega\varepsilon_o}} = n(\omega) \tag{A.44}$$

where $n(\omega)$ is the complex index of refraction. Recall that the Fresnel reflection coefficient for parallel polarization (i.e., **E**-fields in the plane of incidence (the meridians)) is given by [94, p. 632], [173, p. 458]

$$\Gamma_\parallel(\vartheta_i) \equiv \frac{E_{z,R}}{E_{z,i}} = \frac{\sqrt{(\varepsilon_r - jx) - \sin^2\vartheta_i} - (\varepsilon_r - jx)\cos\vartheta_i}{\sqrt{(\varepsilon_r - jx) - \sin^2\vartheta_i} + (\varepsilon_r - jx)\cos\vartheta_i} . \tag{A.45}$$

where θ_i is measured from the normal. As can easily be seen, $\Gamma_\parallel(\theta_i) = 0$, i.e.- the numerator goes to zero, when

$$\theta_i \;=\; \theta_{i,B} \;=\; \mathbf{tan}^{-1}\!\left(\sqrt{\varepsilon_r - j\frac{\sigma}{\omega\varepsilon_o}}\right) \;=\; \mathbf{tan}^{-1} n \tag{A.46}$$

where $\theta_{i,B}$ is the *complex* Brewster angle, and this reduces to the lossless dielectric form of the Brewster angle when $\sigma \to 0$. For this reason, the Zenneck wave is commonly described as being a non-uniform plane wave incident at a complex Brewster angle. The quantity $\Gamma_{\parallel}(\theta_i)$ is generally complex, and for parallel polarization the phase of $\Gamma_{\parallel}(\theta_i)$ approaches 180° for $\theta_i \gg \theta_{i,B}$ and 0° for $\theta_i < \theta_{i,B}$. At $\theta_i = \theta_{i,B}$ the phase of $\Gamma_{\parallel}(\theta_{i,B})$ is equal to 90°.   Magnitude-wise, Barlow and Brown conclude that at this angle,

"... $\Gamma(\theta_{i,B}) = 0$ so that the wave *does not radiate.* Thus we say that *the Zenneck wave is simply* **an inhomogeneous plane wave** *incident onto a flat surface at the [complex] Brewster angle.*" [124, p. 30]

And, Stratton observes that,

"If complex angles of incidence are admitted, the reflection coefficient $\Gamma(\theta)$ can be made to <u>vanish</u> *whatever the nature of the media.*" [2, p. 516]

Any source that can synthesize fields incident at this *complex* angle will launch a Zenneck wave. Stratton also pointed out that a complex angle of incidence can be associated with an *inhomogeneous* (or nonuniform) plane wave incident onto the boundary interface. [2, p. 516] The inhomogeneous wave issue was also discussed by Noether. [136, p. 177-178]

More recently, Kukuskin noted that, *"The near-field of any antenna contains a [wave-number] spectrum of nonuniform plane waves and, therefore, this spectrum contains a wave with a suitable wave-number."* [165] However, the energy in this special spatial component is generally such a small percentage of the total spectral density and the interaction region on the boundary so small that the excitation efficiency for coupling to a Zenneck wave field on natural lossy media is trifling, and will only be observed if the antenna's radiation field is negligible. It appears that this manner of mode-matching [169-172] to natural earth as a Zenneck surface waveguide would be appallingly inefficient.

However, if the excitation could be limited to only a plane wave incident at the complex Brewster angle, as with Hill and Wait's infinite aperture example, [39] there would be *no radiation field* and the solution would consist of only Zenneck's surface wave. On the other hand, as the complex index of refraction, n, becomes much, much greater than unity, $\lambda_s \to k_o$ (in Equation A.12), the pole coalesces with the branch point and the *excitation* of this Zenneck surface wave term diminishes. In that case, the 1909 Sommerfeld solution morphs smoothly back onto the Norton curves shown in Figures 5 and 6 above. (It snaps back immediately with the Norton-Niessen's erroneous "quick fix", which serendipitously nulls out the surface wave integration contribution.) Then one is left only with a radiation field (now called the Norton "ground wave") and no Zenneck surface wave component.

Finally, we reiterate that Zenneck was responsible for moving the the science of radio propagation into the 20th century. It was van der Pol who said that *Zenneck's analysis was the first to demonstrate the influence of terrestrial boundary conditions on propagation.* [111]

A.5. The Burrows-Wise (Decomposable) Formulation

In order to interpret his experimental measurements, Burrows calculated field strength propagation based on the Weyl-Norton prediction and on the Sommerfeld-Rolf prediction. These computations were simplified by means of a collection of series representations and expressions derived from Sommerfeld's 1909 solution developed earlier at AT&T by Dr. W. Howard Wise for use along the surface. Burrows re-expressed Wise's antenna formulae for $e^{+j\omega t}$ time variation and used them to obtain predictions for the *attenuation factor* corresponding to Weyl's solution (W), and the *attenuation factor* corresponding to Sommerfeld's solution

$$(S): \quad W \;=\; A - \frac{B}{2} \;\approx\; C \tag{A.47}$$

$$S \;=\; A + \frac{B}{2} \;\approx\; C + B \tag{A.48}$$

The quantities A, B, and C are from piecewise convergent series expansions derived by Wise [101] and Rice, [105] and re-expressed by Burrows, [106, p. 226], [123], who published them in the February 1937 issue of the Proceedings of the IRE as:

$$A \;=\; 1 + \sum_{n=1}^{\infty} \frac{x^n \, e^{2in\,(\delta + \pi/4)}}{1 \cdot 3 \cdot 5 \,\cdots\, (2n-1)} \tag{A.49}$$

$$B \;=\; \sqrt{2\pi x}\; e^{-\left(\frac{x}{2}\right)\sin 2\delta + i\left(\frac{x}{2}\right)\cos(2\delta + \delta + \pi/4)} \tag{A.50}$$

$$C \;=\; -\sum_{n=1}^{\infty} \frac{1 \cdot 3 \cdot 5 \,\cdots\, (2n-1)}{x^n \, e^{2in\,(\delta - \pi/4)}} \tag{A.51}$$

$$x\,e^{2i\delta} \;=\; \frac{2\pi r/\lambda}{\varepsilon - 2i\sigma l/f} \qquad \text{for} \;\; 0 \le \delta \le \pi/4 \tag{A.52}$$

(Before using these expressions the careful reader should also refer to Burrows' January 1937 BSTJ (Bell System Technical Journal) paper [120, see Appendix I] where greater detail is provided, as well as consulting the original papers by Wise and Rice.) The results in Equations (A.47) and (A.48) were then multiplied by the unattenuated field and plotted in Figure 7 above as curves 2 and 3, respectively. Burrows noticed that on his graph the Weyl (1919) and Sommerfeld (1909) solutions *differ by just the Zenneck surface wave component,* (B), [100], whose contribution vanishes exponentially with distance. In 1936, as a result of his Seneca Lake experiment, Burrows asserted that,

> "...[the] formula given by Norton gives values agreeing with Weyl." [93, p. 284]

And, in 1937 Burrows wrote,

> "[I] found that Norton's results agree with those of Weyl. Since Norton has derived his results from a formula of van der Pol and Niessen, their formulas presumably agree with those of Weyl [too]." [106]

That is (although derived differently), at ranges of interest, numerically one has

$$N \approx W \tag{A.53}$$

where N is the attenuation factor associated with Norton's solution. Stratton observes that,

> "The formulas of van der Pol, Niessen, Burrows and Norton are for the most part extensions of Wey's results. Field strengths calculated by these formulas are in accord with each other and apparently in good agreement with experiment." [2, p. 585]

[The complete Sommerfeld 1909 solution smoothly transitions from the quasi-static near field to a Zenneck-wave dominated region, and then smoothly descends asymptotically down to the Weyl-Norton solution as the knee of the Zenneck wave is passed, after which the contribution from the surface wave component decreases far below the Norton-Weyl field, leaving Sommerfeld and Norton-Weyl equivalent, as illustrated in Figure 6.] Burrows compared his measured data against the predictions of the various formulations, as shown in Figure 7, and concluded that, "... simple antennas do not generate a Sommerfeld surface wave," which was consistent with the analytical results of Norton's "quick fix" of Sommerfeld 1909 to get Weyl's solution.

A.6. Weyl's 1919 Solution[76],[77]

Weyl's formulation is discussed in Noether [136], Stratton [2, pp. 577-582, 584-587], King [174], Bremmer [175], Brekhovskikh [176], and Banos [79, pp. 20-24, 154-155]. As briefly mentioned at the close of Section A.1, above, Weyl's method employed a representation utilizing a bundle (i.e., a superposition) of inhomogeneous plane waves incident on the earth at various complex angles. Weyl employs a *spherical Hankel function* identity (recall the *spherical* Bessel functions), instead of Sommerfeld's cylindrical Bessel function identity for e^{ikr}/r. (Incidentally, Weyl's approach was actually mentioned in Sommerfeld's 1909 paper.) For $e^{-i\omega t}$ time variation and $z > 0$, Weyl found a solution, which, when the antenna is on the interface, transforms to the expression [2, p. 581]:

$$\Pi_o(z > 0) = 2k_E^2 \int_0^\infty \frac{J_0(\lambda \rho)}{k_E^2 \sqrt{\lambda^2 - k_o^2} + k_o^2 \sqrt{\lambda^2 - k_E^2}} e^{-\sqrt{\lambda^2 - k_o^2}\, z} \lambda\, d\lambda . \quad (A.55)$$

This is identical to Sommerfeld's 1926 solution expressed by Equation (A.4). In the general case, where $h_T \neq 0$, Stratton points out that Weyl's solution can also be decomposed into the expression

$$\Pi_W = \frac{e^{ik_o R_o}}{R_o} - \frac{e^{ik_o R_E}}{R_E} + 2k_E^2 \int_0^\infty \frac{J_0(\lambda \rho)}{k_E^2 \sqrt{\lambda^2 - k_o^2} + k_o^2 \sqrt{\lambda^2 - k_E^2}} e^{-\sqrt{\lambda^2 - k_o^2}\,(z + h_T)} \lambda\, d\lambda , \quad (A.56)$$

where h_T is the antenna height above ground, and $R_{o,E}^2 = \rho^2 + (z \mp h_T)^2$. [175] (It is inherently assumed that the Weyl image is at $z = -h_T$ even though the earth is *not* a perfect conductor, and conventional image theory does not apply. Note that an electrical image is not necessarily located at the position of the geometrical image. [177, pp. 27-29]) Also, using the Sommerfeld identity, Equation (A.3), when $k_E \rightarrow \infty$ the integral in Equation (A.54) goes to $2\dfrac{e^{ik_o R_E}}{R_E}$ and one is left with the expression

$$\Pi_W (z > 0) = \frac{e^{ik_o R_o}}{R_o} + \frac{e^{ik_o R_E}}{R_E} , \quad (A.57)$$

which, when $h_T \rightarrow 0$, reduces to Equation (A.16). In reviewing Norton's analyses, Stratton [2, p. 585] pointed out that Norton wrote Weyl's solution as

$$\Pi_W = \frac{e^{ik_o R_o}}{R_o} - \frac{e^{ik_o R_E}}{R_E} + V , \quad (A.58)$$

where V is the integral in the third term of Equation (A.56). For an incremental vertical antenna located on the lossy surface, Norton ascertained an integral approximation which leads to Equation (A.22) with F(p) expressed by Equation (A.19). And, this is what many broadcast consultants used in FCC applications down through the 20th century.

As before, the form of Equation (A.56) implies that Weyl's expressions can also be interpreted like Sommerfeld's 1926 solution, that is, as a direct contribution from the antenna plus a contribution from its geometrical image (which, together form a space wave), and a third term propagating along the interface. However, Stratton notes that,

> "Weyl obtains an asymptotic series representation of the diffracted field by applying the method of "steepest descent" . . . When $k_E \gg k_o$, Weyl's solution also reduces to a form which can be interpreted as the superposition of a space wave and a surface wave, but the Weyl surface [i.e., ground] wave is *not* identical with that of Sommerfeld and Zenneck." [2, p. 584]

The physical nature of *Weyl's ground wave*, which emerges from the steepest descent integration, is distinctly different from the nature of the *Zenneck surface wave* (which is a *discrete mode* surface-guided wave) resulting from the contour integration around a distinct pole on the λ-plane. While Weyl's integration path gives a *ground wave*, it gives no *Zenneck surface wave* since the Zenneck-Sommerfeld pole is not encircled.

Burrows, using the piecewise convergent series solutions of Wise and Rice, concluded that, not just asymptotically but also in close, along the surface of the earth, Weyl's ground wave is identical to Norton's ground wave and was what he was observing experimentally at Seneca Lake. Twenty-five years later, Banos [79, p. 155] pointed out that Burrows' use of the series representations of Wise and Rice, which merely demonstrate that the surface wave disappears in the asymptotic range, does not justify such an interpretation by Burrows. However, this unjustified conclusion, from a pivotal experiment historically identified as being "crucial", wrongly gave widespread confidence in the "sign error" assertion declared by Norton, Niessen, and Rice.

A.7. Other Analyses

There are a large number of other interesting and historical publications which we have not included, and the curious reader is directed toward them. Stratton outlined Van der Pol's solution [2, pp. 582-583], and illustrations of it have been provided by Sarkar et al. [161] Van der Pol started from Equation (A.1) (with (A.2)) and, without the approximations of Sommerfeld or the series representations of Wise and Rice, but with the aid of a *"complex distance"*, obtained the Hertz potential by means of a clever integration resulting in an effective line-source in the lower half-space, *below the geometrical image* of the physical source. [114] Also, Banos [79, pp xii and 24] cited the work of Strutt [178-180] and Kruger [181], who discussed the effect of antenna altitude above lossy ground. Because of its historic commercial significance in broadcasting, there are dozens of papers and reports, and thousands of broadcast license applications all employing Norton's results.

More recently Sarkar et al [160], [161] have pointed back to Schelkunoff [182], [183], who employed the notion of an equivalent transmission line (whose characteristic impedance is a function of incident angle) [183, pp. 251-260], [184], [129, pp. 12-13 and 18-20], [135 pp. 56-59]. Shelkunoff concluded that a vertically polarized electric *current element* (an antenna) disposed above a lossy dielectric cannot produce a Zenneck surface wave, which was contrary to the initial results reached by Sommerfeld. Based on Schelkunoff's integrals [182], Sarkar et al have even developed a new Green's function [185] for the analysis of wave propagation on lossy media, the advantage of which is a much more rapid convergence in numerical computations for the ground wave radiation field.

APPENDIX – B
FLAT EARTH ZENNECK PROPAGATION BEHAVIOR

Equation (A.35) may be used to calculate propagation charts illustrating the Zenneck surface wave field strength at selected frequencies for an oscillating charge distribution of fixed magnitude, Q. These were calculated at frequencies of 10 MHz, 1 MHz, 0.1 MHz and 0.01 MHz, respectively, with constitutive parameters taken as $\varepsilon_r = 15$ and $\sigma = 0.008$ mhos/m, which are from the FCC's R-3 map for Central Ohio. The vertical axis is the RMS field strength (in mV/m) and the horizontal axis is range in kilometers.

From the graphs it can be seen that the lower the frequency, the less the propagation attenuation, and the "knee" (typical of transmission line propagation) moves outward to greater distances. However, consistent with conservation of energy, the energy density (and field strength) decreases with greater "knee" distances. Greater propagation range at low frequencies gives greater "spreading loss", but the Zenneck wave suffers greater exponential *attenuation* at the higher frequencies.

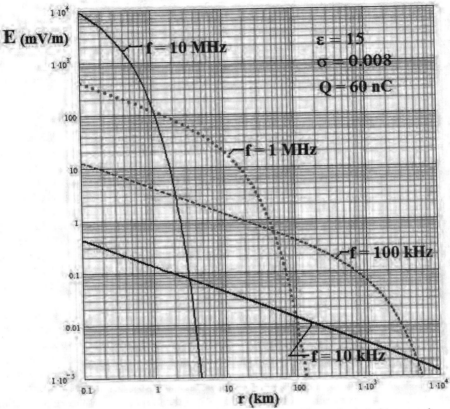

Figure B.1. **Zenneck wave field strength vs. range for selected frequencies. The higher the frequency, the greater the propagation attenuation but the smaller the region over which the energy is spread.**

APPENDIX C
THUMBNAIL BIOSKETCHES OF KEY CONTRIBUTORS TO RADIO SCIENCE
(Gleaned and edited from Wikipedia and other open literature sources.)

Jonathan Adolf Wilhelm Zenneck. (1871 - 1959) Doctorate in 1894 (at age 23).
Zenneck entered the Evangelical-Theological Seminary at Maulbronn in 1885. He then attended a Blaubeuren seminary in 1887, where he learned Latin, Greek, French, and Hebrew. In 1889, Zenneck enrolled in the Tübingen University. At the Tuebingen Seminary, Zenneck studied mathematics and natural sciences. In 1894, Zenneck took the State examination in mathematics and natural sciences and the examination for his doctor's degree. In 1894, Zenneck conducted zoological research at the Natural History Museum, London. Zenneck served in the military between 1894-1895.

In 1895, Zenneck left zoology and turned over to the new field of radio science, becoming assistant to Karl Ferdinand Braun (1850-1918) and lecturer at the "Physikalisches Institut" in Strasbourg, Alsace. His colleagues there were L.I. Mandelshtam (who later established the Institute for Nonlinear Oscillations at Moscow State University) and N.D. Papalexi (also later of Moscow State). It will be recalled that his advisor, Karl Ferdinand Braun, shared the 1909 Nobel Prize with Guglielmo Marconi for contributions to Wireless Telegraphy. Professor Braun invented the Braun tube (a forerunner of the oscilloscope). While at Strassbourg, Zenneck added a transverse (horizontal) deflection and a time-base to Braun's tube and created the modern oscilloscope for graphically displaying electrical waveforms.

Zenneck declared that, during this impressionable time it was Nikola Tesla's published lectures that had introduced him to wireless. Many years later, on June 17, 1931, Professor Zenneck wrote to Tesla from Munich,

> "As a young assistant at Professor F. Braun's Physics Institute – it may have been in the year 1896 – in the library I hit upon your investigations described in the book edited by Thomas C. Martin. [Thomas Commerford Martin, 3rd President of the AIEE (in 1887-1889).] I devoured the book like an intensely exciting novel: it opened to me a new physical universe. Wireless telegraphy came into being a

few years later. I have gotten my copy of the book again, and eagerly studied your high frequency investigations. It is clear to me that your pioneering work in the realm of high-frequency is the best instruction for neophytes wanting to be engaged in wireless telegraphy. And, yet, after many years, I always return again to the book and other publications about your investigations. I am convinced that many concepts, which have emerged in the development of wireless telegraphy, were already established among the abundance of your ideas and research.

The World War caused me to be detained in the United States under difficult circumstances. For me, it was a special event to come in contact with you. As much as I had previously admired you as the original inventor, so enchanted was I that now, to me, you are one of the most gracious people to be acquainted with that I have ever encountered. The hours which I was permitted to spend together with you will forever be among the most beautiful memories of my life."

In 1899, Zenneck started propagation studies of wireless telegraphy, first over land, but then became more interested in the larger ranges that were reached over sea. In 1900 he started ship-to-shore experiments in the North Sea near Cuxhaven, Germany. In 1902 he conducted tests of directional antennas.

Zenneck left Strasbourg in 1905 when he was appointed assistant-professor at the Danzig Technische Hochschule. In 1906 he became professor of experimental physics at the Braunschweig Technische Hochschule. Also in 1906, Zenneck wrote the textbook *Electromagnetic Oscillations and Wireless Telegraphy*, (then the standard textbook on the subject). [A shorter version, *Wireless Telegraphy,* was published in 1908, and a second edition of the latter appeared in 1912. It was subsequently translated into English and published by McGraw-Hill in 1915.] In 1907 Zenneck was the first to obtain an exact solution of the wave equation in the presence of a lossy conducting boundary. [33] This solution of Maxwell's equations had a "surface wave" property, and was thought to provide a possible explanation for how Marconi's 1901 signals were able to get around the curvature of the earth.

In 1909, the year that Sommerfeld's solution appeared, Zenneck joined Badische Anilin und Sodafabrik in Ludwigshafen to experiment with electrical discharges in air to produce bound nitrogen as fertilizer. When World War I broke out, Zenneck went to the front as a Captain in the Marines. However, in 1914, the German government sent him (along with Nobel Laureate Professor Karl Ferdinand Braun from Strassbourg) to the United States as a consultant and technical advisor in a patent case of great importance to Germany. It was April 24, 1915 when the IRE held the celebrated banquet at Luchow's honoring Professors Braun and Zenneck, with the famous photo of Lee de Forest and Nikola Tesla standing next to Zenneck. (In 1914, the IRE had made Zenneck its first 'Fellow'.)

When the United States entered the war Zenneck was declared a Prisoner of War and was interned first at Ellis Island and then at Fort Oglethorpe, Georgia. He returned to Germany in July of 1919 and resumed his duties as Professor of Experimental Physics at the Technische Hochschule in Munich. At that time he resumed propagation studies, now with shortwaves and was the first one in Germany to study the Ionosphere with vertical sounding at his station at Kochel, Bavaria.

Zenneck was awarded the 1928 IRE Medal of Honor for his achievements in basic radio technology research. In 1933 he was elected Vice-President of the IRE. In 1950 Zenneck's student, George Goubau (who was one of the WWII "paperclip" German scientists), published a technique for launching axial cylindrical surface waves (the G-line). Since the 1930s Zenneck directed the Deutsche Museum in Munich and rebuilt it after World War II, devoting his studies back to zoological research. Zenneck passed away on April 8, 1959.

Interestingly, Sommerfeld and Zenneck owned alpine ski huts near each other where they discussed these problems together. Zenneck thanks Sommerfeld in his 1907 paper [33], and Sommerfeld, in his 1909 paper [40], thanks his "colleague" Zenneck "for special expert advice on technical aspects of the problem".

<u>Arnold Johannes Wilhelm Sommerfeld.</u>[*] (1858-1951) Ph.D. in 1891 (at age 23) from the Albertina University at Königsberg, East Prussia. Sommerfeld lectured on mathematical physics at Göttingen in 1895. These lectures on partial differential equations evolved over his teaching career to become Volume VI of his textbook series *Lectures on Theoretical Physics*, under the title *Partial Differential Equations in Physics*. In 1897 he became Chair of Mathematics at the Technische Universität Clausthal. In 1900, Sommerfeld was appointed Chair of Applied Mechanics at the Königliche Technische Hochschule Aachen. Finally, in 1906 Wilhelm Röntgen (Director of the Physics Institute at Munich) selected Sommerfeld as professor of physics and director of the new Theoretical Physics Institute at the University of Munich.

While at Munich, Sommerfeld came in contact with Einstein's 1905 special theory of relativity, which was not yet widely accepted at that time. Sommerfeld's

---

[*] Gleaned from the internet.

mathematical contributions to the theory helped its acceptance by the skeptics. In 1914 he worked with Léon Brillouin on the propagation of electromagnetic waves in dispersive media. He became one of the founders of quantum mechanics. He was nominated a record 81 times for the Nobel Prize, and served as Ph.D. advisor to more Nobel Prize winners in physics than any other supervisor before or since. He introduced the 2nd quantum number (azimuthal quantum number) and the 4th quantum number (spin quantum number). He also introduced the fine-structure constant, and pioneered X-ray theory. His students include Werner Heisenberg, Wolfgang Pauli, Peter Debye, Hans Bethe, Ernst Guillemin (who introduced Green's function as the impulse response and convolution to circuit theory and linear systems), Alfred Landé, and Léon Brillouin, as well as post-graduate students Linus Pauling, Isidor Rabi and Max von Laue. Albert Einstein told Sommerfeld, "What I especially admire about you is that you have, as it were, pounded out of the soil such a large number of young talents." From 1942 to 1951, Sommerfeld worked on putting his lecture notes in order for publication. They were published as the six-volume *Lectures on Theoretical Physics*. And, through these published lectures his influence on young students continues to the present. [186]

Arnold Sommerfeld belongs with Max Plank, Albert Einstein and Niels Bohr as the founders of modern physics. Besides his invaluable contributions to quantum theory, Sommerfeld worked in other fields of physics, such as the classical theory of electromagnetism and *the* fundamental investigation of wave propagation, which is the focus of the present paper. It was Sommerfeld who proposed the basic solution to the problem of a radiating hertzian dipole over a lossy conducting earth, which over the years led to many applications (and had such a broad impact on 20th century civilization!). His Sommerfeld identity and Sommerfeld integrals are still, to the present day, the most common way to solve this kind of problem.

Also, as a mark of the prowess of Sommerfeld's school of theoretical physics and the rise of theoretical physics in the early 1900s, as of 1928 nearly one-third of the ordinarius professors (professors of the highest rank) of theoretical physics in the German-speaking world had been Sommerfeld's students.

Sommerfeld owned an alpine ski hut to which students and colleagues were often invited for discussions of physics as demanding as the sport. Interestingly, Sommerfeld and Zenneck owned alpine ski huts near each other and discussed these problems. In fact, Zenneck thanks Sommerfeld in his 1907 paper, and, in his 1909 paper, Sommerfeld thanks his "colleague" Zenneck "for special expert advice on technical aspects of the problem".

<u>Hermann Klaus Hugo Weyl.</u> (1885-1955) Ph.D at the University of Göttingen under David Hilbert. After teaching for several years at Göttingen he became the chair of mathematics at the ETH (Eidgenössche Technische Hochschule – Federal Institute of Technology) Zürich, where he was a colleague of Albert Einstein, who was then working on the general theory of relativity. Weyl published technical and some general works on space, time, matter, geometry, philosophy, logic, symmetry and the history of mathematics.

Weyl published the first paper attempting to combine general relativity with the laws of electromagnetism (Gravitation und Elektrizität, 1918) in which he employed a symmetric affine connection. The theory failed, but opened a decades-long search for a unified field theory. His book Raum, Zeit Materie (Space, Time, Matter) was published in 1918, and went through 5 editions by 1923. Weyl introduced the concept of the vierbein (tetrad) into general relativity in 1929.n 1919, Hermann Weyl turned his attention, briefly, to the propagation of electromagnetic waves and developed a plane wave spectrum solution complimenting Sommerfeld's 1909 paper. Weyl's solution is valid at large distances and fails to capture the Zenneck pole, leaving only what later became known as the Norton ground wave.

Weyl left Zürich in 1930 to become David Hilbert's successor at Göttingen, leaving when the Nazis assumed power in 1933. The events persuaded him to move to the new Institute for Advanced Study in Princeton, New Jersey. He remained there until his retirement in 1951. His research has had major significance for theoretical physics as well as purely mathematical disciplines including number theory. He was one of the most influential mathematicians of the twentieth century, and an important member of the Princeton Institute for Advanced Study during its early years.

REFERENCES

1.   C. R. Burrows, "Radio Gain," *IEEE Transactions on Antennas and Propagation*, **AP-15**, No. 3, May 1967, pp. 404-410. See footnote 6 on p. 408.
2.   J. A. Stratton, *Electromagnetic Theory*, McGraw-hill, 1941, p. 585.
3.   H. Hertz, *Untersuchungen über die Ausbreitung der Elektrischen Kraft*," translated as *Electric Waves*, edited by D. E. Jones, Macmillan, 1893; republished by Dover, 1962.
4.   J. H. Bryant, *Heinrich Hertz: The beginning of Microwaves*, IEEE, 1988 IEEE/MTT-S Hertz Centennial Celebration, 1988.
5.   N. Tesla, "The True Wireless," *Electrical Experimenter*, May, 1919, pp. 28-30, 61-63, 87.
6.   G. Marconi, "Wireless Telegraphic Communication," Nobel Lecture, Dec.11, 1909.
7.   G. Marconi, "Signals Across the Atlantic," *Electrical World*, **38**, Dec. 21, 1901, pp. 1023-1025.
8.   J. A. Ratcliffe, "Marconi: Reactions to His Transatlantic Radio Experiment," *Electronics and Power*, May 2, 1974, pp. 320-323.
9.   J. R. Strutt, Lord Rayleigh, "On the bending of Waves round a Spherical Obstacle," *Proceedings of the Royal Society (London)*, **72**, May 28, 1903, pp. 40-41.
10. M. Macdonald, "The Bending of Electric Waves round a Conducting Obstacle," *Proceedings of the Royal Society (London)*, **71**, January 21, 1903, pp. 251-258.
11. C. R. Burrows, "The History of Radio Wave Propagation Up to the End of World War I," *Proceedings of the IRE*, **50**, May 1962, pp. 682-684.
12. A. E. Kennelly, "On the Elevation of the Electrically Conducting Strata of the Earth's Atmosphere," *Electrical World and Engineer*, **39**, March 15, 1902, p. 473.
13. O. Heaviside, "Theory of Electric Telegraphy," Encyclopedia Britannica, 10th edition, December 19, 1902, Vol. 33, pp. 213-218. (See p. 215.)
14. K. Davies, *Ionospheric Radio Propagation*, National Bureau of Standards Monograph 80, 1966; republished by Dover, 1966, p. 1.
15. G. W. Gardner, "Origin of the Term Ionosphere," *Nature*, **224**, December 13, 1969, p. 1096.
16. "Bibliography of the Kennelly-Heaviside Layer," *Proceedings of the IRE*, **19**, No. 6, June 1931, pp. 1066-1071.
17. G. N. Watson, "The Diffraction of Electric Waves by the Earth," *Proceedings of the Royal Society (London)*, **A95**, October 1918, pp. 83-99.
18. G. N. Watson, "The Transmission of Electric Waves Around the Earth," *Proceedings of the Royal Society (London)*, **A95**, July 15, 1919, pp. 546-563.
19. B. Vvedensky, "The Diffractive Propagation of Radio Waves," *Technical Physics of the USSR*, **2**, 1935, pp. 623-639.
20. B. Vvedensky, "The Diffractive Propagation of Radio Waves: Part II – Elevated Transmitter and Receiver," *Technical Physics of the USSR*, **3**, 1936, pp. 915-925.

21. B. Vvedensky, "The Diffractive Propagation of Radio Waves: Part III – The Case of Dielectric Ground," *Technical Physics of the USSR*, **4**, 1937, pp. 579-591.

22. C. R. Burrows, "Radio Propagation Over Spherical Earth," *Proc. IRE*, **23**, No. 5, May 1935, pp. 470-480. (Reprinted in *Bell Systems Technical Journal*, **14**, July 1935, pp. 477-488.)

23. B. van der Pol, and H. Bremmer, "The Diffraction of Electromagnetic Waves from an Electrical Point Source Round a Finitely Conducting Sphere, with Applications to Radiotelegraphy and the Theory of the Rainbow - Part I," *Philosophical Magazine*, Serial 7, **24**, July 1937, pp. 141-175.

24. B. van der Pol, and H. Bremmer, "The Diffraction of Electromagnetic Waves from an Electrical Point Source Round a Finitely Conducting Sphere, with Applications to Radiotelegraphy and the Theory of the Rainbow - Part II," *Philosophical Magazine*, Serial 7, **24**, No. 164, Supplement, Nov. 1937, pp. 825-864.

25. B. van der Pol, and H. Bremmer, "Ergebnisse einer Theorie über die Fortpflanzung electromagnetischer Wellen über eine Kugel endlicher Leitfähigkeit," *Hochfrequenztechnik und Elektroakustitik*, **51**, Heft 6, June 1938, pp. 181-188.

26. B. van der Pol, and H. Bremmer, "The Propagation of Radio Waves over a Finitely Conducting Spherical Earth," *Philosophical Magazine*, Series 7, **25**, No. 171, Supplement, June 1938, pp. 817-834.

27. B. van der Pol, and H. Bremmer, "Further Note on the Propagation of Radio Waves over a Finitely Conducting Spherical Earth," *Philosophical Magazine*, Series 7, **27**, No. 182, March 1939, pp. 261-275.

28. B. van der Pol, and H. Bremmer, "The Propagation of Wireless Waves Round the Earth," *Phillips Technical Review*, **4**, No. 9, September 1939, pp. 245-253.

29. H. J. Aitken, *Syntony and Spark - The Origins of Radio*, Princeton Univ. Press, 1985, pp. 195, 286.

30. J. Zenneck, *Wireless Telegraphy*, McGraw-Hill, 1915, p. 421. (See Note #219.)

31. Balth. Van der Pol, "Über die Ausbreitung elektromagnetischer Wellen," *Zeitschrift* für *Hochfrequenztechnik*, **37**, April 1931, pp. 152-156.

32. S. Boksen, *Nikola Tesla und Sein Werk*, Deutscher Verlag für Jugend und Volk, 1932, p. 286.

33. J. Zenneck, "Über die Fortpflanzung ebener elektromagnetischer Wellen längs einer ebenen Leiterfläche und ihre Beziehung zur drahtlosen Telegraphie," (On the propagation of plane electromagnetic waves along a flat conducting surface and their relation to wireless telegraphy), *Annalen der Physik*, Serial 4, **23**, September 20, 1907, pp. 846-866.

34. F. Hack, "Die Ausbreitung ebener elektromagnetischer Wellen länges eines geschichteten Leiters, besonders in den Fällen der drahtlosen Telegraphie," *Annalen der Physik*, **27**, 1908, pp. 43-63.

35. A. Blondel, A., *Association Francaise pour de l'Académie des Sciences Comptes Rendu Congres de Nantes*, 1898, p. 212. [According to Zenneck, himself, Blondel was probably the first to conceive of radio waves as surface waves and is cited in Marconi's Nobel lecture.]

36. E. Lecher, "Ueber draghtlosen Telegraphie," *Physikalischer Zeitschrift.*, **3**, April 1902, p. 274.

37.   A. L. Cullen, "The Excitation of Plane Surface Waves," *Proceedings of the IEE* (British), **101**, Part IV, (Monograph No. 93R), August 1954, pp. 225-234. [See Section (5).]

38.   R. P. Kerr, "Gravitational Field of a Spinning Mass as an Example of Algebraically Special Metrics," *Physical Review Letters*, **11**, 1963, pp. 237-238.

39.   D. A. Hill, and J. R. Wait, "Excitation of the Zenneck Surface Wave by a Vertical Aperture," *Radio Science*, **13**, No. 6, November-December 1978, pp. 969-977.

40.   A. Sommerfeld, "Über die Ausbreitung der Wellen in der Drahtlosen Telegraphie," *Annalen der Physik*, **28**, 1909, pp. 665-695.

41.   A. Sommerfeld, "Problems of Radio," published as Chapter VI in *Partial Differential Equations in Physics – Lectures on Theoretical Physics: Volume VI*, Academic Press, 1949, pp. 236-257, 296, 312-317, 326-328.

42.   A. Sommerfeld, "Über die Ausbreitung der Wellen in der Drahtlosen Telegraphie," *Annalen der Physik*, **81**, December 1926, pp. 1135-1153.

43.   Sommerfeld, A., and F. Renner, "Strahlungsenergie und Erdabsorption bei Dipolantennen (Radiation Energy and Earth Absorption with Dipole Antennas)," *Annalen der Physik*, **41**, 1942, No. 1, pp. 1-36

44.   Sommerfeld, A., and F. Renner, "Strahlungsenergie und Erdabsorption bei Dipolantennen (Radiation Energy and Earth Absorption with Dipole Antennas)," *Hochfrequenz Technik und Elektroakustik*, **59**, 1942, pp. 168-173.

45.   A. Sommerfeld, "Ausbreitung der Wellen in der Drahtlosen Telegraphie. Einfluss der Bodenbeschaffenheit auf gerichtete und ungerichtete Wellenzüge (Spreading of Waves in Wireless Telegraphy – Influence of the Properties of the Soil)," *Jahrbuch für Drahtlosen Telegraphie und Telephoney*, **4**, December 1910, pp. 157-176.

46.   A. Sommerfeld, "Ueber die Fortpflanzung elektrodynamischer Wellen längs eines Drahtes," *Annalen der Physik*, **67**, 1899, pp. 233-290.

47.   A. Sommerfeld, *Electrodynamics*, (Lectures on Theoretical Physics: Vol. III), Academic Press, 1964, p. 178.

48.   O. Heaviside, "Practice Versus Theory – Electromagnetic Waves," *The Electrician*, Oct. 19, 1888, p. 772; reprinted in *Electrical Papers*, 1892, Vol. II, p. 490.

49.   E. Lecher, "Eine Studie über elektrische Resonanzerscheinungen," (A Study of Electrical Resonance Phenomena), *Annalen der Physik*, **41**, 1890, pp. 850-871.

50.   N. Tesla, "Experiments with Alternate Currents of Very High Frequency and Their Application to Methods of Artificial Illumination," *The Electrical World*, **XVIII**, No. 2, July 11, 1891, pp. 19-27. (Publication of a lecture delivered before the AIEE at Columbia University on May 20, 1891.)

51.   N. Tesla, "Experiments with Alternate Currents of Very High Frequency and Their Application to Methods of Artificial Illumination," a republication of the previous reference in *Inventions, Researches and Writings of Nikola Tesla*, by T.C. Martin, 2nd edition, Barnes and Noble, 1992, pp 145-197. (See pp. 192-195, and Figures 126, 127, and 128 on pages 192, 193, and 194.)

52.   N. Tesla, "On Light and Other High Frequency Phenomena," (a lecture delivered before the Franklin Institute, Philadelphia, February 1893, and before the National Electric Light Association, St. Louis, March 1893), published in *Inventions,*

Researches and Writings of Nikola Tesla, by T. C. Martin, 2nd edition, Barnes and Noble, 1992, pp. 294-373. (See Fig. 183, p. 339, and Fig. 186, p. 351.)

53. O. J. Lodge, Lightning Conductors, Whittaker and Co., 1892, pp. 108-115, 311-312.

54. G. Mie, "Elektrische Wellen an zwei parallen Drähten," Annalen der Physik, 2, 1900, pp. 201-249.

55. D. Hondros, "Über die elektromagnetische Drahtwellen," Annalen der Physik, 30, 1909, pp. 905-950.

56. Terman, F.E., "Resonant Lines in Radio Circuits," Electrical Engineering, July 1934, pp. 1046-1053.

57. S. A. Schelkunoff, "Anatomy of a 'Surface Wave'," IRE Transactions on Antennas and Propagation, 7, No. 5, December 1959, pp. S133-S139.

58. K. A. Norton, "The Physical Reality of Space and Surface Waves in the Radiation Field of Radio Antennas," Proc. IRE, 25, No. 9, September 1937, pp. 1192-1202.

59. "Obituary," Electrical Engineering, 62, No. 1, January 1943, p. 76.

60. H. Pratt, "Nikola Tesla," Proceedings of the IRE, 44, No. 9, September 1956, pp. 1106-1108.

61. Photo-reproduced letter from Zenneck (Technische Hochschule at Munich) to Tesla, dated June 17, 1931. Published in Tribute to Nikola Tesla, Nikola Tesla Museum, V. Popovich, editor, Beograd, Yugoslavia, 1961, pp. LS 46-47.

62. N. Tesla, Nikola Tesla On His Work With Alternating Currents and Their Application to Wireless Telegraphy, Telephony, and Transmission of Power, L. I. Anderson, editor, Sun Publishing Co., 1992, p. 75.

63. "Fritz Lowenstein: Inventor of Radio Devices Died with Praises Unsung," Philadelphia Public Ledger, Nov. 16, 1922.

64. F. Lowenstein, "The Mechanism of Radiation and Propagation in Radio Propagation," Proceedings of the IRE, 4, No. 3, June 1916, pp. 271-281.

65. B. F. Miessner, On the Early History of Radio Guidance, San Fran Cisco, Press, 1964, p. 6.

66. L. I. Anderson, Nikola Tesla: Guided Weapons and Computer Technology, Twenty-First Century Books, 1998, pp. 141-146.

67. J. H. Hammond, and E. S. Purington, "A History of Some Foundations of Modern Radio-Electronic Technology," Proceedings of the IRE, 45, No. 9, September 1957, pp. 1191-1208. (See p. 1198.)

68. F. Lowenstein, "Telephone-Relay," US Patent # 1,231,764 (Filed: April 24, 1912; Issued: July 3, 1917).

69. L. Espenschied, "Critique of - A History of Some Foundations of Modern Radio-Electronic Technology," Proceedings of the IRE, 47, No. 7, July 1959, pp. 1253-1258.

70. J. H. Hammond, and E.S. Purington, "Rebuttal to the Critique and Comments by Espenschied," Proceedings of the IRE, 47, No. 7, July 1959 pp. 1258–1268.

71. F. Lowenstein, "Variable Electrical Apparatus," US Patent # 1,258,423 (Filed: August 13, 1916; Issued: March 5, 1918).

72. F. Lowenstein, "Capacities," Proceedings of the IRE, 4, No. 1, February 1916, pp. 17-32.

73. F. E. Terman, Radio Engineers' Handbook, McGraw-Hill, 1943, p. 121.

74.   L. E. Whittemore, "The Institute of Radio Engineers – Fifty Years of Service," published in *Turning Points in American Electrical History*, J. E. Brittain, editor, IEEE Press, 1976, pp. 94-100.

75.   H. Weyl, "Gravitation und Elektrizität," *Sitzungsberichte der Preussischen Akad. der Wissenschaften*, 1918, pp. 465-480. (Republished in *The Principle of Relativity*, Dover, 1952, pp. 199-216.)

76.   H. Weyl, "Ausbreitung elektromagnetischer Wellen über einem ebenen Leiter (Propagation of Electromagnetic Waves Over a Plane Conductor)," *Annalen der Physik*, **60**, November 1919, pp. 481-500.

77.   H. Weyl, "Erwiderung auf Herrn Sommerfelds Bemerkungen über die Ausbreitung der Wellen in der drahtlosen Telegraphie," *Annalen der Physik*, **62**, 1920, pp. 482-484.

78.   J. R. Wait, "Theory of Ground Wave Propagation," published as chapter 5 in *Electromagnetic Probing in Geophysics*, Golem Press, 1971, pp. 163-207. (See p. 164.)

79.   A. Banos, *Dipole Radiation in the Presence of a Conducting Half-Space*, Pergamon Press, 1966, pp. 20-24, 154-155.

80.   T. Kahan, and G. Eckart, "On the Electromagnetic Surface Wave of Sommerfeld," *Physical Review*, **76**, No. 3, August 1, 1949, pp. 406-410.

81.   R. E. Collin, *Field Theory of Guided Waves*, McGraw-Hill, 1960, p. 491.

82.   B. Friedman, *Principles and Techniques of Applied Mathematics*, Wiley, 1956, pp. 214, 283-286, 290, 298-300.

83.   B. Rolf, "Graphs to Professor Sommerfeld's Attenuation Formula for Radio Waves," *Proceedings of the IRE*, **18**, March 1930, pp. 391-402.

84.   W. H. Wise, "Note on the Accuracy of Rolf's Graphs of Sommerfeld's Attenuation Formula," *Proceedings of the IRE*, **18**, November 1930, pp. 1971-1972.

85.   S. S. Kirby and K. A. Norton, "Field Intensity Measurements at Frequencies from 285 to 5400 Kilocycles per Second, *Proceedings of the IRE*, **80**, No. 2, May 1932, pp. 841-862.

86.   C. B. Feldman, "The Optical Behavior of the Ground for Short Radio Waves," *Proceedings of the IRE*, **21**, June 1933, pp. 764-801.

87.   C. N. Anderson, "Attenuation of Overland radio Transmission in the Frequency Range 1.5 to 3.5 Megacycles per Second," *Proceedings of the IRE*, **21**, No. 10, October 1933, pp. 1447-1462.

88.   C. R. Burrows, A. Decino, and L.E. Hunt, "Ultra-Short-Wave Propagation Over Land," *Proceedings of the IRE*, **23**, No. 12, December 1935, pp. 1507-1535.

89.   W. F. Snyder and C. L. Bragaw, *Achievement in Radio: Seventy Years of Radio Science, Technology, Standards, and Measurement at the National Bureau of Standards*, National Bureau of Standards, Special Publication 555, 1986, p. 112.

90.   J. R. Wait, "Electromagnetic Surface Waves," published in *Advances In Radio Research*, J. A. Saxton, editor, Academic Press, 1964, Vol. 1, pp. 157-217. "Corrections," *Radio Science*, **69D**, No. 7, 1965, pp. 969-975.

91.   J. R. Wait, "Characteristics of Antennas over Lossy Earth," published as Chapter 23 in *Antenna Theory: Part 2*, edited by R. E. Collin and F. J. Zucker, Mcgraw-Hill, 1969, pp. 386-437. (See pp. 390-391.)

92.   K. A. Norton, "Propagation of Radio Waves Over a Plane Earth," *Nature*, **135**, June 8, 1935, pp. 934-935.

93.   C. R. Burrows, "Existence of a Surface Wave in Radio Propagation," *Nature*, **138**, August 15, 1936, p. 284.

94.   E. C. Jordan, and K. G. Balmain, *Electromagnetic Waves and Radiating Systems*, Prentice-Hall, 1968, p. 632.

95.   K. A. Norton, "The Propagation of Radio Waves over the Surface of the Earth and in the Upper Atmosphere – Part I," *Proceedings of the IRE*, **24**, 1936, pp. 1367-1387.

96.   FCC Rules and Regulations, Volume III, Part 73, US Government Printing Office, March 1980. See §73.184.§ 73.184, Graph 20, "Relative Field Intensity F(p)/p vs. Numerical Distance p".

97.   C. R. Burrows, "Radio Propagation Over Plane earth – Field Strength Curves," *Bell System Technical Journal*, **16**, No. 1, January 1937, pp. 45-75. "Addendum," *BSTJ*, Vol. 16, #4, October 1937, pp. 574-577. [See p. 74 of the January article.]

98.   C. R. Burrows, "Addendum," *Bell System Technical Journal*, **16**, #4, Oct. 1937, pp. 574-577.

99.   C. R. Burrows, "The Surface Wave in Radio Propagation over Plane Earth," *Proc. Radio Club of America*, **14**, No. 2, August 1937, pp. 15-18. Available at: http://www.jumpjet.info/PioneeringWireless/eJournals/RCA/1937-02.pdf

100.   W. H. Wise, "The Physical Reality of Zenneck's Surface Wave," *Bell System Technical Journal*, **16**, No. 1, January 1937, pp. 35-44.

101.   W.H. Wise, "Asymptotic Dipole radiation Formulas," Bell System Technical Journal, **8**, October 1929, pp. 662-667.

102.   W. H. Wise, "The Grounded Condenser Antenna Radiation Formula," *Proceedings of the IRE*, **19**, No. 8, September 1931, pp. 1684-1689.

103.   W. H. Wise, "Note on Dipole Radiation Theory," *Physics*, **4**, October 1933, pp. 354-358.

104.   H. Ott, "Bemerkung zum Beweis von W. H. Wise über die Nichtexistenz der Zenneckschen Oberflächenwelle im Antennenfeld," Zeitschrift für Naturforschung, A, Astrophysik, Physik, und physikalische Chemie, **8**, 1953, 100-103.

105.   S. O. Rice, "Series for the Wave Functions of a Radiating Dipole at the Earth's Surface," *Bell System Technical Journal*, **16**, No. 1, January 1937, pp. 101-109.

106.   C. R. Burrows, "The Surface Wave in Radio Propagation over Plane Earth," *Proc. IRE*, **25**, No. 2, February 1937, pp. 219-229.

107.   C. R. Burrows, "The Surface Wave in Radio Transmission," *Bell Laboratories Record*, **15**, June 1937, p. 321-324. http://www.americanradiohistory.com/Archive-Bell-Laboratories-Record/30s/Bell-Laboratories-Record-1937-06.pdf

108.   R. E. Collin, "Hertzian Dipole Radiating Over a Lossy Earth or Sea: Some Early and Late 20th Century Controversies," *IEEE Antennas and Propagation Magazine*, **46**, No. 2, April 2004, pp. 64-79.

109.   B. van der Pol and K. F. Niessen, "Über die Ausbreitung Elektromagnetischer Wellen üeber eine Ebene Erde," *Annalen der Physik*, Ser. 5, **6**, August 22, 1930, pp. 273-294.

110. B. van der Pol and K. F. Niessen, "Über die Raum Wellen von einem vertikalen Dipolesender auf Ebene Erde," *Annalen der Physik*, Ser. 5, **10**, July 21, 1931, pp. 485-510.

111. B. van der Pol, "Über die Ausbreitung Elektromagnetischer Wellen," *Jahrbuch für Drahtlosen Telegraphie/ Zeitschrift fuer Hochfrequenztechnik*, **37**, April 1931, pp. 152-156.

112. K. F. Niessen, "Bemerkung zu einer Arbeit von Murry und einer Arbeit von van der Pol und Niessen uber die Ausbreitung elektromagnitischen Wellen," *Annalen der Physik*, Series 5, **16**, April 3, 1933, pp. 810-820.

113. K. F. Niessen, "Über die Entferntun Raumwellen eines vertikalen Dipolesenders oberhalb einer ebenen Erde von beliebiger Dielektrizitätskonstante und beliebiger Lightfähigkeit," *Annalen der Physik*, Series 5, **18**, December 24, 1933, pp. 893-912.

114. B. van der Pol, "Theory of the reflection of the light from a point source by a finitely conducting flat mirror, with an application to radiotelegraphy," *Physica*, **2**, August 1935, pp. 843-853.

115. K. F. Niessen, "Zur Entscheidung zwischen den Beiden Sommerfeldschen Formeln für die Fortpflanzeng von Drahtlosen Wellen," *Annalen der Physik*, **29**, 1937, pp. 585-596.

116. A. Ishimaru, *Electromagnetic Wave Propagation, Radiation and Scattering*, Prentice-Hall, 1991, pp. 609-610.

117. J. R. Wait, *Electromagnetic Wave Theory*, Harper and Row, 1985, p. 254.

118. J. R. Wait, "The Ancient and Modern History of EM Ground-Wave Propagation," *IEEE Antennas and Prop. Magazine*, **40**, No. 5, Oct. 1998, pp. 7-24; Correction: **40**, No. 6, December 1998, p. 22.

119. T. P. Sarkar, R. J. Mailoux, A. A. Olner, M. Salazar-Palma, and D. L. Sengupta, *History of Wireless*, Wiley-Interscience, 2006.

120. C. R. Burrows, "Radio Propagation Over Plane Earth – Field Strength Curves," *Bell System Technical Journal*, **16**, No. 1, January 1937, pp. 45-75. "Addendum," *BSTJ*, **16**, #4, October 1937, pp. 574-577.

121. L. I. Mandelstam and N. D. Papalexi, "Noveishie Issledovaniya Rasprostraneniya Radiovolin Vdol' Zemnoi Poverkhnosti. Sb. Statei (Latest Research of Radio Wave Propagation Along Earth's Surface)," Collected Papers, Moscow-Leningrad: Gostekhizdat, 1945.

122. K. A. Norton, "The Propagation of Radio Waves over the Surface of the Earth and in the Upper Atmosphere: Part II," *Proceedings of the IRE*, **25**, 1937, pp. 1203-1236. [See: Corrections," by R. J. King, *Radio Science*, **4**, No. 3, March 1969, p. 267.]

123. T. Kahan, and G. Eckhart, "On the Existence of a Surface Wave in Dipole Radiation over a Plane Earth," *Proceedings of the IRE*, **38**, July 1950, pp. 807-812. (See Fig. 6.)

124. H. M. Barlow and J. Brown, *Radio Surface Waves*, Oxford University Press, 1962, pp. 12-13, 25, 82.

125. G. Goubau, "Surface Waves and Their Application to Transmission Lines," *Journal of Applied Physics*, **21**, November 1950, pp. 1119-1128.

126. J. F. Corum and K. L. Corum, "RF Coils, Helical Resonators and Voltage Magnification by Coherent Spatial Modes," *Microwave Review*, **7**, No. 2,

Corum & Corum: The Radio Surface Wave Propagation Experiment

September 2001, pp. 36-45.   http://www.mwr.mediaris.net/pdf/Vol7No2-07-Jcorum.pdf

127. S. Ramo, W.R. Whinnery, and T. Van Duzer, *Fields and Waves in Communication Electronics*, Wiley, 2nd edition, 1984, p. 460.

128. T. Milligan, *Modern Antenna Design*, McGraw-Hill, 1st edition, 1985, pp. 8-9.

129. J. R. Wait, *Electromagnetic Waves in Stratified Media*, IEEE Press, 1996, p. 35.

130. A. Sommefeld, "Drahtlose Telegraphie," published in *Differentialgleichungen der Physik*, Vol. II, P. Frank and R. von Mises, Braunschweig, 1935, Dover reprint, 1961, Vol. 2, p. 918-952. [See p. 928.]

131. L. Felsen and N. Marcuvitz, *Radiation and Scattering of Waves*, Prentice-Hall, 1973, pp. 261, 511.

132. D. S. Jones, *Theory of Electromagnetism*, Pergamon Press, 1964, p. 370.

133. R. E. Collin, *Antennas and Radiowave Propagation*, McGraw-Hill, 1985, p. 377-385.

134. "Summary of the Normal Mode Theory Symposium (July 1955)," *IRE Transactions on Antennas and Propagation*, **AP-4**, No. 1, January 1956, p. 92.

135. J. R. Wait, *Lectures on Wave Propagation Theory*, Pergamon Pr., 1981, pp. 71-72.

136. F. Noether, "Spreading of Electric Waves Along the Earth," published in the book translation *Theory of Functions As Applied To Engineering Problems*, Technology Press, MIT, 1942, Part 2, Section E, pp. 167-184. [Originally published by Springer, Berlin, in 1931 under the title *Funktionentheorie und Ihre Anwendung in der Technik*, Part II, R. Rothe, F. Ollendorf, and K. Pohlhausen, editors.] See pp. 177-178.

137. P. S. Epstein, "Radio-Wave Propagation and Electromagnetic Surface Waves," *Proceedings of the National Academy of Sciences*, **33**, 1947, pp. 195-199.

138. H. Ott, "Reflexion und Brechung von Kugelwellen: Effekte 2. Ordnung," *Annalen der Physik*, Series 5, **41**, 1942, pp. 443-466.

139. H. Ott, "Die Saddlepunkte in der Umgebung eines Pols Mit Anwendungen auf die Wellenoptik und Akustik," *Annalen der Physik*, Series 5, **43**, 1943, pp. 393-403.

140. J. D. Jackson, *Classical Electrodynamics*, Wiley, 1st edition, 1962, pp. 183-185.

141. G. Tyras, *Radiation and Propagation of Electromagnetic Waves*, Academic Press, 1969, pp. 102-105.

142. H. Bateman, *Electrical and Optical Wave-Motion,* Cambridge University Press, 1915, pp. 73-75.

143. R. F. Harrington, *Time-Harmonic Fields*, McGraw-Hill, 1961, pp. 130, 202.

144. J. R. Wait and W. C. G. Fraser, "Radiation from a Vertical Dipole over a Stratified Ground (Part II)," *IRE Transactions on Antennas and Propagation*, **AP-3**, No. 4, October 1954, pp. 144-146.

145. M. V. Shuleikin, "Rasprostranenie elektromagnitoni enerhii [Propagation of Electromagnetic Energy]," First Russian Radio Press, 1923.

146. C. E. Smith, "Standard Broadcast Antenna Systems," published as chapter 9 in the *National Association of Broadcasters Engineering Handbook*, G.W. Bartlet, editor, 6th ed., 1975, pp. 190-191.

147. M. Dolukhanov, *Propagation of Radio Waves*, Mir Publishers, Moscow, 1971, pp. 23, 74.

148. C. E. Smith, *Directional Antennas*, Cleveland Institute of Electronics, 1st ed., 1946, p. 1.9. (See Table I.)

149. R. Li, "The Accuracy of Norton's Empirical Approximations for Ground Wave Attenuation," *IEEE Trans. Ant. and Prop.*, **AP-31**, No. 4, July 1983, pp. 624-628.

150. W. R. Burrows, *VHF Radio Wave Propagation in the Troposphere*, Intertext Books, 1968, pp. 28-39.

151. K. A. Norton, "The Calculation of Ground-Wave Field Intensity Over a Finitely Conducting Spherical Earth," Proceedings of the IRE, **29**, No. 12, December 1941, pp. 623-639.

152. C. R. Burrows, "Radio Propagation Over Spherical Earth," *Proceedings of the IRE*, **23**, No. 5, May 1935, pp. 470-480. Reprinted in *Bell System Technical Journal*, **14**, July 1935, pp. 477-485.

153. H. Bremmer, *Terrestrial Radio Waves*, Elsevier Publishing Co., 1949, pp. 105-124.

154. D. A. Hill and J. R. Wait, "Ground Wave Attenuation Function for a Spherical Earth with Arbitrary Surface Impedance," *Radio Science*, **15**, No. 3, May-June 1980, pp. 637-643.

155. S. Rotheram, "Ground-Wave Propagation. Part I - Theory for Short Distances," *Proceedings of the IEE* (London), **128**, Part F, October 1981, pp. 275-284.

156. S. Rotheram, "Ground-Wave Propagation. Part II – Theory for Medium and Long Distances and Reference Propagation Curves," *Proceedings of the IEE* (London), **128**, Part F, October 1981, pp. 285-295.

157. R. P. Eckert, "Modern Methods for Calculating Ground-Wave Field Strength Over A Smooth Spherical Earth," FCC Office of Engineering and Technology report: FCC/OET R86-1, February 1986, 43 pages.

158. N. DeMinco, "Propagation Prediction Techniques and Antenna Modeling (150 to 1705 kHz) for Intelligent Transportation Systems (ITS) Broadcast Applications," *IEEE Antennas and Propagation Magazine*, **42**, No.4, August 2000, pp. 9-34.

159. V. Trainotti, "Simplified Calculation of Coverage Area for MF AM Broadcast Station," *IEEE Antennas and Propagation Magazine*, **32**, June 1990, pp. 41-44.

160. T. K. Sarkar, et al, "Electromagnetic Macro Modeling of Propagation in Mobile Wireless Communication: Theory and Experiment," *IEEE Antennas and Propagation Magazine*, **54**, No. 6, December 2012, pp. 17-43.

161. T. K. Sarkar, et al, "The Physics of propagation in a Cellular Wireless Communication Environment," *Radio Sci. Bulletin*, No. 343, Dec. 2012, pp. 5-21.

162. L. E. Vogler, "Point-to-Point Communication on the Moon by Ground Wave Propagation," *Journal of Research of the National Bureau of Standards, Section D , Radio Propagation*, **67D**, No. 1, January/February, 1963, pp. 5-21

163. H. M. Barlow and A. L. Cullen, "Surface Waves," *Proc. IEE (London)*, pt. III, **100**, November 1953, pp. 329-341.

164. J. A. Fleming, *The Principles of Electric Wave Telegraphy and Telephoney*, Longmans, Green and Co., 2nd ed. 1910, pp. 729-744.

165. A. V. Kukushkin, "On the Existence and Physical Meaning of the Zenneck Wave," *Russian Physics – Uspekhi*, **52**, No. 7, 2009, pp. 755-756.

166. G. Goubau, "Ueber die Zenneck Bodenwelle," *Zeitschrift für angewandt Physik*, **3**, 1951, pp. 103-107. Translation at: http://nedyn.com/Goubau_1951-X.pdf

167. L. J. Chu and W.L. Barrow, "Electromagnetic Waves in Hollow Metal Tubes of Rectangular Cross-Section," *Proceedings of the IRE*, Vol. 26, No. 12, December 1938, pp. 1520-1555.

168.   H. J. Reich, P. F. Ordnung, H. L. Krauss, and J. G. Skalnik, *Microwave Theory and Techniques*, Van Nostrand, 1953, pp. 291-293.

169.   G. Goubau, "Waves on Surfaces," *IRE Transactions on Antennas and Propagation*, **AP-7**, No. 5, December 1959, pp. S140-S146. (See p. S145.)

170.   G. Goubau, "On the Excitation of Surface Waves," *Proceedings of the IRE*, **40**, July 1952, pp. 865-868.

171.   E. A. Wolff, *Antenna Analysis*, Wiley, 1966, p. 33. [See Equation (3.66).]

172.   M. Abramowitz and I. A. Stegun, *Handbook of Mathematical Functions*, Dover reprint, 9th edition, 1972, p. 360, #9.1.9.

173.   J. D. Kraus and K. R. Carver, *Electromagnetics*, McGraw-Hill, 1973, p. 458.

174.   R. W. P. King, *The Theory of Linear Antennas*, Harvard University Press, 1956, Chapter 7.

175.   H. Bremmer, "Propagation of Electromagnetic Waves," chapter in *Handbuch der Physik*, S. Flugge, editor, Springer, Vol. 16, 1958, pp. 434-639. (See pp. 521-522.)

176.   L. M. Brekhovskikh, *Waves in Layered Media*, Academic Press, 1960, pp. 259-270.

177.   J. R. Wait, "Complex Image Theory - Revisited," *IEEE Antennas and Propagation Magazine*, **33**, No. 4, August 1991, pp. 27-29.

178.   M. J. O. Strutt, "Strahlung von Antennen unter dem influss der Erdboden-eigenschaften. A. Elektrische Antennen; B. Magnetische Antennen.," *Annalen der Physik*, **1**, 1929, pp. 721-772.

179.   M. J. O. Strutt, "Strahlung von Antennen unter dem influss der Erdboden-eigenschaften. C. Rechnung in zweiter Näherung," *Ann. der Phys.*, **4**, 1930, p. 1-16.

180.   M. J. O. Strutt, "Strahlung von Antennen unter dem influss der Erdboden-eigenschaften. D. Strahlungsmessungen mit Antennen," *Annalen der Physik*, **9**, 1931, pp. 67-91.

181.   M. Krueger, ""Die Theorie der in endlicher Entfernung von der Trennungsebene zweier Medien erregten Kugelwelle für endliche Brechung Indizes," *Zeitschrift für Physik*, **121**, 1943, pp. 377-438.

182.   S. A. Schelkunoff, "Modified Sommerfeld's Integral and It's application," *Proceedings of the IRE*, 24, No. 10, October 1936, pp. 1388-1398.

183.   S. A. Schelkunoff, *Electromagnetic Waves*, Van Nostrand, 1943, pp. 428-435.

184.   R. B. Adler, L. J. Chu, and R. M. Fano, *Electromagnetic Energy Transmission and Radiation*, Wiley, 1960, pp. 358-369.

185.   W. M. Dyab, T. K. Sarkar and M. Salazar-Palma, "A Physics-Based Green's Function for Analysis of Vertical Electric Dipole Radiation Over an Imperfict Ground Plane," *IEEE Transactions on Antennas and Propagation*, **61**, No. 8, August 2013, p. 4148-4157.

186.   M. Eckert, *Arnold Sommerfeld: Science, Life and Turbulent Times, 1868-1951*, Springer, 2013, 471 pages.

AUTHOR BIOSKETCHES

Dr. James F. Corum (Ph.D. in Electrical Engineering, Ohio State 1974) is a Life Senior

Member of the IEEE, an Emeritus Life Member of the American Association of Physics Teachers, and listed in *Who's Who in Engineering* and *American Men and Women of Science*. He served in academia for 18 years on the engineering/physics faculties at several universities, and as Senior Scientist, Research Leader, and Chief Scientist for a number of technology corporations.

He has consulted for industry and dozens of government agencies, and is currently President of National Electrodynamics. Dr. Corum is the authentic inventor of the contra-wound and cross-wound Toroidal Helix Antenna technology (1972, 1981, 1986, 1988) [which evolved from the Multi-Turn Loop antenna invented at Ohio State, plus the Cubical Quad antenna conceived at HCJB in Quito, Ecuador], the 60 Hz Ring Power Multiplier (1998, 2003, 2009, 2013) [which evolved from the microwave RPM and surface waves on helices], and the Polyphase Surface Wave Probe (2013, 2014, 2015), and holds several dozen patents. He has authored over 125 technical papers and reports, and his principal publications are on Antennas, Microwaves, and Relativistic Electrodynamics. While study manager for the DARPA National Panel of Radar Experts in 1990, he was cited as a "National Treasure" by the Office of the Secretary of Defense.

===

Kenneth L. Corum (B.A. in Physics, Gordon College 1976; Engineering graduate studies at U. Mass.) Mr. Corum taught digital techniques, and software engineering for Compugraphics, ATEX/Kodak, Hewlett-Packard and Sun Microsystems. He was Director of the Commercial Satellite Division of Pinzone Communications in Cleveland, OH. He was Staff Consultant for Sun Microsystems (now Oracle) in Burlington, MA and taught industrial software courses in England, France, Germany, Latvia, Switzerland, the Netherlands, Russia, India, China, and Chile, as well as across the US and Canada. Mr. Corum is currently Chief Scientist for Texzon Technologies in Red Oak, Texas. Mr. Corum holds several domestic and foreign patents. He is the discoverer of the *modulated common-mode Radar-backscatter phenomenon from baseband differential-mode nonlinear systems*. He also discovered and documented the Trichel pulse excited VCO (Voltage-Controlled-Oscillator), and the 2-frequency RF-injected parametric regenerative mixer/detector technique employed by Mahlon Loomis in his 1865 RF experiments. His fundamental work on slow-wave helical resonators and Tesla's laboratory generation of ball lightning was published in *Uspekhi* by the Russian Academy of Sciences in 1990. This electric fire-ball

phenomenon was recently experimentally replicated by the Russian Academy of Sciences in Moscow and Troitsk in 2012. His most recent activity involved a modern replication of C.R. Burrows' 'crucial' Seneca Lake measurements and the experimental verification and NIST-traceable documentation of Tesla's 1897-1899 (Zenneck mode) surface wave propagation phenomenon. He spoke by invitation at Belgrade and Novi Sad as a guest of the Serbian Academy of Sciences and Arts in 1993. He is listed in *American Men and Women of Science*, and other dictionaries. He received the 1915 *Tesla White Dove Award* at the International Global Forum in Serbia. Over 100 patents are now pending or applied for on the new Texzon surface wave technology of which he is co-inventor.

AFTERWORD[*]

THE SENECA LAKE EXPERIMENT REPEATED[†]

After writing the above report, back in 2013, on the 'crucial experiment' that was used to vindicate Weyl's radiowave propagation theory over Sommerfeld's and justify Norton's mythical sign error (which was taught as scientific fact in the foremost textbooks, eminent technical journals, and academia around the world for the better part of the 20th century), it occurred to us that repeating Burrows' Seneca Lake experiment was in order (especially with the advent of Texzon's innovative surface wave technology, now Patent Pending).

The repeat experiment was performed on September 4, 2014 near the very spot where Burrows and his associates from Bell Labs had conducted the 'crucial experiment' in 1936. We described the event at a recent symposium on wireless, at Baylor University in Waco, Texas.[‡]

"We thought it would be constructive to repeat Burrows' experiment at Seneca Lake. When we used a conventional vertical half-wave dipole we obtained Norton's groundwave radiation curve for the constitutive parameters that we measured in situ at 52 MHz ($\varepsilon_r = 82.5$, $\sigma = 0.067$).[§] However, when we repeated the experiment with a field-matched surface-wave probe (as disclosed in Texzon's patent applications) we observed just the opposite effect, as seen in Figures D.1, D.2, and D.3. The dominant field contribution was consistent with that predicted by Zenneck theory *not* Norton. Our experiment was conducted on September 4, 2014 from the docks at "The Anchor Inn and Boat Rental" on Salt Point Road, two miles north of Watkins Glen on the west shore of Seneca Lake. The transmitter and antennas were set up on the Inn's wharf, and a rented motor-boat was used from which to perform field strength measurements out on the lake, just as Burrows

---

[*]   Prepared February 29, 2016.
[†]   The Texzon technology described herin is Patent Pending.
[‡]   The 2016 Texas Symposium on Wireless & Microwave Circuits & Systems, Baylor University, Waco, Texas, March 31-April 1, 2015.
[§]   Burrows reported that $\varepsilon_r = 82.2$ and $\sigma = 0.045$ mhos/m were measured for Seneca Lake water in the summer of 1936. Today the US Coast Guard asserts that $\sigma = 0.065$ near the north-end of the lake.

had done in 1936. Location identification was made with a GPS receiver. We took data all over the Southern end of the lake."[1]

The attached photos and graphs document the measured results and should convey the environmental conditions under which the experiment was performed. The theoretical Zenneck mode fields may be calculated from Equations (A.35) and (A.43) above. In our experiment, f_o = 52 MHz, V = 7.07 (RMS), and C = 3 pF, with an RF probe current of I = ωq = 0.007 A (RMS). The graphs in Figures D.1 and D.2 show the predicted and measured field strengths in RMS mV/m and dBµV/m, respectively. The latter may be compared against Burrows' plot directly.

Photo 1. "To make a test under these crucial conditions, an experiment on the propagation of 2-meter waves has been conducted over Seneca Lake, NY." Charles R. Burrows, 1936*

---

* Burrows, C.R., "Existence of a Surface Wave in Radio Propagation," *Nature*, Aug. 15, 1936, p. 284.

Photo 2. Initial field strength measurements being performed on Seneca Lake.

Photo 3. Field strength measurement performed out at 2.05 miles. The FIM-71 registers 17µV

Photo 4. (Left) Field strength measurement on Seneca Lake.

Photo 5. (Right) Photo showing the Zenneck probe (the small golden doublet, left side) mounted at six feet with common-mode chokes, and the conventional 52 MHz vertical λ/2 dipole with a Guanella balun mounted at 10 feet (right side). Our 2014 data is shown in Figures D.1 and D.2, below. Construction details and procedures for the Zenneck probe were disclosed in several patent applications. The Texzon technology described herein is Patent Pending.

The New Seneca Lake Measurements

The half-wave dipole was resonant and impedance-matched. The Zenneck probe was field-matched and the Texzon technology is Patent Pending. The results are presented in our Figures D.1 and D.2 below. Figure D.2 is to be compared with Burrows' Figure 1 published in the Bell Labs Record,[2] and reproduced below as our Figure D.3.

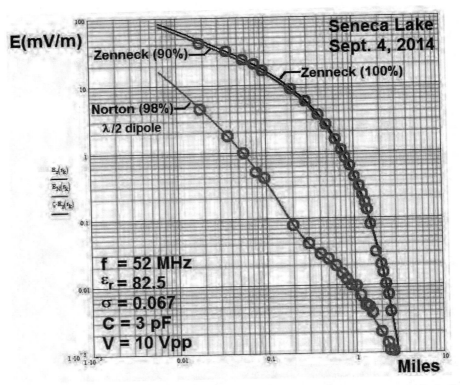

Fig. D.1. Theoretical <u>predictions</u> (curves) and <u>measured data</u> (circles) for the Seneca Lake experiment. The units are mV/m vs. miles. We performed the experiment at a wavelength of 5.77 meters (52 MHz), while Burrows operated at a wavelength of 2 meters (150 MHz). The anomaly at the 3 mile point was probably due to shallower water and ledges at the far shore. The constitutive parameters aren't really uniform, but are functions of the water temperature, which varies with water depth, and would locally bend the propagation curve. The Texzon technology described herein is Patent Pending.

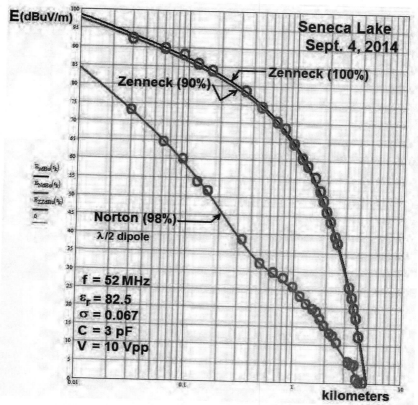

Figure D.2. Theoretical <u>predictions</u> (curves) and <u>measured data</u> (circles) for the vertically polarized Seneca Lake experiment. The units are dBμV/m vs. kilometers. [dBμV/m = 20 log $E_{mV/m}$ + 60 dB] We performed the experiment at a wavelength of 5.77 meters (52 MHz), while Burrows operated at a wavelength of 2 meters (150 MHz). The range of our data collection was, essentially, from 10 m out to 4 km. Burrows' range was from 1 m out to 2 km. The Texzon technology described herein is Patent Pending.

Our Figure D.2 is to be compared with Burrows results in Figure D.3, below. Professor Banos once wrote, "Burrows made careful measurements ...[He] concludes that his experimental results *prove conclusively that simple antennas do not generate a Sommerfeld surface wave ...*"[3] We strongly challenge that supposition and assert that it has now been *conclusively demonstrated* that even simple structures may be adjusted to launch a Zenneck/Sommerfeld surface wave.

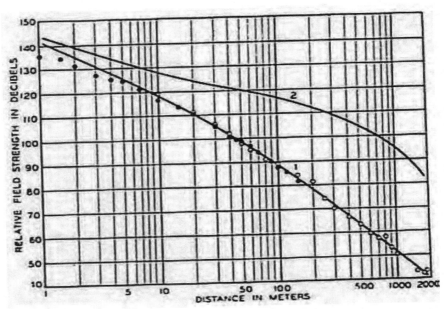

Figure D.3. Reproduction of Burrows' 1937 graph from Reference 2. His caption reads, *"Experimental points to show the actual field strength. They agree with Curve 1, which applies if there is no surface wave. Curve 2 gives the calculated values of what the field strength would be if there were a surface wave."* **(In Ref. 4, Burrows gives ε_r = 82.1 and σ = 0.045 mhos/m for Seneca Lake at 150 MHz.[4]) It has been stated that, "The discrepancy between the experimental points and [the upper] curve, which is a plot of Sommerfeld's formula, is so great that** *there can be no doubt* **as to the incorrectness of the latter."[5] Burrows wrote, "The two formulas predict field strengths that differ enormously." [Ref. 4, p. 221.] And, "At 1800 meters Sommerfeld gives about 1000 times Weyl [60 dB]." [Ref. 2.] And, "The discrepancy between the experimental data and Curve 2 is so great that** *there can be no doubt as to the incorrectness of the latter."* **[Ref. 4, p. 226.] (Unfortunately, Burrows[6] did not go out far enough to see the intersection of the Sommerfeld curve with the Weyl/Norton curve.)**

Finally, note that the Norton field strength over water varies inversely with distance as 1/r, whereas the Zenneck surface-guided field strength varies as $e^{-\alpha r}/\sqrt{r}$, where α depends on the medium's conductivity. As demonstrated in Figures D.1 and D.2 one may, *at will*, swap back and forth between Burrows' lower curve and upper curve!

James Wait, whose judgement we all deeply respect, asserted in a number of places that Burrows' 1936 experiments justified 20th century conventional radiowave propagation thought. Burrows, himself, concluded the Bell Labs investigation with these carefully crafted words,

> "... these tests *prove conclusively* that simple antennas do not generate a surface wave and that this time-honored concept must be given up ..."[7]

A century of wave propagation theory notwithstanding, the experimental evidence would now appear to challenge that assertion.

Photo 6. Seneca Lake is a long, slender, very deep (mean depth: 291 feet; maximum depth: 618 feet) fresh water lake.

Photo 7. This is the vertical ledge on the eastern shore line.

REGIONAL MAP

Fig. D.4. Seneca Lake is one of the 'Finger Lakes' in the State of New York.

Aside: Incidentally, the calculation above can be used to estimate the performance of Tesla's Colorado Springs transmitter. For example, from his <u>Colorado Springs Notes</u>, in the Fall of 1899, we have $f_o \sim 100$ kHz ($\lambda = 3000$ m), $C_T \sim 100$ pF, V ~ 10 MV.[8,9] The FCC R3 map gives the constitutive parameters as $\varepsilon_r = 15$ and $\sigma = 0.015$ mhos/m. From Equation (A.35) above one calculates $E_{RMS} = 8.7$ V/m at r = 26 miles (41.8 km)! [This is a surface wave mode, *not* a radiation field.] Assuming a ground bed resistance of, say, 2 Ω and a receiver effective height of, say, 33 m, and isolated (not bound) capacitance on the order of 100 pF, the Thevenin equivalent source at the receiver would be a 287 volt source in series with a 2 Ω resistance and a series capacitive reactance. This source could deliver about 10 kW to a conjugate-matched load. The possibility throws new light on the O'Neill anecdote,[10] which we all know didn't really happen (*wink, wink*).

. . .

We believe that this is only a start. We've all seen that amazing December 17, 1903 photograph[*] of Orville Wright stretched out on the lower wing of their rickety bi-plane taking off at Kitty Hawk, with Wilber Wright running along beside. The first powered flight only got 20 feet above the ground, lasted 12 seconds, and traveled a distance of 120 feet.[†] Yet, within less than 66 years men walked on the moon and public jet transportation circumvented the globe with thousands of people on a daily basis! Think of it …! While so far we have measured and documented Texzon Technologies' *surface wave* sources (the upper curve in Figures D.1-D.2) only out to 200 miles, we harbor no doubts that we will see the entire planet aglow with Tesla's energy distribution technology in a whole lot less than 66 years.

"Isn't it astonishing that all of these secrets have been preserved for so many years…"

Orville Wright, 1901

REFERENCES

[1] Corum, K.L., M.W. Miller, and J.F. Corum, "Surface Waves and the Crucial Propagation Experiment," Proceedings of the 2016 Texas Symposium on Wireless & Microwave Circuits & Systems, Baylor University, Waco, Texas, March 31-April 1, 2016.

[2] C. R. Burrows, "The Surface Wave in Radio Transmission," *Bell Laboratories Record*, Vol. 15, June 1937, p. 321-324. The *Bell Labs Record* was an in-house publication, and is *not* the BSTJ and is available on the internet at: http://www.americanradiohistory.com/Archive-Bell-Laboratories-Record/30s/Bell-Laboratories-Record-1937-06.pdf

[3] Banos, A., Dipole Radiation in the Presence of a Conducting Half-Space, Pergamon Pr., 1966, p. 155.

[4] C. R. Burrows, "The Surface Wave in Radio Propagation over Plane Earth," *Proc. IRE*, Vol. 25, No. 2, February 1937, pp. 219-229.

[*] The picture was taken by John T. Daniels of Nags Head, NC.
[†] By the end of the day (in the 4th flight), Wilber flew 59 seconds to a distance of 852 feet!

5   T. Kahan, and G. Eckhart, "On the Existence of a Surface Wave in Dipole Radiation over a Plane Earth," *Proceedings of the IRE,* **38**, July 1950, pp. 807-812. (See Fig. 6.)

6   Burrows, C.R., "The Surface Wave in Radio Propagation over Plane Earth," *Proc. Radio Club of America*, vol. 14, No. 2, Aug. 1937, pp. 15-18. At: http://www.jumpjet.info/Pioneering-Wireless/eJournals/RCA/1937-02.pdf

7   C. R. Burrows, "The Surface Wave in Radio Transmission," *Bell Laboratories Record*, vol. 15, June 1937, p. 321-324.

8   Corum, J.F., and K.L. Corum, "The Application of Transmission Line Resonators to High Voltage RF Power Processing: History, Analysis and Experiment," Proceedings of the 19th Southeastern Symposium on System Theory, Clemson University, Clemson, SC, March 15-17, 1987, pp. 45-49.

9   Corum, J.F., and K.L. Corum, Vacuum Tube Tesla Coils, Corum & Associates, 1987, p. 48.

10   O'Neill, J.J., Prodigal Genius, Ives-Washburn, 1944, p. 193.

INDEX

AC circuit theory, ... 61
AC field 179
AC generator 49, 60
AC induction motor **36, 41, 212**
AC machines 59, 61, 213
AC systems........................... 59, 60
AC voltages 60
Adams Plant **22, 40**
air core transformer 61
alternating current 15, 36, 40, 44, 45,
    46, 47, 48, 49, 56, 58, 133, 158,
    249, 333, 389
Alternating current........................ **47**
alternating current................. **46, 48**
alternating-current machines...... 132
alternating-current motor ... 46, 250
alternating-current power systems 59
alternator......92, 115, 116, 117, 120,
    122
ambient medium 113
American Institute of Electrical
    Engineers...53, 56, 111, 113, 115,
    206, 225
Anne Morgan............................. 51
annular distortion....................... 158
antenna ..62, 82, 110, 129, 133, 135,
    136, 137, 138, 139, 150, 216,
    217, 218, 221, 222, 232, 243,
    244, 249, 250, 251, 252, 262,
    266, 334, 350, 359, 366, 367,
    392, 393, 408, 410, 411, 412,
    414, 428, 431, 432, 433, 435,
    437, 439, 443, 446, 447, 448,
    456, 457, 459, 460, 461, 480
antipodes................... 158, 215, 398

Astor 64, 66, 72, 231
atmospheric conduction 145, 174,
    221, 389, 391
attenuation... 25, 133, 139, 140, 203,
    218, 220, 235, 237, 392, 404,
    418, 419, 424, 428, 435, 437,
    438, 446, 448, 450, 452, 455,
    457, 458, 461, 462
Baldor motor............................256
ball lightning........ 55, 211, 217, 481
Bell Labs .. 4, 11, 81, 392, 407, 408,
    417, 420, 424, 425, 430, 431,
    432, 440, 442, 482, 487, 490, 493
bladeless turbine 67
Boldt 87, 90
capacitor.... 151, 159, 160, 161, 176,
    181, 201, 212, 215, 219, 244,
    245, 248, 262, 275, 276, 277,
    281, 299, 300, 312, 314, 333,
    346, 365, 366, 388, 399, 400, 416
Capacitor 159, 315, 357
capacitor dielectric 244
carbon arc 125
cavity resonant frequencies 220, 232,
    233
charge carriers.... 187, 188, 366, 367
charge oscillation 135
Closed circuit 174
Colorado Springs 20, 22, 44, 53, 64,
    65, 72, 83, 136, 137, 140, 141,
    145, 150, 154, 155, 157, 162,
    171, 172, 186, 190, 191, 193,
    197, 198, 200, 204, 209, 211,
    214, 215, 216, 217, 219, 221,
    224, 225, 226, 230, 231, 387,

388, 395, 396, 399, 401, 402, 415, 491

condensers 222
constant dielectricity 169
constant permeability 169
constant velocity 171
Corum 4, 10, 11, 24, 25, 187, 188, 191, 211, 217, 218, 219, 220, 225, 226, 227, 229, 392, 396, 398, 407, 476, 480, 481, 493
cosmic rays **52, 68, 219**
Coulomb field 167
coupled Transmitter 179
cryogenic 66
Crystal Radio 242
currents propagated thru the ground 130
cyclotron 44, 52
earth currents. 132
Earth Resonance 158, 162, 394
Earth-Ionosphere 237
earth-ionosphere cavity. 23, 25, 211, 212, 217, 218, 221
Edison 14, 15, 42, 45, 46, 47, 48, 49, 53, 54, 56, 59, 60, 62, 67, 75, 76, 86, 87, 88, 99, 216, 227, 251, 397
Einstein. 53, 184, 412, 416, 466, 467
electric charge 134, 167, 174, 385
Electric Displacement 178
Electric Pierce Arrow 252
electric strength 167
Electrical discharges 151
electrical oscillators 116, 160
electrical resonance 61, 115
electrical resonance 61
electrically coupled elevated terminals 181
electromagnetic field 133, 166, 167, 171
electromagnetic fields 150
electromagnetic radiations 189
electromagnetic spectrum 61

electromagnetic waves .. 20, 23, 133, 137, 138, 140, 142, 150, 169, 171, 187, 188, 232, 354, 392, 393, 395, 466, 467, 470
electromagnetic waves. 20, 133, 393
electromagnetism 148, 182, 185, 189, 416, 443, 466, 467
electron ..44, 52, 184, 185, 187, 188, 189, 190, 363
electron microscope 44, 52
electronic star-wars 86
electrosphere 212, 219, 222, 223, 224
electrostatic. 24, 131, 158, 170, 178, 181, 214, 219, 220, 385, 390, 431
elevated capacity 120, 122
elevated terminal 122, 129, 201, 385, 388, 391, 401, 402
ELF 133, 138, 139, 140, 141, 142, 152, 154, 155, 156, 163, 169, 181, 217, 218, 219, 220, 223, 225, 226, 228, 230, 232, 235, 393, 437
ELF antennas 139
ELF waves 140, 394
EM funnel 179
ether ... 112, 113, 184, 185, 186, 187, 189
ether vibrations 112
ether vortex model 184
excitation of longitudinal-type waves 145
extraterrestrial signals 65
extremely low frequency ... 132, 133, 141, 211, 230, 394
Faraday cage 361, 367, 368
Farragut Letter 93
fluorescent lamps 62
fluorescent lights 44, 213, 266
Franklin D. Roosevelt 83
Franklin Institute 105, 115, 118, 134, 213, 225, 405, 472